Composite Truss Bridges With Concrete-filled Rectangular Steel Tube Members

Composite Truss Bridges With Concrete-filled Rectangular Steel Tube Members

Yongjian LIU
Chongqing University, P.R. China
Chang'an University, P.R. China

Xuhong ZHOU
Chongqing University, P.R. China

ELSEVIER

Elsevier
Radarweg 29, PO Box 211, 1000 AE Amsterdam, Netherlands
125 London Wall, London EC2Y 5AS, United Kingdom
50 Hampshire Street, 5th Floor, Cambridge, MA 02139, United States

ISBN: 978-0-443-44011-3

Publisher: Matthew Deans
Acquisitions Editor: Glyn Jones
Editorial Project Manager: Juwairiya Ali
Production Project Manager: Maria Bernard
Cover Designer: Raman Kumar

Typeset by MPS Limited, Chennai, India

Contents

Preface *ix*
Main symbols *xiii*

1. Introduction **1**

1.1 The proposal of concrete-filled steel tube composite truss girder bridge 1
1.2 Characteristics of the bridges 2
1.3 Characteristics and components of the bridges 7
1.4 About this book 11
Extended reading 13

2. General design and common structural details **15**

2.1 Overview 15
2.2 Structural system and general layout 15
2.3 Main truss 22
2.4 Bracing system 28
2.5 Deck system 29
2.6 Construction methods 32
2.7 Summary 44
Extended reading 44

3. Basic mechanical properties of concrete-filled rectangular steel tubular trusses **45**

3.1 Overview 45
3.2 Mechanical behavior and flexural capacity of concrete-filled steel tubular trusses 45
3.3 Basic mechanical properties of shaped steel tubes and concrete-filled steel tubular truss joints 53
3.4 Rectangular concrete-filled steel tubular truss tests 79
3.5 Summary 91
Extended reading 91

4. Static performance of concrete-filled rectangular steel tubular truss joints **93**

4.1 Overview 93
4.2 Experimental study on the compressive performance of concrete-filled rectangular steel tubular truss joints 93
4.3 Experimental study on the tensile and bending performance of concrete-filled rectangular steel tubular truss joints 119
4.4 Experimental study on T-type, Y-type, and K-type joints of concrete-filled steel tubular trusses 145
4.5 Bearing capacity calculation formula for concrete-filled rectangular steel tubular truss joints 183
4.6 Summary 193
Extended reading 194

5. Fatigue performance of joints in concrete-filled rectangular steel tubular trusses **197**

5.1 Overview 197
5.2 Experimental study on stress concentration in joints of CFRST trusses 197
5.3 Fatigue test study on joints of CFRST trusses 227
5.4 Parametric analysis of SCF for CFRST truss joints 247
5.5 Conclusion 278
Extended Reading 279

6. Flexural performance of concrete-filled rectangular steel tube composite truss girders **281**

6.1 Overview 281
6.2 Experiment on composite truss girders with concrete-filled rectangular steel tube in negative moment region 281
6.3 Full-scale test on a concrete-filled rectangular steel tube composite truss girder bridge 297
6.4 Simplified calculation of flexural capacity for concrete-filled steel tube composite truss girders 312
6.5 Summary 322
Extended reading 323

7. Calculation methods for concrete-filled rectangular steel tubular composite truss bridges **325**

7.1 Overview 325
7.2 Detailing requirements for concrete-filled rectangular steel tubular composite truss girders 325

7.3 Calculation of internal forces in concrete-filled steel tubular composite truss girders 328
7.4 Calculation method for bearing capacity of concrete-filled rectangular steel tubular composite truss girders 333
7.5 Summary 346
Extended reading 347

8. Continuous rigid frame bridge with concrete-filled rectangular steel tubular composite truss girders 349

8.1 Overview 349
8.2 Project overview 349
8.3 Structural design 350
8.4 Structural verification 361
8.5 Economic analysis 378
8.6 Summary 381
Extended reading 381

9. Prefabricated concrete-filled rectangular steel tubular composite truss bridge 383

9.1 Overview 383
9.2 Design overview 383
9.3 Structural design 384
9.4 Structural verification 396
9.5 Economic analysis 420
9.6 Summary 426
Extended reading 428

Index *429*

Preface

Concretefilled steel tube (CFST) structures possess superior mechanical properties, construction performance, and economic efficiency. They have been widely applied in CFST arch bridges and have become a prominent hallmark of Chinese bridge engineering and construction. Truss structures primarily bear axial forces in their members: the bending load is resisted by the couple formed by the axial forces of the upper and lower chords, while the shear load is resisted by the axial forces of the brace members. Such structures are characterized by high stiffness and high material utilization efficiency in member cross-sections. The CFST composite truss girder bridge, derived from the combination of CFST structures and truss structures, employs CFST members in the truss. This fully leverages the excellent compressive performance of CFST, reducing the thickness of steel tube walls, simplifying the configuration of stiffening ribs and truss joints, and enhancing the static and fatigue performance of the joints. Furthermore, when the chords of the CFST truss are connected to the concrete deck slab via shear connectors to form a composite truss girder, the CFST composite truss girder bridge becomes a double composite structure, with its flexural performance further improved.

Compared with traditional concrete box girder bridges, CFST composite truss girder bridges use brace members and lower chords to replace the braces and bottom slabs of concrete box girders. This not only improves material utilization but also significantly reduces the self-weight of the structure. Based on existing engineering application cases of CFST composite truss girder bridges, compared with concrete box girder bridge schemes, the steel consumption of CFST composite truss girder bridges is roughly the same or slightly higher (by approximately 10%), while the concrete consumption is significantly reduced (by approximately 50%), resulting in prominent economic advantages. In contrast to circular-section CFST structures, CFST composite truss girder bridges with rectangular-section CFST chords feature simple joint configurations, flexible arrangement of member cross-sectional dimensions, and easy direct placement of deck slabs on the concretefilled rectangular steel tube (CFRST) chords. They also facilitate prefabricated construction and have a high degree of standardization. Therefore, the new truss girder bridge type proposed in this book is a high-performance bridge structural system with high bearing capacity, high stiffness, excellent economic efficiency, and efficient construction, boasting broad application prospects.

The research work presented in this book originated at Hunan University. In 1998, with the support of the Hunan Provincial Natural Science Foundation Project (98JJY2042) and the Hunan Provincial Key Laboratory Project (99JZY2130), preliminary studies were conducted on the performance and design methods of high-strength CFST joints, as well as experimental research on CFST composite structure joints. The authors also participated in the compilation of the Technical Specification for Rectangular Concretefilled Steel Tube Structures (CECS 159:2004). Since 2004, research on the structural design theory and application of CFRST composite truss girder bridges has been carried out at Chang'an University. This research has successively received funding from the Shaanxi Provincial Natural Science Foundation (2005E215), the Transportation Construction Science and Technology Project (2013318812410), the Scientific Research Project of Shaanxi Provincial Department of Transportation (13-26K, 13-20K), the National Natural Science Foundation of China (51378068), and the National Key R&D Program of China (2016YFC0701202). This book summarizes some of the authors' achievements over more than 20 years in the theoretical research and engineering practice of CFRST and tube-structured bridges.

The book consists of nine chapters in total. Chapter 1 (Introduction) introduces the composition and characteristics of CFRST composite truss girder bridges. Chapter 2 elaborates on the structural layout and type selection of such bridges. Chapter 3 presents the basic mechanical properties of CFRST trusses and analyzes the influence of concrete filled in chord tubes on the mechanical properties of the trusses. Through experimental research, Chapters 4 and 5 discuss the static and fatigue performance of CFRST truss joints, and introduce the joint failure modes, bearing capacity calculation methods, and the hot spot stress method for fatigue performance evaluation. Chapter 6 conducts experimental and theoretical research on CFRST composite truss girder structures with concrete deck slabs. Chapter 7 summarizes existing theoretical research results and proposes a calculation method for CFRST composite truss girders. Chapter 8 provides engineering examples of CFRST composite truss continuous rigid frame bridges. Chapter 9 introduces the general drawings of prefabricated CFRST composite truss continuous beam bridges with spans ranging from 50 m to 80 m.

During the writing of this book, Dr. Ma Yinping from Chongqing University participated in the writing of Chapters 3 and 8; Dr. Jiang Lei participated in the writing of Chapter 5; Dr. Long Gang from Xi'an Municipal Design and Research Institute Co., Ltd. participated in the writing of Chapter 8; and Dr. Liu Bin from China Northwest Municipal Engineering Design and Research Institute Co., Ltd. participated in the writing of Chapter 9. PhD students Wu Wenxiang and Zhang Yingjie participated in the collation and typesetting of the manuscript. The authors hereby express their sincere gratitude to all of them.

Due to the authors' limited knowledge and capabilities, there may be shortcomings in this book. The authors sincerely welcome comments and corrections from readers.

Main symbols

A_b	Concrete actual bearing area or local bearing area
A_c	Cross-sectional area of composite truss girder concrete deck slab
A_d	Concrete spread bearing area or calculated base area
A_{cc}	Cross-sectional area of concrete inside the compressed chord of the concrete-filled steel tube truss
A_{st}	Cross-sectional area of the steel tube of the tension chord of the concrete-filled steel tube truss
A_{sb}	Cross-sectional area of the brace member of the concrete-filled steel tube truss
A_{sc}	Cross-sectional area of the compressed steel tube of the concrete-filled steel tube truss
A_{dc}	Cross-sectional area of the compressed concrete deck slab
b_0	Width of the main tube of the joint or the chord of the truss
b_1, b_2	Width of the branch tube of the joint or the brace member of the truss
$b_{e,p}$	Effective width of the branch tube of the joint or the brace member of the truss
e	Eccentricity of the joint
E_c	Elastic modulus of concrete
E_s	Elastic modulus of steel
f_c	Axial compressive strength of concrete
$f_{c'}$	Compressive strength of concrete cylinder specimen
f_{cu}	Compressive strength of concrete cube specimen
f_u	Tensile and compressive ultimate strength of steel
f_y	Yield strength of steel
h_0	Height of the main tube of the joint or the chord of the truss
h_1, h_2	Height of the branch tube of the joint or the brace member of the truss
g	Gap of K-type joint
k_c	Local bearing strength enhancement coefficient of plain concrete
k_{cs}	Transverse local bearing strength enhancement coefficient of rectangular concrete-filled steel tube
k_s	Coefficient of steel tube participation in transverse local bearing under transverse local pressure load of rectangular concrete-filled steel tube
k_L	Length influence coefficient of concrete inside the tube
$l_{p,eff}$	Length of punching-shear line
L	Calculated span of the truss
L_c	Length of concrete inside the tube of the joint
m_s	Plastic yield moment of steel tube brace under horizontal load
M_u	Flexural bearing capacity (ultimate moment) of the truss
n	Number of segments between load action and adjacent support
N	Total number of panel points of the truss
N_t	Axial force of the tension chord of the truss
N_c	Axial force of the compressed chord of the truss
N_{ucs}	Transverse local bearing strength of rectangular concrete-filled steel tube
N_b	Axial force of the branch tube of the joint or the brace member of the truss
P_Q	Punching-shear failure bearing capacity of rectangular steel tube joint

P_Y	Yield-line failure bearing capacity of rectangular steel tube joint
P_{YQ}	Combined yield-line and punching-shear failure bearing capacity
S_{hs}	Total hot-spot stress amplitude
SCF	Stress concentration factor
t_0	Wall thickness of the main tube of the joint or the chord of the truss
t_1, t_2	Wall thickness of the branch tube of the joint or the brace member of the truss
θ	Angle between branch tube and main tube of the joint or between brace member and chord of the truss
β	Width ratio of branch tube to main tube of the joint
τ	Aspect ratio of rectangular steel tube
2γ	Width-to-thickness ratio of rectangular steel tube
ν_c	Poisson's ratio of concrete
ν_s	Poisson's ratio of steel
η_{eff}	Reduction coefficient of brace member resistance considering brace member failure and joint failure
$k_{c/h}$	Ratio of elastic stiffness of tension joint of rectangular concrete-filled steel tube to simplified model of tension joint of rectangular steel tube

CHAPTER 1

Introduction

1.1 The proposal of concrete-filled steel tube composite truss girder bridge

Concrete box girders are currently the mainstream bridge structure form in China. Their application is extremely extensive, ranging from prefabricated small box girders for medium and small spans to prestressed concrete continuous girder bridges and continuous rigid frame bridges for long spans. However, due to the large self-weight of concrete box girders, they are associated with a long construction period, high costs of construction equipment, and temporary facilities, which affects the technical and economic benefits. Therefore exploring the optimization of their structural form holds significant theoretical significance and practical engineering value.

Concrete exhibits excellent compressive performance but poor tensile performance, being prone to cracking and resulting in low material utilization efficiency. In contrast, steel possesses the characteristics of light weight, high strength, and isotropy. To address the limitations of concrete box girders, the brace and bottom slab of the concrete box girder can be replaced with steel tubes and concrete-filled steel tube structures, respectively, forming a composite truss girder bridge composed of concrete filled steel tubular truss and a concrete deck slab (Fig. 1.1). In the optimized bridge type, steel structures are used instead of concrete structures in areas where tensile stress may occur, leading to a significant improvement in material utilization efficiency and economic performance.

According to the cross-sectional form, the truss members of concrete-filled steel tube (CFST) are mainly classified into circular cross section and rectangular cross section. CFST with circular cross section exhibits a significant confining effect and high ductility, and its mechanical properties have also been thoroughly studied. Composite truss girder bridges with circular cross section CFST have been applied in such projects as Xiangjiaba Bridge in Zigui County, Hubei Province; Zidong Bridge in Nanhai District, Guangdong Province; Ganhaizi Bridge in Sichuan Province; Wanzhou Bridge in Chongqing Municipality; and Kegu Bridge on Wenchuan-Maerkang Expressway.

Composite Truss Bridges With Concrete-filled Rectangular Steel Tube Members
DOI: https://doi.org/10.1016/B978-0-443-44011-3.00002-0

Figure 1.1 Steel tubular concrete composite truss bridge.

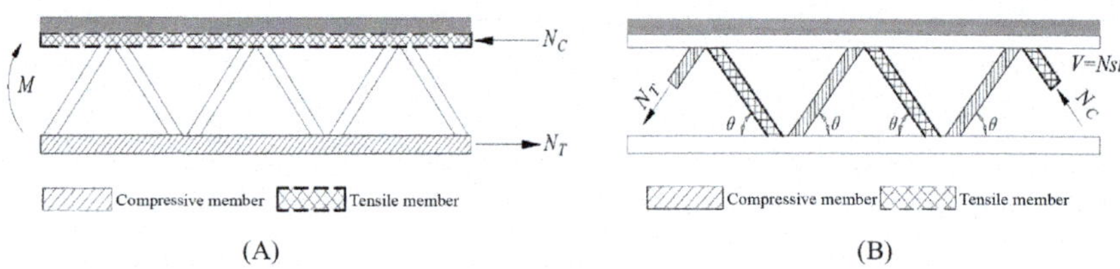

Figure 1.2 Force transmission of composite truss girder bridge. (A) Flexural load path. (B) Shear load path.

1.2 Characteristics of the bridges

From the perspectives of structural characteristics and engineering application effects, the CFST composite truss girder bridge is a high-performance bridge structural system, which is mainly reflected in three aspects: high structural performance, high construction performance, and high economic performance.

1.2.1 Structural high performance

The high structural performance refers to that the force transmission system of the structure can give full play to the mechanical properties of the structural components themselves, and the mechanical properties of the structural materials are fully utilized. The high structural performance of the CFST composite truss girder bridge is reflected in two aspects:

1. The basic force-transmitting units for bending resistance and shear resistance of the composite truss girder are axial loaded members. The stress distribution on the cross-section of the members is uniform (Fig. 1.2), resulting in high material utilization efficiency.

2. The composite action is fully utilized. The composite action of CFST composite truss girder bridges can be divided into two categories. The first is the composite action of the CFST structure formed by the steel tubes of the truss chords and the concrete filled in the tubes. This composite action can be fully utilized when the chord members are under compression; meanwhile, it can also improve the mechanical properties of joints and avoid local failure of truss joints. The second is the slab-truss composite action, which is formed when the concrete deck slab and the CFST truss are connected to each other as a whole through direct embedding or shear connectors. This composite action can significantly enhance the flexural stiffness and bearing capacity of the CFST composite truss girder. When a CFST composite truss girder bridge possesses both of the above-mentioned composite actions, it can be referred to as a composite truss girder bridge with dual composite actions.

1.2.2 High construction performance

The construction methods for the main truss of CFST composite truss girder bridges are flexible. They can adopt either the "disassembled assembly method" for members (Fig. 1.3A), which follows the principle of "breaking the whole bridge into members," or the segmental assembly method (see Fig. 1.3B) and full-span erection method (see Fig. 1.3C) that follow the principle of "integrating members into a whole." Specifically, the truss members are assembled into small segments or full-span girder segments first, and then transported to the bridge site for installation.

CFST composite truss girder bridges can be applied to different bridge locations such as canyon mountainous areas, river-crossing and sea-crossing sites, and can meet the requirements of different transportation conditions,

(A) (B) (C)

Figure 1.3 Construction methods of the main truss. (A) Piece-by-piece assembly. (B) Segmental construction. (C) Full span launching method.

hoisting capacities, and construction methods. During the construction of the concrete filled in the steel tubes and the deck slab, the steel tube truss can also serve as a construction platform, which reduces the use of formwork and lowers the costs of construction measures.

1.2.3 High economic performance

CFST composite truss girder bridges feature high material utilization efficiency and allow flexible selection of construction methods based on on-site conditions, resulting in superior corresponding economic performance. To analyze the economic efficiency of CFST composite truss girder bridges, several typical CFST composite truss girder bridges are selected below, and a comparison is made between these bridges and the concrete box girder bridge scheme.

1.2.3.1 Nanhai Zidong Bridge

The main bridge of Nanhai Zidong Bridge is a double-pylon, three-span cable-stayed bridge with a single cable plane, and its span arrangement is (69 + 140 + 69) m. Both the cable towers and the main girder adopt CFST structures. The main girder has a height of 3.0 m and a width of 25.5 m, with an inverted trapezoidal cross section with cantilevers, and the node spacing is 3 m. Concrete is poured into the lower chord members, brace members, and lower lateral bracing steel tubes, while the bridge deck is a cast-in-place prestressed concrete slab. The original design scheme of this bridge was a prestressed concrete continuous rigid frame bridge.

Table 1.1 lists the material consumption of the main girder design scheme of Zidong Bridge, its original design scheme, and other bridges of the same type. It can be seen from the table that compared with the

Table 1.1 Main material quantity table.

Scheme Materials	Zidong Bridge	Existing concrete continuous rigid frame bridge	Ratio
Steel plate and steel tube (kg/m^2)	150.1	0.19	/
Steel bar (kg/m^2)	70.67	203.21	
Total steel (kg/m^2)	220.77	203.4	1.09
Total concrete (m^3/m^2)	0.759	1.359	0.56

concrete continuous rigid frame bridge, although the steel consumption of Zidong Bridge increases by 9%, the concrete consumption decreases by 45% (Fig. 1.4).

1.2.3.2 Xiangjiaba Bridge

Xiangjiaba Bridge in Zigui County, Hubei Province, is a CFST continuous rigid frame bridge, as shown in Fig. 1.5. The bridge adopts a three-span continuous rigid frame structure, with a span arrangement of (43.3 + 72.2 + 43.3) m. The main girder is a CFST composite truss girder, with a height of 3.55 m, a width of 9.5 m, and a height-span ratio of 1/20. Its cross section is in the shape of a quadrangular pyramid, and a spatial truss is adopted; the brace members are arranged in a triangular pattern, with a longitudinal and transverse node spacing of 3.8 m.

Table 1.2 presents the comparison of material consumption and cost between Xiangjiaba Bridge and a prestressed reinforced concrete box girder

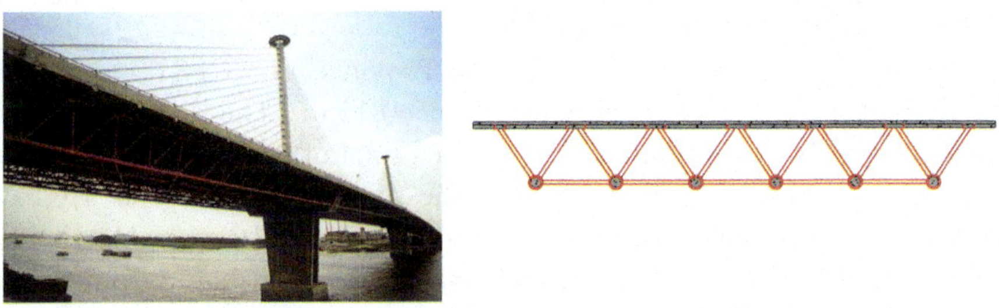

(A) (B)

Figure 1.4 Nanhai Zidong Bridge. (A) Full-scale bridge. (B) Schematic cross section.

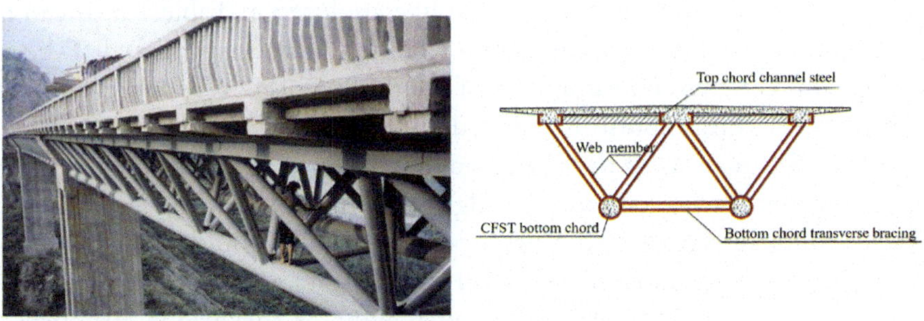

(A) (B)

Figure 1.5 Xiangjiaba Bridge. (A) Full-Scale bridge. (B) Schematic cross section.

Table 1.2 Comparison of material usage and cost.

Scheme Materials	Xiangjiaba Bridge	Prestressed concrete box girder bridge	Ratio
Steel plate and steel tube (kg/m^2)	119.5	/	/
Steel bar (kg/m^2)	34.8	142.1	
Total steel (kg/m^2)	154.3	142.1	1.09
Total concrete (m^3/m^2)	0.3	0.66	0.45

bridge with a similar span. It can be seen from the table that compared with the prestressed reinforced concrete continuous box girder bridge, although the steel consumption of Xiangjiaba Bridge increases by 9%, the concrete consumption decreases by 55%.

1.2.3.3 Ganhaizi Bridge

Ganhaizi Bridge in Ya'an City, Sichuan Province, is a CFST continuous beam bridge with a total span length of 1811 m. The entire bridge is divided into three continuous units, with a total of 36 spans, and the span lengths are mainly 44.5 and 62.5 m. The superstructure adopts a CFST composite truss girder with a constant cross section and circular steel tubes, with a girder height of 4.4 m and a height-span ratio ranging from 1/9 to 1/14. The lower chord members are CFST components, while the brace members are hollow steel tube components. The substructure adopts CFST lattice piers.

The main material indexes of the original simply supported T-beam bridge design scheme and Ganhaizi Bridge are listed in Table 1.3. It can be seen from the table that compared with the simply supported T-beam bridge, the concrete consumption of Ganhaizi Bridge is reduced by 64%, and the steel consumption is reduced by 24%.

From the analysis of the above engineering cases, it can be concluded that compared with concrete bridges of similar span, the adoption of CFST composite truss girder bridges significantly reduces the consumption of concrete and prestressed steel bars, despite a slight increase in steel consumption. Consequently, the self-weight of the bridge structure is also reduced accordingly, demonstrating superior economic efficiency (Fig. 1.6).

Table 1.3 Comparison of main material indicators of Ganhaizi Bridge.

Scheme / Materials	Ganhaizi Bridge	Existing t-beam bridge	Ratio
Steel plate and steel tube (kg/m^2)	8057	/	/
Steel bar (kg/m^2)	7015	19,664	
Total steel (kg/m^2)	15,072	19,664	0.76
Total concrete (m^3/m^2)	52,207	14,4307	0.36

Figure 1.6 Ganhaizi Bridge. (A) Full-scale bridge. (B) Schematic cross section.

1.3 Characteristics and components of the bridges

1.3.1 Structural characteristics of concrete-filled rectangular steel tube members

As discussed in Section 1.2, CFST composite truss bridges represent a high-performance structural system. Compared with circular CFST composite truss bridges, which are more commonly applied in engineering practice, concrete-filled rectangular steel tube (CFRST) composite truss bridges possess the following distinctive characteristics:

First, in terms of connections and joint configuration, circular steel tube joints are typically spatially intersecting, resulting in complex connection details and greater challenges in controlling welding quality. By contrast, rectangular steel tube joints feature simple construction, straight member cutting, and reduced welding difficulty (Fig. 1.7).

Second, the cross-sectional dimensions of rectangular steel tubes can be set with greater flexibility, allowing for different forms to be adopted

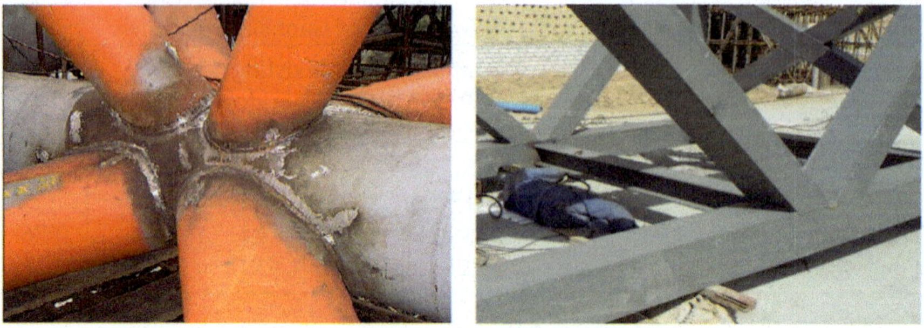

Figure 1.7 Node structure comparison.

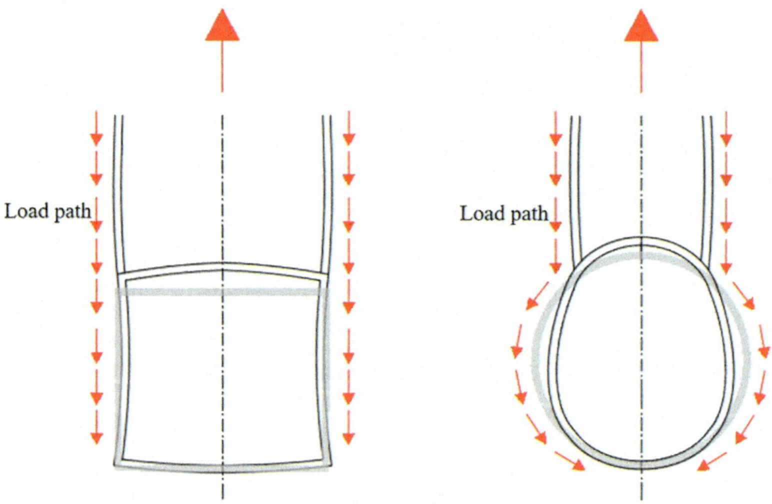

Figure 1.8 Comparison of node force transmission paths.

according to specific requirements. Under equal material consumption, CFRST members can achieve improved mechanical performance through adjustments in cross-sectional dimensions. Moreover, the brace members and chord members of CFRST trusses can be arranged with equal width, which alters the load-transfer mechanism at the truss joints and significantly enhances the bending capacity and stiffness of the truss (Fig. 1.8).

Third, the top surface of the CFRST truss chord is flat, which facilitates the installation of shear connectors, while the deck slab can also be directly placed on the chord (Fig. 1.9). This configuration makes CFRST

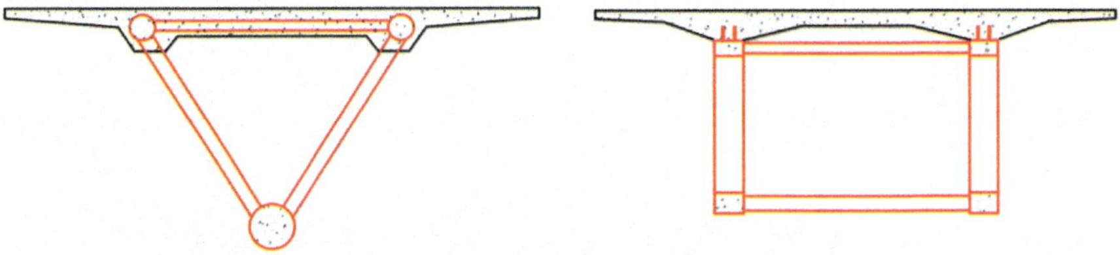

Figure 1.9 Comparison of bridge deck and main truss connection.

Figure 1.10 Storage comparison of steel pipes.

composite truss bridges more suitable for the use of precast deck slabs, thereby achieving a higher level of prefabrication.

In addition, rectangular steel tubes, unlike circular ones, can be stacked neatly, making transportation, storage, handling, and installation more convenient (Fig. 1.10).

In summary, compared with circular CFST composite truss bridges, CFRST composite truss bridges offer certain advantages in terms of structural configureuration and construction, serving as a complementary alternative to their circular counterparts. Research on CFRST composite truss bridges can enrich the structural forms of CFST bridges and is expected to demonstrate promising prospects for practical engineering applications.

1.3.2 Components of the bridges

A CFRST composite truss bridge generally consists of three main parts: the CFRST main truss, the bracing system, and the deck system, as illustrated in Fig. 1.11.

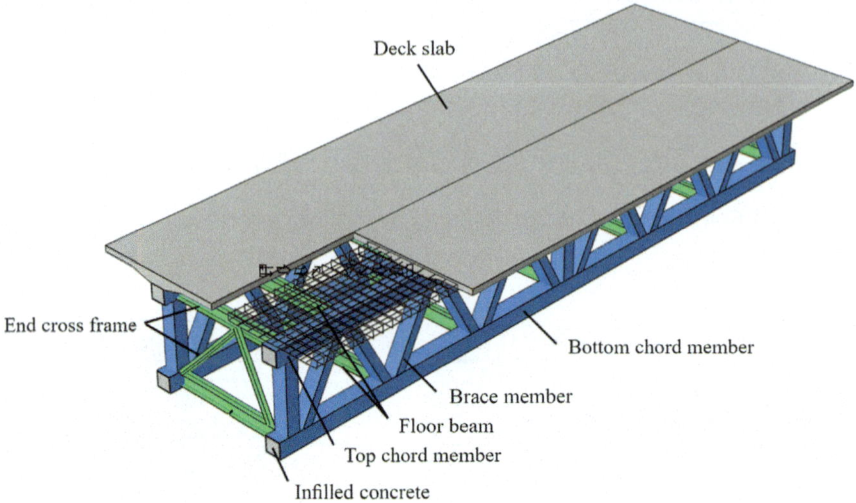

Figure 1.11 Composition of rectangular steel tube concrete composite truss bridge.

The CFRST main truss serves as the primary vertical load-bearing structure. Its chord members are constructed with CFRST sections, while its brace members are made of hollow rectangular steel tubes.

The bracing system is divided into longitudinal and transverse bracing, abbreviated as longitudinal bracing and transverse bracing. Longitudinal bracing is arranged within the upper and lower chord planes of the main truss: that located in the upper chord plane is referred to as the upper longitudinal bracing, and that in the lower chord plane as the lower longitudinal bracing. The main function of longitudinal bracing is to resist horizontal loads and provide lateral support for the main truss. Transverse bracing is arranged in the transverse planes of the bridge span structure, where those located within the span are called intermediate transverse bracing, and those at both ends are called end transverse bracing. The main function of transverse bracing is to make the truss laterally form a geometrically invariable system, thereby enhancing the torsional resistance of the section. In addition, the end transverse bracing can transfer the horizontal loads borne by the longitudinal bracing to the supports. When the transverse spacing between the main trusses is relatively small, the upper and lower longitudinal bracing can be simplified into cross beams (as shown in Fig. 1.11).

The deck system is composed of cross beams, longitudinal beams, and deck slabs. Cross beams are arranged transversely and supported at the nodes of the main truss; longitudinal beams are arranged longitudinally and supported on the cross beams; the deck slab is supported on both longitudinal and cross beams, directly carrying live loads. When the spacing between main trusses or the panel length is small, the longitudinal and cross beams of the deck system can be omitted, with the deck slab directly connected to the truss chords. In Fig. 1.11, the upper and lower longitudinal bracing are simplified into cross beams, while the longitudinal and cross beams of the deck system are also simplified, with the deck slab directly connected to the chords.

The connection between the deck slab and the truss chords of composite truss bridges can be classified into two types: embedded connections and seated connections. The former is applicable to cast-in-place deck slabs, in which the slab and the chord are cast together to form an integral structure that bears loads jointly. The latter is suitable for both cast-in-place and precast deck slabs, where shear connectors are installed on the surface of the rectangular chord members to connect with the concrete deck slab. Both methods ensure that the composite action between the deck slab and the main truss is fully realized.

1.4 About this book

1.4.1 Definitions of key terms

In this book, bridges that adopt a truss structure as the main load-bearing system are collectively referred to as truss girder bridges. Bridges in which the chords of the main truss are filled with concrete are collectively referred to as CFST composite truss bridges. Where the context is clear, they may also be abbreviated as composite truss bridges. Based on the cross-sectional form of the members, CFST composite truss bridges can be classified into CFRST composite truss bridges and circular CFST composite truss bridges.

When all or part of the truss chords are filled with concrete, the structure is termed a CFST truss. When a CFST truss is integrated with a concrete deck slab to act as a single structural system, it is referred to as a CFST composite truss girder.

Throughout this book, the term “joint” refers to truss joints, all of which are assumed to be fully welded. When the research focus is on joints, the terms branch member and main member correspond, respectively, to the brace member and chord member in truss studies.

In this book, the longitudinal direction of the bridge refers to the spanwise direction, the transverse direction refers to the direction perpendicular to the longitudinal axis within the horizontal plane, and the vertical direction refers to the direction perpendicular to the horizontal plane.

1.4.2 Chapter arrangements

This book is divided into nine chapters:

Chapter 1 Introduction analyzes the structural, constructional, and economic characteristics of CFST composite truss bridges in comparison with other bridge types. It introduces the concept of CFRST composite truss bridges and presents their distinctive features and components.

Chapter 2 General Design and Common Structural Details describes the overall design principles of CFRST composite truss bridges. Based on the characteristics of rectangular tube sections, it further discusses member cross-sectional forms, joint types, and connection methods among different components.

Chapter 3 Fundamental Mechanical Behavior of CFRST Trusses analyzes the load-bearing characteristics of CFRST trusses, the basic mechanical behavior of rectangular steel tube joints and CFST truss joints, and finally investigates the relationship between truss failure modes and joint performance through truss tests.

Chapters 4 and 5 Static Performance of CFRST Truss Joints and Fatigue Performance of CFRST Truss Joints present experimental studies on the static and fatigue performance of joints, examining their failure modes and proposing design calculation methods for joint bearing capacity and fatigue resistance.

Chapter 6 Flexural Behavior of CFRST Composite Trusses reports experimental research on both model specimens and full-scale bridges, establishing formulas for determining failure modes and corresponding flexural capacity while accounting for slab-truss composite action.

Chapter 7 Calculation Methods for CFRST Composite Truss Bridges summarizes existing theoretical research and proposes calculation methods for the flexural capacity of CFST composite trusses, as well as for the bearing capacity and fatigue performance of CFRST joints.

Chapters 8 and 9 Continuous Rigid-Frame Bridges with CFRST Composite Trusses and Prefabricated CFRST Composite Truss Bridges provide two engineering applications. The first is a completed and operational project on the Huangyan Expressway in Shaanxi Province. The second presents a standard design catalog for prefabricated CFRST composite truss bridges with spans of 50–80 m, developed based on the research outcomes. These chapters are intended to serve as references for the engineering application and promotion of CFRST composite truss bridges.

Extended reading

Gao YM, Liu YJ, Zhou XH, et al. High-performance CFST composite truss bridge[J]. China Journal of Highway and Transport 2018;31(12):174–87.

Liu B, Liu YJ, Yang YH, et al. Design of PBL-stiffened steel truss-concrete composite girder bridge with truss made of rectangular steel tubes[J]. Bridge Construction 2019;49(05):97–102.

Liu B, Liu YJ, Zhou XH, et al. Design of mid-span fabricated RCFST composite truss bridge[J]. Journal of Traffic and Transportation Engineering 2017;17(04):20–31.

Liu YJ, Gao YM, Zhou XH, et al. Technical and economic analysis in steel-concrete composite girder bridges with small and medium span[J]. China Journal of Highway and Transport 2017;30(3):1–13.

Liu YJ, Zhang N, Zhang JG. Review on mechanical behavior research of concrete filled rectangular hollow section tube stiffened with PBL[J]. Journal of Architecture and Civil Engineering 2012;29(4):13–7.

Tian ZJ, Liu YJ, Jiang L, et al. A review on application of composite truss bridges composed of hollow structural section members[J]. Journal of Traffic and Transportation Engineering (English Edition 2019;6(1):94–108.

Zhang LY, Li ZS, Cheng MF, et al. Composite beam structure with concrete-filled steel tubular spatial truss [M]. Beijing: China Communications Press Co., Ltd; 1999.

Zhou XH, Liu YJ. Steel bridges[M]. Beijing: China Communications Press Co., Ltd; 2020.

Zhou XH, Liu YJ, Jiang L, et al. Review on mechanical behavior research of concrete filled rectangular hollow section tube stiffened with PBL[J]. China Journal of Highway and Transport 2017;30(11):45–62.

CHAPTER 2

General design and common structural details

2.1 Overview

Rectangular concrete-filled steel tube (CFST) composite truss bridges exhibit distinctive features in terms of member cross-sectional forms, joint types, and connection methods between components. Based on these characteristics, this chapter introduces the overall design principles and common structural details of this new bridge type.

2.2 Structural system and general layout

2.2.1 Structural forms and classification

According to the relative position of the deck system, truss girder bridges can be classified into through truss bridges, deck truss bridges, half-through truss bridges, and double-deck truss bridges, as illustrated in Fig. 2.1. In deck truss bridges, cantilevered deck slabs can be arranged to reduce the spacing between the main trusses, thereby lowering material consumption. When the clearance beneath the bridge is limited, a through truss bridge may be adopted; in cases of shorter spans with relatively small truss heights, a half-through truss bridge can also be considered. Where bridge site resources are constrained, a double-deck truss bridge may serve as an alternative.

The use of rectangular CFST chords not only exploits their superior compressive capacity and allows for reduced truss height but also takes advantage of the flexibility of rectangular section dimensions. By increasing the width-to-height ratio of the section, the out-of-plane moment of inertia of the main truss members can be enhanced, thereby improving the out-of-plane stability of the truss.

2.2.2 General layout

The general layout of a truss girder bridge mainly involves span arrangement, elevation configuration, and cross-sectional layout. A rational overall layout must not only satisfy alignment and clearance requirements but also take into

Composite Truss Bridges With Concrete-filled Rectangular Steel Tube Members
DOI: https://doi.org/10.1016/B978-0-443-44011-3.00003-2

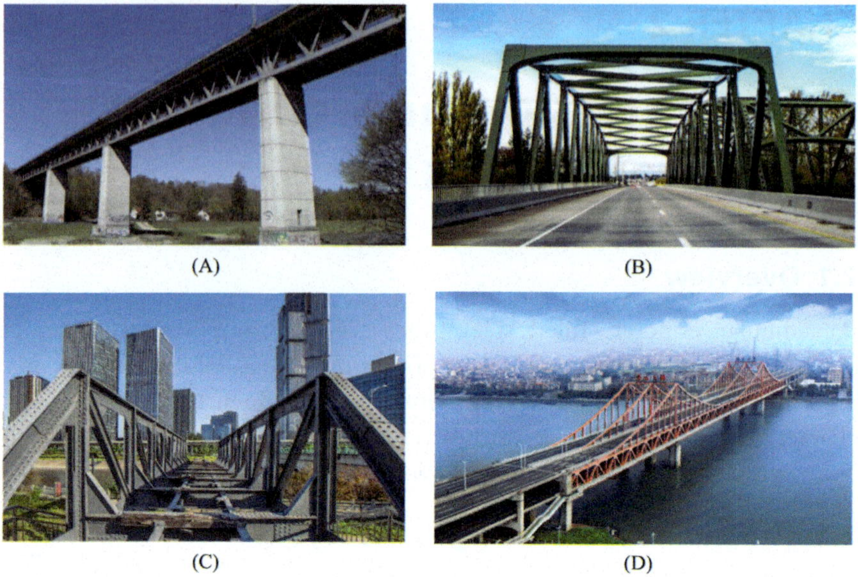

Figure 2.1 Truss bridges with different deck arrangements. (A) Deck truss bridge. (B) Through truss bridge. (C) Half-through truss bridge. (D) Double-deck truss bridge.

account structural performance, construction feasibility, economy, and aesthetics. Existing engineering examples of truss girder bridges can serve as valuable reference.

2.2.2.1 Span arrangement

For continuous truss girder bridges, the side-to-main span ratio is a key parameter in the overall layout, as it directly affects structural performance and material consumption. From a structural standpoint, adopting unequal span arrangements—with a side-to-main span ratio in the range of 0.6–0.8—results in more comparable mid-span bending moments between the side and main spans, leading to a more uniform moment distribution and a more rational overall force response. As shown in Fig. 2.2, the main span lengths of continuous truss girder bridges generally fall within the range of 40–180 m, with flexible options for configuring the side spans. The side-to-main span ratio commonly lies between 0.5 and 1.0.

The truss height, defined as the distance between the centerlines of the upper and lower chords, is closely related to the internal forces of truss members and the flexural stiffness of the structure. In truss girder bridges, the

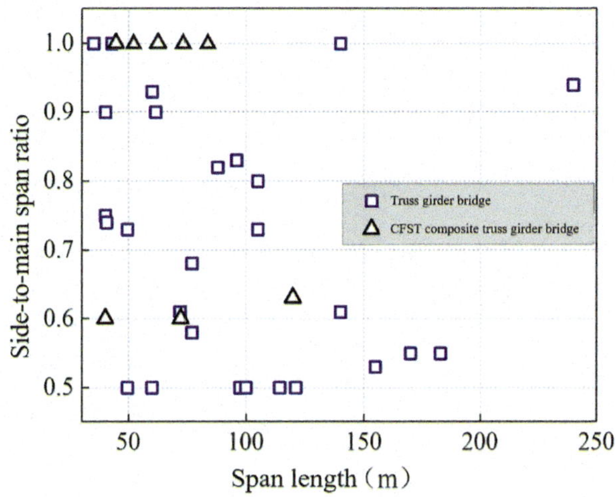

Figure 2.2 Span arrangement of continuous truss bridges.

upper and lower chords form a force couple to resist bending moments, with the truss height representing the lever arm length of this couple. Consequently, under the same loading conditions, a greater truss height reduces the axial forces in the chords, allowing for smaller cross-sectional dimensions. However, this also leads to longer brace members. Therefore, a comprehensive consideration is required to achieve an optimal balance in material consumption for the entire truss. The span-to-depth ratio of a truss is defined as the ratio between its truss height and the design span. Fig. 2.3 presents the span-to-depth ratios of several constructed CFST composite truss girder bridges. As shown, the span-to-depth ratio of continuous CFST composite truss girder bridges generally falls within the range of 1/12 to 1/16, while that of continuous rigid-frame bridges is around 1/22. For spans exceeding 150 m, it is advisable to adopt variable truss heights or to employ a continuous truss reinforced with a third chord.

Fig. 2.4 illustrates the relationship between the steel consumption and the span-to-depth ratio for a standard 4 × 70 m span composite truss girder bridge (with a bridge width of 12.75 m). As shown in the figure, with an increase in the span-to-depth ratio and under comparable stress levels of the members, the steel consumption of the chord members tends to decrease, while that of the brace members tends to increase. Since the rate of reduction in chord steel consumption is greater than the rate of increase in brace steel consumption, the total steel consumption exhibits

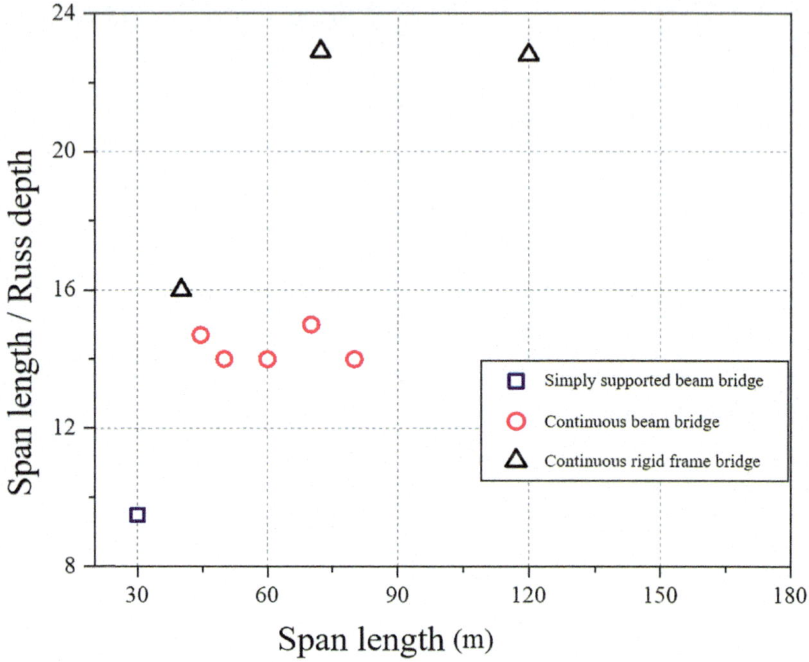

Figure 2.3 Height-to-span ratio of steel tube concrete composite truss bridge.

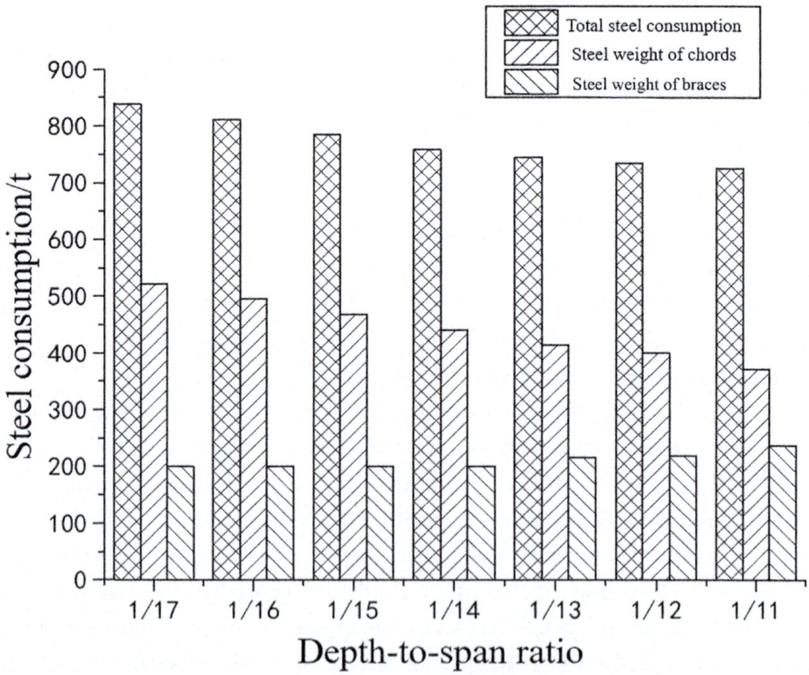

Figure 2.4 Steel consumption of 4×70 m rectangular steel tube concrete composite truss continuous beam bridge.

an overall decreasing trend. However, when the span-to-depth ratio exceeds 1/14, the downward trend in total steel consumption becomes less pronounced.

2.2.2.2 Elevation configuration

The main trusses of truss girder bridges can be designed with either parallel chords or curved chords. In curved-chord trusses, the section height is relatively large near the supports and reduced at mid-span, aligning with the bending moment distribution pattern of continuous girder bridges. This configuration leads to more rational internal force distribution and greater spanning capacity.

Parallel-chord trusses, by contrast, have a constant truss height and relatively uniform member specifications, making them well-suited for standardized fabrication and installation. Fig. 2.5 illustrates a truss girder bridge with a parallel-chord arrangement. Curved-chord trusses, with varying truss heights along the span, may employ either an upper or lower curved chord depending on the deck system's position. The chord curvature generally approximates the bending moment envelope of the structure, resulting in a more uniform internal force distribution, more efficient load transfer, and enhanced spanning capacity. An example of such a bridge is shown in Fig. 2.6.

Building on the advantages of both parallel-chord and curved-chord truss girder bridges, a further development is the stiffened-chord truss girder bridge, also referred to as the third-chord truss girder bridge. The main structure is a parallel-chord truss, with an additional stiffening chord typically installed in the negative bending moment regions near the piers. This configuration effectively increases the truss height in the stiffened

Figure 2.5 Parallel chord girder bridge.

(A) (B)

Figure 2.6 Curved chord truss bridge. (A) Curved bottom chord arrangement. (B) Curved top chord arrangement.

(A) (B)

Figure 2.7 Truss bridge with stiffened chords. (A) Stiffening chord arrangement at top chord. (B) Stiffening chord arrangement at bottom chord.

region, thereby reducing the stress level of the chords. Stiffened-chord trusses are commonly employed in long-span truss girder bridges.

The stiffening chord can be arranged either at the upper chord or at the lower chord, as illustrated in Fig. 2.7. In addition to incorporating stiffening chords, a combined beam–arch truss girder bridge system may also be adopted. In such cases, arch ribs are introduced at mid-span to share loads with the truss, thereby further enhancing the spanning capacity of the truss girder bridge (Fig. 2.8).

2.2.2.3 Cross-sectional layout

The cross section of CFST trusses generally adopts two basic forms: the tri-chord section and the quad-chord section. Spatial trusses often employ the tri-chord configuration (Fig. 2.9A), consisting of two upper chords and one lower chord. In this case, the brace members connected to the chords do not lie in the same plane as the chords, hence the designation spatial truss. In contrast, planar trusses have equal numbers of upper and lower chords, with the brace members and chords lying in the same plane.

Figure 2.8 Truss bridge using a beam-arch combination system.

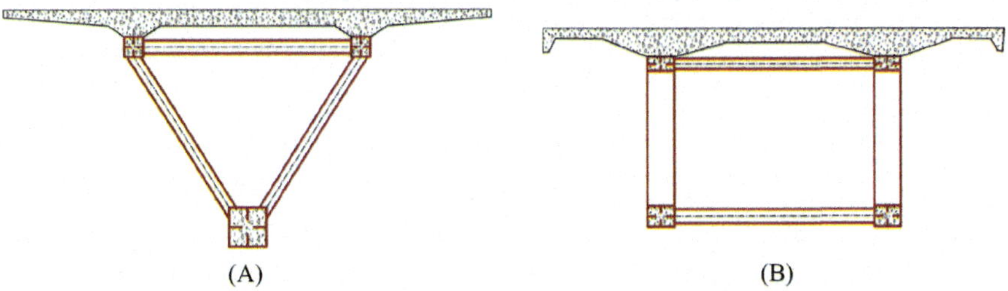

Figure 2.9 Common cross-sectional forms of composite truss bridges. (A) Tri-chord cross-section. (B) Multicell cross section.

Multiple parallel planar trusses, when connected by bracing systems, form a spatially stable structure. A typical arrangement is the quad-chord configuration, formed by two planar trusses with identical parameters (see Fig. 2.9B).

Compared with the tri-chord section, the quad-chord section offers the following advantages: (1) The additional lower chord increases the axial load-carrying capacity. Planar truss members lie within the same plane, making layout and installation more convenient. (2) Chords and brace members can be designed with equal widths, significantly enhancing the bearing capacity and stiffness of joints, thereby fully exploiting the advantages of rectangular members. (3) With two lower chords, the fatigue stress amplitude of chords and joints under live loads is markedly reduced compared to that of the tri-chord section, which is beneficial for fatigue design. (4) Therefore, the quad-chord cross-sectional arrangement is recommended for rectangular CFST composite truss girder bridges.

Based on the different cross-sectional configurations shown in Fig. 2.9, together with practical requirements for bridge width in engineering applications, rectangular CFST composite truss girder bridges can adopt a wider variety of cross-sectional forms. These are detailed in Tables 2.1 and 2.2.

2.3 Main truss

2.3.1 Brace member arrangements

The brace member arrangement of the main truss (also referred to as the truss pattern) includes triangular (Warren-type) trusses, modified triangular

Table 2.1 Several combination forms based on three-limb cross section.

Cross-sectional configurations	Name
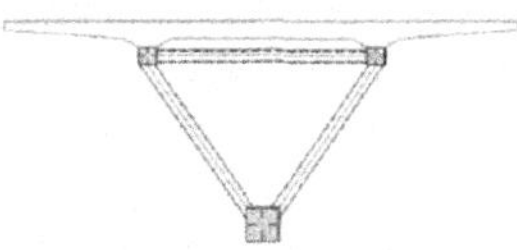	Single main truss cross section
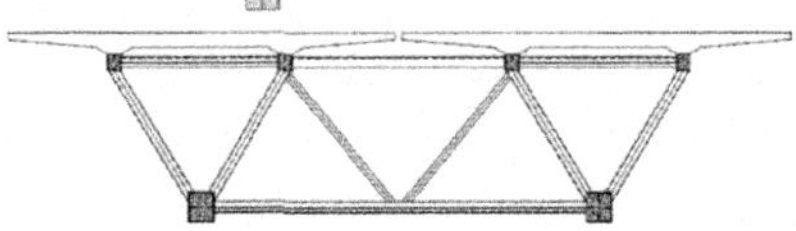	Double main trusses cross section
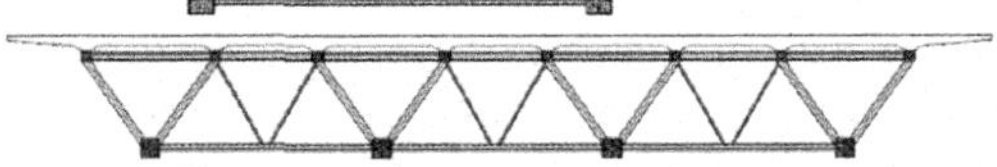	Multi-main trusses cross section

Table 2.2 Several combination forms based on four-limb cross-section.

Cross-sectional configurations	Name
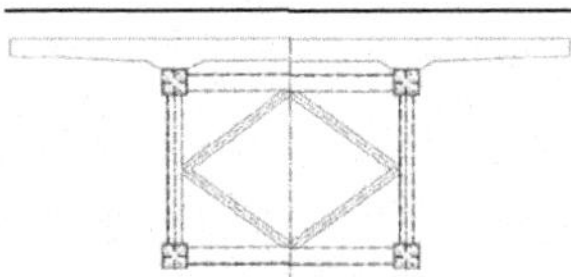	Single main truss cross section
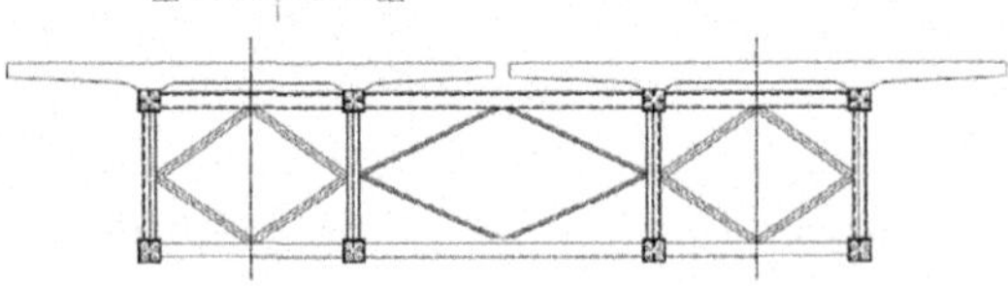	Double main trusses cross section
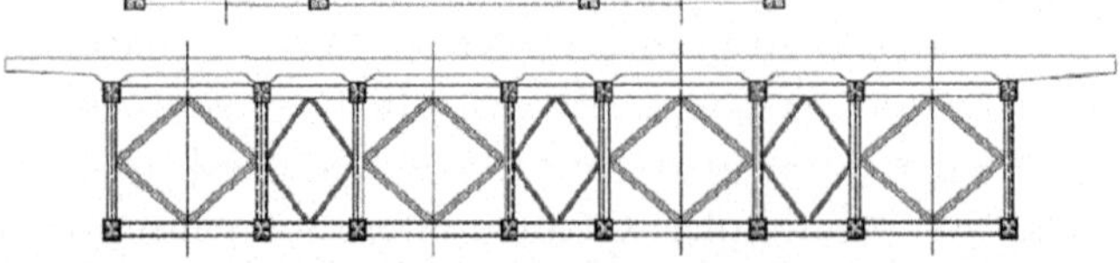	Multimain trusses cross section

trusses, single-diagonal (Pratt-type) trusses, diamond-pattern brace trusses, and K-type brace trusses, as illustrated in Fig. 2.10.

The triangular brace truss, shown in Fig. 2.10A, is the most commonly used form. This type of truss has a clear force-transfer mechanism, a simple configuration, and relatively few member and joint types, making it convenient for standardized design, fabrication, and installation. It is frequently adopted in truss girder bridges of relatively short spans. As the bridge span increases, the truss height must also be increased. To maintain a reasonable panel length and diagonal inclination angle, vertical brace members may be added, as shown in Fig. 2.10B. Vertical brace members are designed as compression members in deck-type truss bridges, and as tension members in through-type truss bridges.

Another option is the single-diagonal truss, as shown in Fig. 2.10C. In this form, the diagonal brace members of adjacent panels on one side of the truss are arranged in parallel. Under uniform distributed loading and midspan concentrated loads, the vertical members (short members) act as compression members, while the diagonal members (long members) act as tension members. This arrangement results in rational force distribution and efficient material utilization.

As bridge spans continue to increase, the truss height must also increase to meet strength and stiffness requirements, which can lead to excessively long diagonal brace members. In such cases, diamond-pattern trusses (see Fig. 2.10D) or K-type trusses (see Fig. 2.10E) can be employed.

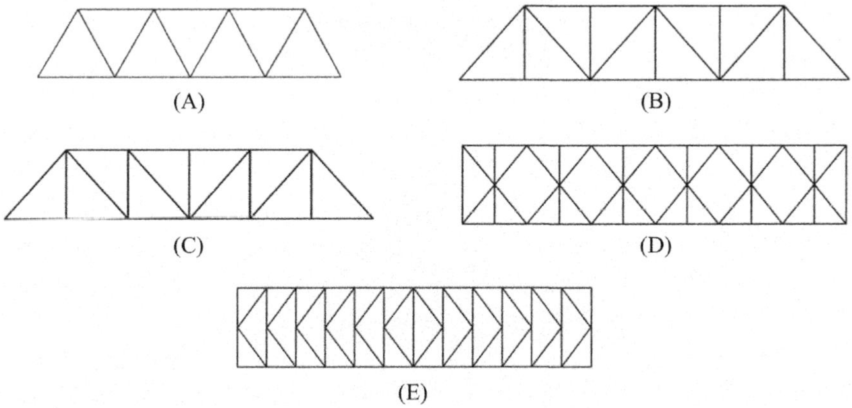

Figure 2.10 Brace arrangement. (A) Triangle truss. (B) Modified triangular truss. (C) Single-diagonal truss. (D) Diamond brace member truss. (E) K-braced truss.

These arrangements shorten the effective length of brace members, allow for reasonable adjustment of panel lengths, and better accommodate the span of longitudinal beams in the deck system. Compared with triangular and single-diagonal trusses, however, diamond-pattern and K-type trusses involve more member types and joint types, which increases the complexity of fabrication and assembly.

2.3.2 Types of main truss members

In professional English terminology, rectangular hollow sections (RHS) is usually referred to as a rectangular hollow section, and it can also be called a box section. At present, finished rectangular steel tubes are mainly manufactured using the three processing methods shown in Fig. 2.11.

(A)

(B)

(C)

Figure 2.11 Finished rectangular steel pipe processing method. (A) Fabrication method 1. (B) Fabrication method 2. (C) Fabrication method 3.

1. Forming from circular sections: A steel plate is first rolled into a circular section and closed by high-frequency welding. The section is then reshaped into a rectangular hollow section using a sizing mill.
2. Direct rectangular forming: A steel plate is directly rolled into a rectangular shape, and the open seam is closed by high-frequency welding to form an RHS.
3. Half-section welding: Steel plates are rolled separately into two half-rectangular sections. A splice plate is welded at the joint of the top and bottom plates, and finally, the section is closed using arc welding. Commercial RHS members are produced in fixed standard dimensions. The cross-sectional size of square hollow sections typically ranges from 20 × 20 ×2 to 400 × 400 ×1 6 mm, while that of rectangular hollow sections generally ranges from 40 × 25 × 2 to 500 ×300 × 16 mm.

Factory-produced RHS can generally meet the sectional requirements of members in short-span bridges. However, for medium-span and long-span bridges, the available standard RHS dimensions are often insufficient. In such cases, large cross-sectional dimensions can be achieved by welding together four individual plates to form a box-shaped section, as illustrated in Fig. 2.12. In professional English terminology, this type of section is referred to as a built-up box section. In this book, both categories mentioned above will be uniformly referred to as RHS.

To enhance the load-bearing capacity of the compression chords in the main truss, rectangular CFST sections may be adopted (Fig. 2.13B). For tension chords, either RHS (see Fig. 2.13A) or rectangular CFST sections can be employed. Filling the interior of tension chords with concrete can effectively improve the load-bearing capacity of truss joints, while the application of prestressing further enhances their tensile performance.

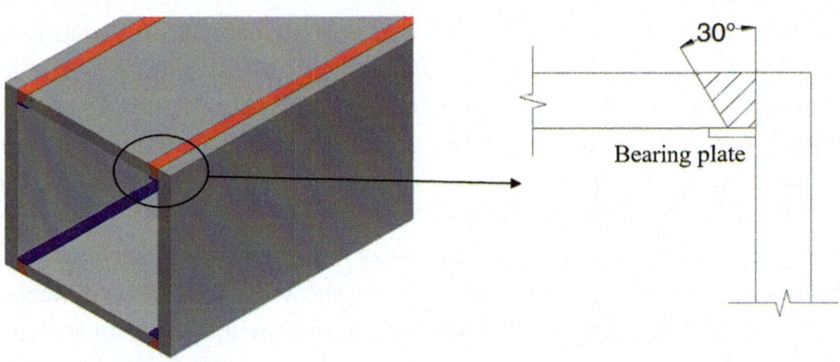

Figure 2.12 Rectangular steel tube section formed by welding four plates.

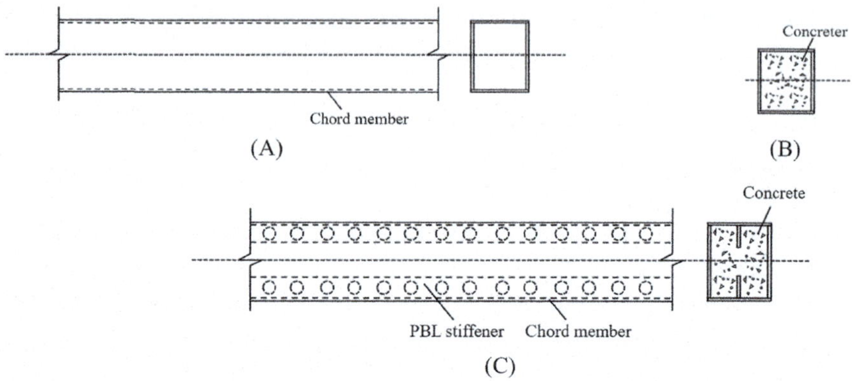

Figure 2.13 Main truss chord cross section. (A) Rectangular steel tube. (B) Concrete-filled rectangular steel tube. (C) PBL-stiffened concrete-filled rectangular steel tube.

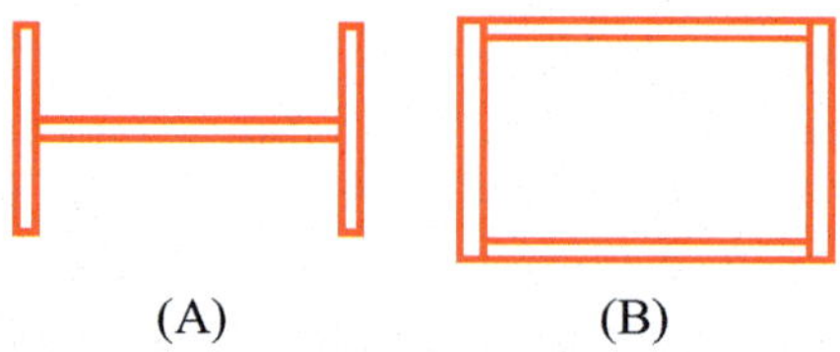

Figure 2.14 Main truss brace cross section. (A) H-shaped cross section. (B) Rectangular cross section.

As the cross-sectional dimensions of the chords increase, compression chords may encounter local stability issues, in which case longitudinal stiffeners can be installed inside the chord. For CFST structures, the compaction of the infilled concrete significantly affects their compressive performance. The stiffeners may be further optimized into perforated steel plate ribs (Perfobond Leister ribs, abbreviated as PBL ribs). This configuration is referred to as a PBL-stiffened rectangular CFST structure (see Fig. 2.13C). The incorporation of PBL ribs not only improves the local buckling resistance of the steel tube but also enhances the bond strength at the steel–concrete interface, preventing debonding and effectively transferring the loads from the tube wall to the core concrete. Thus the PBL ribs serve the dual functions of stiffeners and connectors.

The brace members of the truss may adopt the cross-sectional forms illustrated in Fig. 2.14. The H-shaped section shown in Fig. 2.14A requires less material; however, when the compressive forces in the brace members are relatively large, its stability is somewhat inferior to that of the section form depicted in Fig. 2.14B.

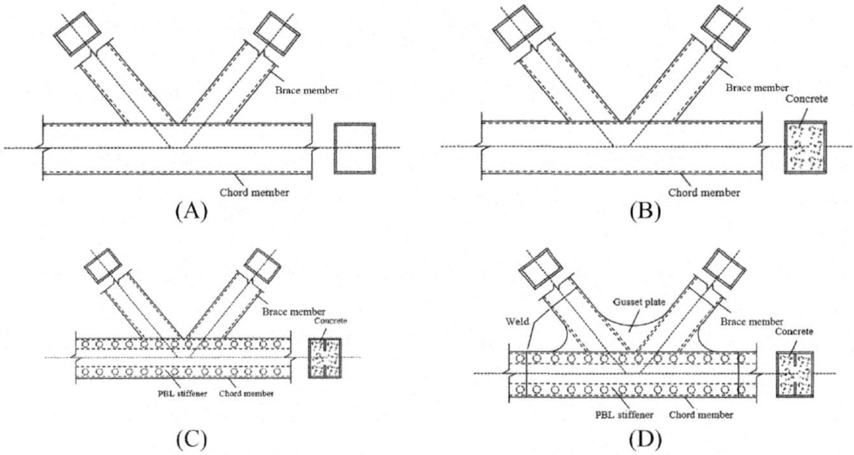

Figure 2.15 Joint forms of rectangular CFST trusses. (A) Rectangular tubular joint. (B) Concrete-filled rectangular steel tubular joint. (C) PBL-stiffened concrete-filled rectangular steel tubular joint. (D) Gusset-plated PBL-stiffened concrete-filled rectangular steel tubular joint.

2.3.3 Types of main truss joints

Joints are critical components of truss girder bridges. The static and fatigue performance of joints are key issues that must be addressed in design. With advancements in welding technology, fully welded joints have been widely adopted in the design and construction of truss girder bridges. Most welding work is carried out in prefabrication yards, ensuring high welding quality, while only minimal and simple member connections are required on-site. This approach better aligns with the requirements of prefabricated and standardized construction.

For rectangular CFST truss girder bridges, the commonly used joint types correspond to the chord section types shown in Fig. 2.13, as illustrated in Fig. 2.15. Rectangular hollow section joints (see Fig. 2.15A) feature a simple configuration with relatively low fabrication and installation difficulty. However, when the branch member is subjected to compression, the side wall of the chord may be crushed or experience local buckling. In K-type tubular joints, the chord may also undergo shear failure under combined tensile and compressive forces from the branch members. When the chord is filled with concrete to form a rectangular CFST joint (see Fig. 2.15B), such failure modes can be effectively prevented. Nevertheless, the improvement in joint capacity under tensile branch forces is not significant compared with rectangular hollow section joints.

PBL-stiffened rectangular CFST joints (see Fig. 2.15C) form concrete dowels at the openings of the stiffeners within the chord, which significantly enhance the joint's capacity to resist tensile forces from branch members. Additionally, the horizontal component of the branch axial force can be more rapidly transferred to the chord through the stiffeners, thereby reducing the load transfer length and improving transfer efficiency. For rectangular CFST joints where the chord and branch have equal widths, joint plates can also be introduced to enhance fatigue performance. In this configuration, the side plates of both the chord and the brace member within the joint region are fabricated from a single steel plate, allowing welds at the joint to be staggered over a certain distance. The edges of the joint plate are designed with smooth circular transitions, and full-penetration butt welds are used to connect with adjacent plates (see Fig. 2.15D).

2.4 Bracing system

In truss girder bridges, longitudinal bracings are typically arranged within the horizontal planes of both the upper and lower chords. When the bridge deck system is positioned within one of these longitudinal planes, bracing in that plane may be omitted; however, temporary longitudinal bracing should still be provided during the construction stage. Common forms of longitudinal bracing in steel truss girders include cross-shaped, diamond-shaped, and K-shaped arrangements, as illustrated in Fig. 2.16. When the spacing between the main trusses is relatively small, the longitudinal bracing can be simplified into a transverse beam configuration, as shown in Fig. 2.17.

For through-type steel truss girder bridges, the arrangement of transverse bracing is not restricted by deck clearance and can be flexibly configured according to structural detailing and load-bearing requirements. Typical structural forms are illustrated in Fig. 2.18A.For deck-type steel truss girder bridges, however, the arrangement of transverse bracing must account for clearance requirements of the bridge deck. Common structural forms of transverse bracing and portal frames are shown in Fig. 2.18B.

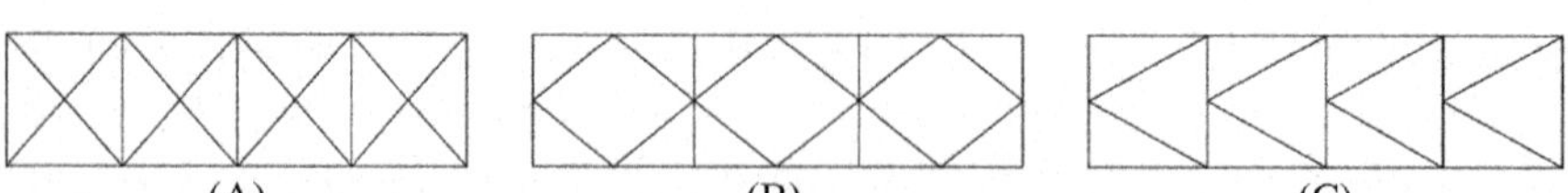

Figure 2.16 Vertical connection form. (A) Crisscross configuration. (B) Diamond configuration. (C) K-shaped configuration.

Figure 2.17 Vertical connection form. (A) Schematic of simplified longitudinal bracing. (B) Existing bridge example.

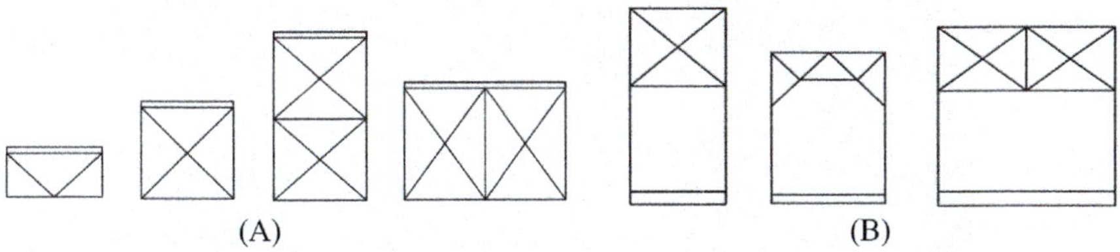

Figure 2.18 Common cross-link structures. (A) Transverse bracing system of deck truss bridge. (B) Transverse bracing system of through truss bridge.

2.5 Deck system

The deck system of rectangular CFST truss girder bridges consists of longitudinal beams, transverse beams, and deck slabs. In conventional steel truss girder bridges, the deck slabs rest on the longitudinal and transverse beams, while the transverse beams are connected to the main truss chords. For CFST truss girders, when the spacing between the main trusses is relatively small, the longitudinal and transverse beams of the deck system may be omitted, allowing the deck slab to connect directly to the main truss chords.

The connection between the deck slab and the main truss can be realized in two ways: embedded connection and bearing-type connection. For circular CFST chords, whose surfaces are curved, the connection is typically achieved by embedding the chord into the concrete deck slab, as shown in Fig. 2.19A. For rectangular CFST chords, whose top surfaces are flat, the deck slab can be directly supported on the chord and connected through shear connectors. In the case of bearing-type connections, the deck slabs may be prefabricated in segments and assembled on-site. This construction

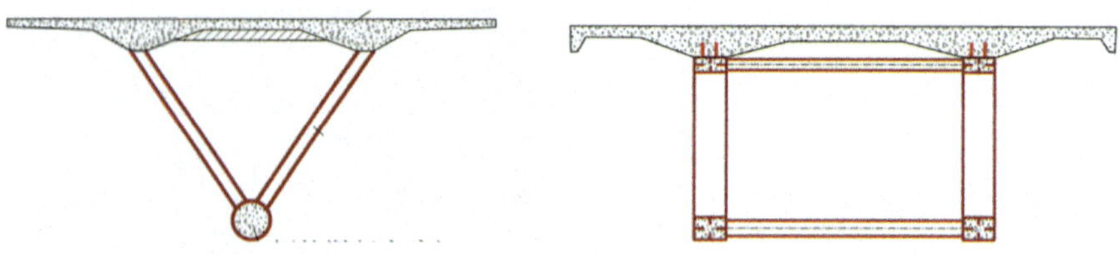

Figure 2.19 Plate truss connection method. (A) Embedded. (B) Simply supported.

Figure 2.20 Cast-in-place concrete bridge deck.

method can significantly reduce on-site workload and enhance the degree of prefabrication. Therefore, for rectangular CFST truss girder bridges, the bearing-type slab–truss connection is recommended.

Deck slabs can be categorized into two types: precast slabs and cast-in-place slabs (Figs. 2.20 and 2.21). The construction of cast-in-place slabs is similar to that of conventional steel–concrete composite bridges. The use of precast slabs can mitigate the effects of concrete shrinkage and hydration heat on the structure, while also enhancing the degree of prefabrication and improving construction efficiency for truss girder bridges.

The segmentation of precast concrete deck slabs is generally divided into two approaches: full-width segmentation and longitudinal–transverse segmentation. The full-width segmentation method is suitable for bridge deck widths ranging from 12 to 16 m, with the longitudinal length of each deck slab

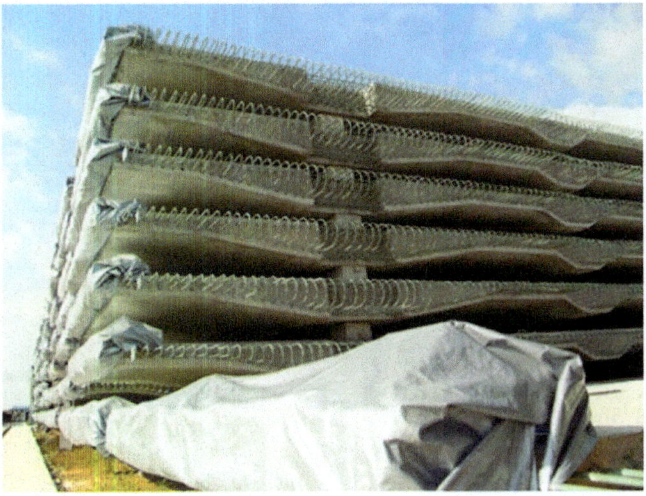

Figure 2.21 Precast concrete bridge deck.

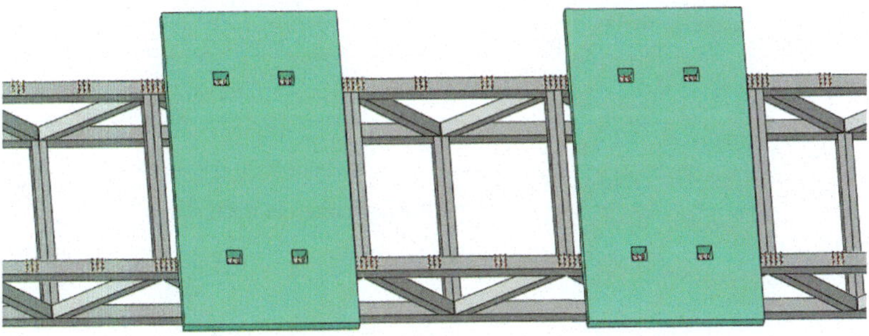

Figure 2.22 Precast concrete bridge deck.

segment typically between 4 and 6 m. In this case, continuous transverse wet joints must be provided across the bridge deck (Fig. 2.22), and shear grooves must be installed along the longitudinal direction at the top chords of the main trusses (Fig. 2.23). The longitudinal–transverse segmentation method, on the other hand, is applicable to wider bridge decks, where the segment lengths in both longitudinal and transverse directions range from 4 to 8 m. For this segmentation approach, continuous wet joints are required in both longitudinal and transverse directions.

To ensure that the composite action is fully developed, a sufficient number of shear connectors should be installed at the wet joints and at the shear grooves reserved in the deck slabs (Fig. 2.24).

Figure 2.23 Shear trough structure.

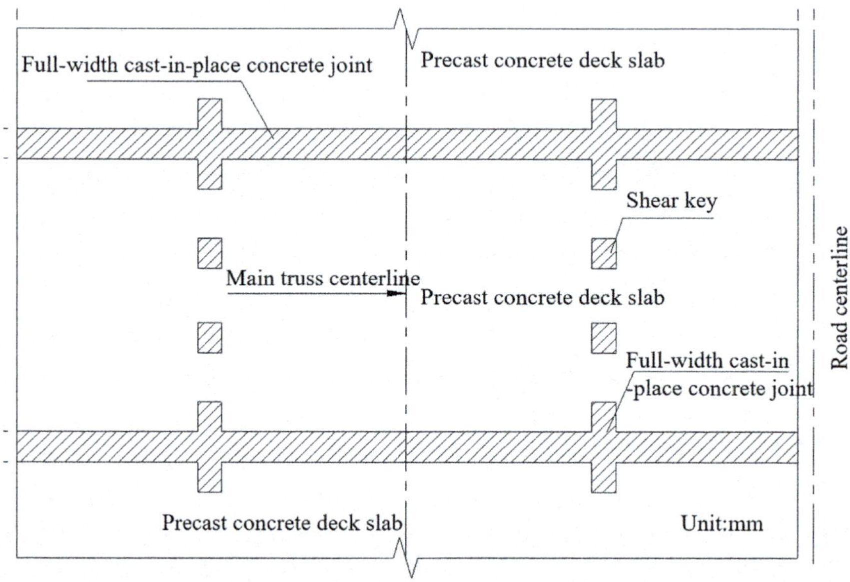

Figure 2.24 Wet joints and shear groove arrangement.

2.6 Construction methods

For CFST truss girder bridges, the typical construction sequence involves first installing the steel truss at the bridge site, followed by the casting of concrete into the tubular members and the construction of the deck slab. However, if site conditions are favorable and transportation and erection capacities are sufficient, the concrete may be cast into the tubular members prior to installing the CFST truss. Furthermore, an even higher degree of prefabrication can be achieved by completing both the CFST truss and the concrete deck slab in a fabrication yard, assembling them into composite truss girder segments, and subsequently installing these segments directly at the bridge site. This approach can effectively enhance construction efficiency and improve the overall level of prefabrication for CFST truss girder bridges.

2.6.1 Construction of steel trusses

The construction of steel trusses generally employs methods such as incremental launching, segmental hoisting, and cantilever erection, as illustrated in Fig. 2.25. The incremental launching method has wide applicability, making full use of the advantages of steel girders, including their relatively light self-weight and strong tensile and compressive capacities. The segmental hoisting method is suitable for medium- and small-span bridges with favorable on-site transportation conditions; for construction over water, floating cranes may be employed depending on site conditions. The cantilever erection method is typically used for long-span bridges where incremental launching or large-segment hoisting is constrained.

2.6.1.1 Incremental lunching construction

Depending on the structural form of the rectangular CFST truss, the launched structure may be either an empty steel tube truss or a CFST truss. According to the characteristics of incremental launching and the equipment employed, the launching system may be configured with or without a launching nose. The launching supports can be arranged either continuously along the lower chord of the truss or solely at the truss joints. Unlike incremental launching of steel plate girders or steel box girders, the construction of steel trusses requires careful consideration of the stress states between the members and joints of the main truss. Based on the supporting arrangement during launching, the presence or absence of a launching nose, and whether the truss is filled with concrete during launching, the incremental launching schemes can be classified, as illustrated in Fig. 2.26.

The key points of incremental launching construction include setting up the fabrication platform, segmental assembly on the platform, determination of the launching nose, selection of the launching method, girder launching, and

Figure 2.25 Steel truss construction method. (A) Incremental launching method. (B) Segment erection method. (C) Cantilever segment erection method.

girder lowering. Fig. 2.27 illustrates the construction process of one-sided incremental launching of a steel truss girder. It can be seen that, in general, special temporary equipment is installed and corresponding construction

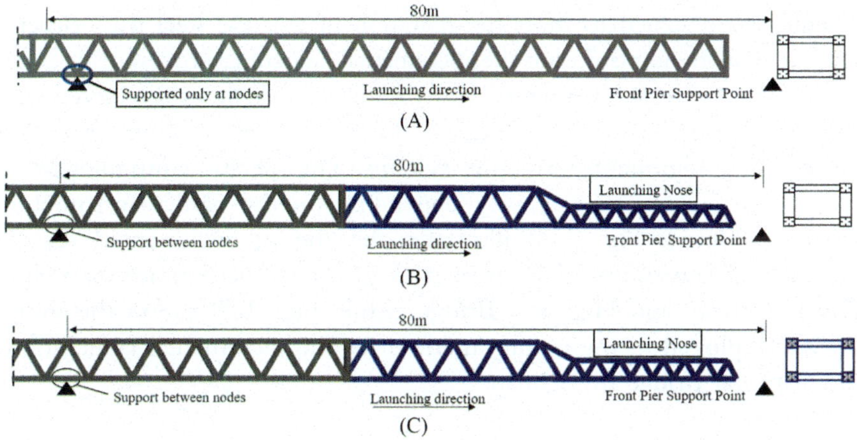

Figure 2.26 Steel truss girder jacking scheme. (A) Incremental launching of steel tubular truss with nodal-point supports. (B) Incremental launching of steel tubular truss with launching nose. (C) Incremental launching of CFST truss with launching nose.

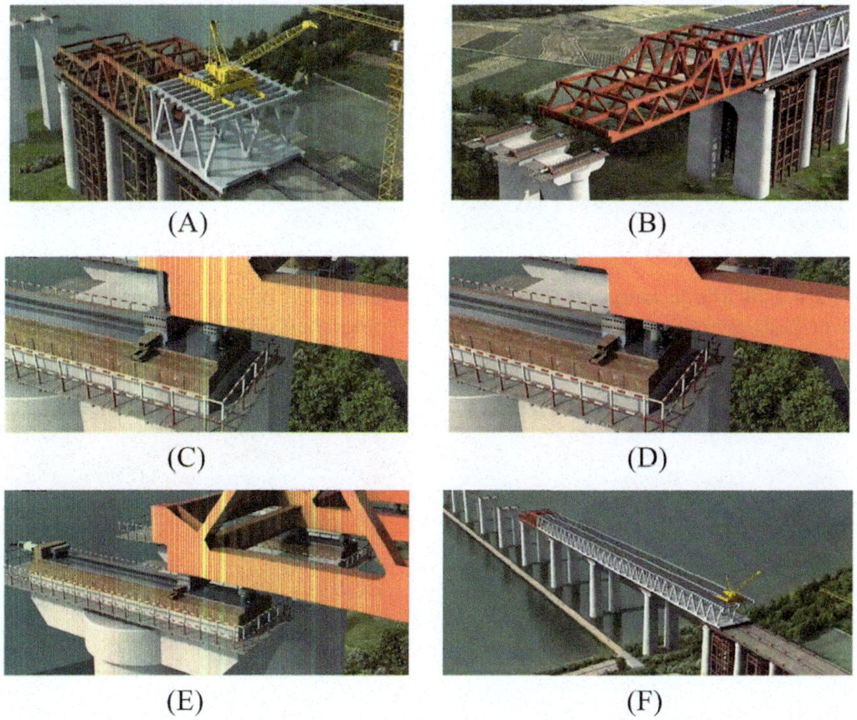

Figure 2.27 Schematic diagram of steel truss beam jacking construction. (A) Launching nose-to-segment docking. (B) Segment advancing. (C) Launching nose pier contact. (D) Jacking shoe at truss joint. (E) Forward movement of sliding shoe by horizontal jack. (F) Next segment launch length.

schemes are designed to meet the requirement that the steel truss girder is supported at the joints during the incremental launching process.

In addition to the launching nose, temporary facilities for incremental launching construction also include sliding devices, jacking or pulling devices, and alignment correction systems. The specific configurations of sliding and jacking devices vary depending on the particular launching scheme adopted. Fig. 2.28 illustrates the sliding device used for the incremental launching of the steel truss girder in the Sanmenxia Yellow River Railway and Highway Bridge, while Fig. 2.29 shows the sliding device employed in the pulling method for the incremental launching of the Ganhaizi Bridge in Sichuan.

2.6.1.2 Segmental erection by lifting

The construction process of steel truss girder erection by lifting generally consists of the following steps: installation of temporary supports on piers

(A) (B)

Figure 2.28 The jacking and sliding device of steel truss beams of Sanmenxia Yellow River highway-railway dual-use bridge. (A) Sliding shoe. (B) Operation of horizontal jack with sliding shoe.

(A) (B)

Figure 2.29 The jacking and sliding device of steel truss beams of Sanmenxia Yellow River highway-railway dual-use bridge. (A) Roller. (B) Launching plate.

(or abutments) and placement of permanent vertical bearings; transportation of the truss segments to the bridge site; hoisting the segments with lifting equipment; lateral adjustment of adjacent segments for alignment, followed by temporary locking; welding of the adjacent segments into an integral unit; lowering the steel truss girder to the design elevation by means of vertical jacks on the temporary supports; removal of temporary supports from the lower chord; installation of permanent longitudinal and transverse restraining devices; and locking of the completed segment prior to lifting the next one.

For standard-span steel truss girder erection, the primary influencing factors are the lifting equipment and the site-specific lifting conditions. For land-based transportation and lifting, when the elevation difference between the bridge deck and the ground is relatively small, mobile cranes are typically employed. Commonly used mobile cranes range from 8-ton to 260-ton units, as well as 300-ton and 500-ton fully hydraulic cranes, providing a wide lifting capacity. However, considerations such as lifting height and the maximum bearing capacity of the subgrade must also be taken into account. Fig. 2.30 illustrates the erection of a steel truss girder segment in the engineering case study of Chapter 8, where only one to two mobile cranes were required for lifting.

For large-scale erection of steel truss girder components over water, lifting operations are typically carried out directly from barges positioned within the waterway. As illustrated in Fig. 2.31, the erection of the Padma Bridge in Bangladesh adopted an integral-span lifting scheme, in which the floating crane Tianyi successfully hoisted and installed a steel truss girder weighing up to 3200 tons onto the bridge piers. Similarly, for the approach

(A) (B)

Figure 2.30 Huangyan Expressway overpass segmented hoisting construction. (A) Side span erection. (B) Main span erection.

Figure 2.31 The entire span of the steel truss of the Padma Bridge in Bangladesh was hoisted.

Figure 2.32 The entire span of the Øresund Bridge approach bridge in Denmark was hoisted.

spans of the Øresund Road and Rail Bridge in Denmark, integral-span erection was employed, as shown in Fig. 2.32. In this project, the large offshore floating crane Svanen was utilized, with segment lengths ranging from 120 to 140 m.

In summary, the segmental erection of steel truss girders can be carried out using various schemes, including the lifting of bare steel tubular trusses, steel tubular concrete trusses, or composite steel trusses integrated with bridge decks.

2.6.1.3 Cantilever erection

Cantilever erection is generally carried out using two methods: single-sided cantilever erection and symmetrical cantilever erection, as illustrated in Figs. 2.33 and 2.34. In the single-sided method, counterweights are placed at the abutment or side span to balance the overturning moment at the cantilever end, enabling incremental extension on one side. In the symmetrical method, erection starts from a pier within the bridge span, and segments are assembled symmetrically in both directions based on the principle of approximate balance.

The construction scheme of the Dongjiang Bridge in Dongguan adopted a combined method of support erection and single-sided cantilever erection. Specifically, the side-span truss was assembled on platform scaffolding, and upon completion of the side span, the truss members of the main span were erected in a cantilever manner from the side-span steel truss. For this bridge, fabrication units, transportation units, and erection units were all organized on a member-by-member basis, including chord members, brace members, and others. During on-site cantilever erection, general-purpose step-type gantry cranes were employed. These cranes are capable of walking step by step during relocation and provide a lifting capacity ranging from 170 to 400 tons. The erection sequence of steel truss members followed the order: lower chords → brace members → upper chords → lateral bracing. The overall construction process of the bridge is illustrated in Fig. 2.35.

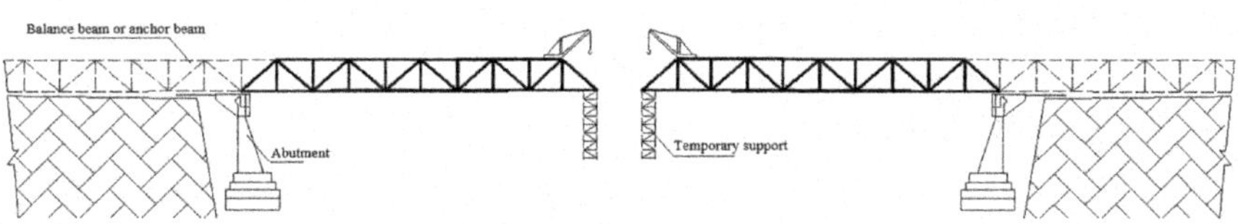

Figure 2.33 Single-sided cantilever assembly.

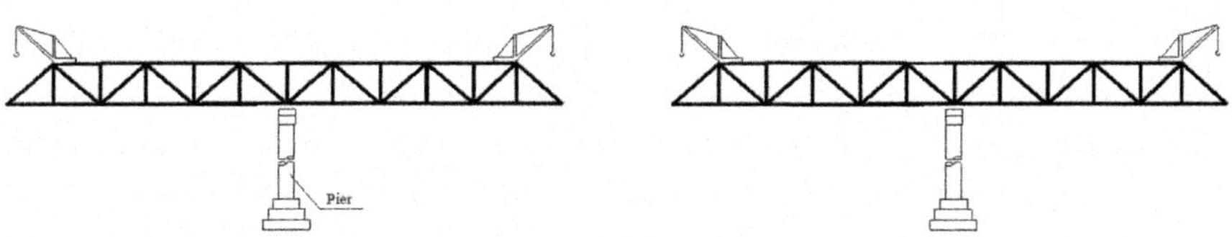

Figure 2.34 Symmetrical cantilever assembly.

Figure 2.35 Cantilever construction of Dongjiang Bridge. (A) Erection gantry crane for the side span. (B) Installation of the bottom chord in the side span. (C) Installation of the brace members and top chord in the side span. (D) Construction by falsework method. (E) Cantilever erection method. (F) Main span closure.

Compared with the former scheme of assembling individual members, the main bridge of the Ulla High-Speed Railway Bridge in Spain adopted a cantilever erection method using spatial segments. The maximum weight of a single erection unit reached 300 tons, as illustrated in Fig. 2.36.

In comparison between the two cantilever erection methods, the segmental cantilever assembly adopted for the Ulla High-Speed Railway Bridge in Spain embodies a more advanced concept of prefabricated construction than the member-by-member cantilever assembly used for the Dongjiang Bridge in Dongguan. This approach provides greater reference value for the cantilever erection of steel truss girder segmental units in prefabricated bridge construction.

2.6.2 Construction of infilled concrete

The construction method in which the steel truss is first erected followed by on-site infilling of concrete into the steel tubes imposes relatively lower demands on transportation and lifting capacity, and may also result in lower stress levels in the truss during launching operations. However, empty steel tube trusses exhibit lower stability during construction, necessitating the installation of temporary supports. In addition, on-site infilling of concrete presents greater construction difficulties and makes it more challenging to ensure quality control. Conversely, the method of infilling the concrete

(A)

(B)

Figure 2.36 Cantilever construction of Ulla Bridge in Spain. (A) Segment unit for cantilever erection. (B) Segment to be lifted.

first to form a concrete-filled steel tube truss prior to erection requires higher-capacity transportation and lifting equipment, but it reduces the volume of on-site construction work, ensures better concrete quality, and improves overall construction efficiency.

For rectangular steel tubes, concrete infilling is preferably carried out using the tremie method, though pumping or staged placement methods may also be adopted. Continuous pouring of concrete is recommended.

2.6.3 Construction of prefabricated deck panels

Prefabricated deck panels are manufactured in precast yards as concrete slab segments, which are subsequently installed on top of the steel truss girders. Finally, concrete is cast into the shear key grooves and the joints between the precast panels, thereby integrating the deck panels with the steel truss girder to form a composite section that resists loads jointly. The length of full-width precast concrete deck panels generally ranges from 2.5 to 6.0 m, with unit weights varying from 15 to 60 tons depending on their dimensions, as illustrated in Fig. 2.37.

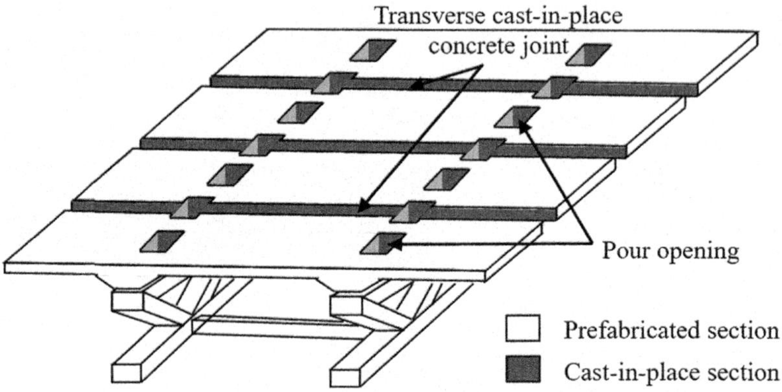

Figure 2.37 Full-width prefabricated units.

After the prefabricated deck panels are hoisted into place, concrete casting can be carried out immediately. Particular attention should be given to the connection between the prefabricated units and the steel truss girder, wherein the connectors are concentrated in designated zones on the concrete slab to form connection grooves. These grooves are typically filled on site with low-shrinkage, microexpansive concrete, ensuring complete compaction and a compressive strength higher than that of the deck panel concrete. In addition, attention must also be paid to the compactness of the contact surface between the full-width prefabricated deck panel and the steel truss girder, excluding the reserved groove areas. Fig. 2.38 illustrates the construction of prefabricated deck panels for Section CB05 of the Hong Kong–Zhuhai–Macao Bridge.

When transportation and lifting capacities are sufficient, the construction method of prefabricating entire steel–concrete truss girder segments can be adopted. In this approach, both the infill concrete of the tubular members and the deck slab are completed at the prefabrication site, after which the steel–concrete composite truss girder segments are directly hoisted into position. Since the major structural work is executed at the prefabrication yard, construction quality is less affected by site conditions. Furthermore, the presence of infill concrete and deck slabs significantly enhances the stability of the composite truss girder segments, thereby reducing the demand for temporary support systems and markedly improving both construction efficiency and the degree of prefabrication of steel–concrete composite truss girder bridges.

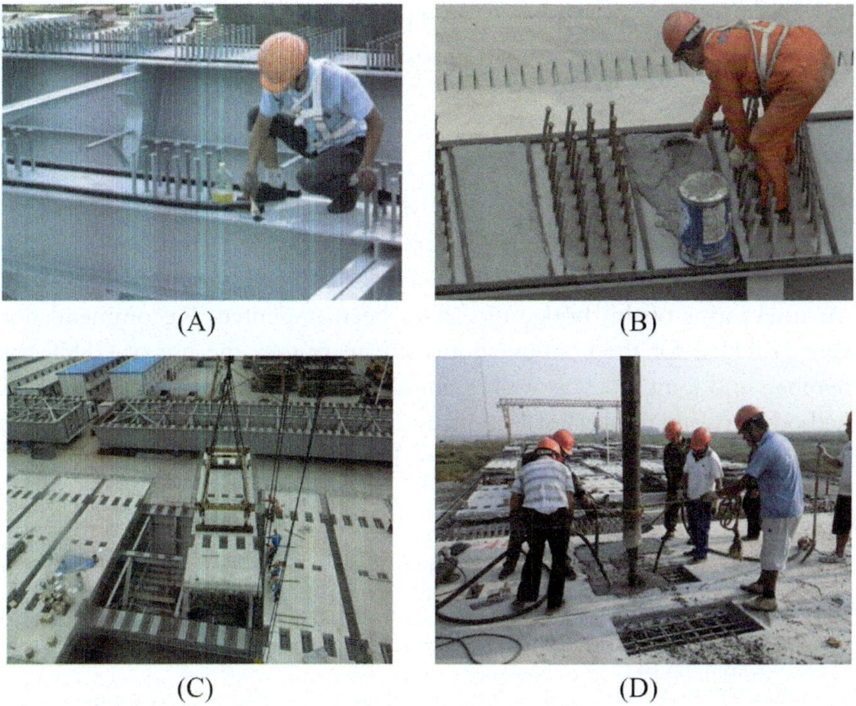

Figure 2.38 Hong Kong-Zhuhai-Macao Bridge CB05 Section prefabricated bridge deck construction. (A) Application of corrosion-resistant rubber strip. (B) Epoxy mortar application. (C) Erection of precast deck slabs. (D) Concreting of shear keys and joints.

Figure 2.39 Installation of steel truss girder erection machine for Wenchuan Keku Bridge.

Fig. 2.39 shows the erection of a full-span steel truss girder of the Wenchuan Keku Bridge on the Wenma Expressway. The truss main girder of this bridge adopts a prestressed steel–concrete truss structure with a span of 30 or 40 m, and a deck slab in composite form with a steel bottom plate.

The bridge has a width of 12.25 m. The erected segment corresponds to a full-span steel–concrete composite truss girder, weighing approximately 220 tons.

2.7 Summary

This chapter has elaborated on the overall design and general structural configuration of rectangular CFST composite truss girder bridges. From the perspectives of main trusses, bracing systems, and deck systems, the applicable structural forms of this bridge type have been presented. Recommendations were provided for the configuration of main trusses, the use of CFST truss members and joint types, as well as simplified bracing systems and connection methods between the deck slab and truss chords. Finally, construction methods for CFST composite truss girders were discussed.

Extended reading

Gao YM, Liu YJ, Zhou XH, et al. High-performance CFST composite truss bridge[J]. China Journal of Highway and Transport 2018;31(12):174–87.

Liu B, Liu YJ, Yang BH, et al. Design of PBL-stiffened steel truss-concrete composite girder bridge with truss made of rectangular steel tubes[J]. Bridge Construction 2019;49(05):97–102.

Liu B, Liu YJ, Zhou XH, et al. Design of mid-span fabricated RCFST composite truss bridge[J]. Journal of Traffic and Transportation Engineering 2017;17(04):20–31.

Liu Y.J., Cheng G., Long G., et al. A steel truss-concrete slab composite beam based on shear connection structure: China, 201310744155. 1[P]. 2014-03-19.

Liu YJ, Gao YM, Zhou XH, et al. Technical and economic analysis in steel-concrete composite girder bridges with small and medium span[J. China Journal of Highway and Transport 2017;30(3):1–13.

Liu Y.J., Liu B., Dai L.X., et al. Fabricated rectangular concrete-filled steel tube composite truss bridge and construction method[P]. Shaanxi： CN106223183A [P]. 2016-12-14.

Liu Y.J., Yuan Z.Y., Cheng G., et al. Continuous rigid frame bridge based on steel truss-concrete slab composite beam: China, 201310745945. 1[P]. 2014-03-26.

Liu YJ, Zhang N, Zhang JG. Review on mechanical behavior research of concrete filled rectangular hollow section tube stiffened with PBL[J]. Journal of Architecture and Civil Engineering 2012;29(4):13–7.

Tian ZJ, Liu YJ, Jiang L, et al. A review on application of composite truss bridges composed of hollow structural section members[J]. Journal of Traffic and Transportation Engineering (English Edition 2019;6(1):94–108.

Zhang GJ, Liu YJ, Hou BB, et al. Parameter analysis of seismic performance of hollow thin-walled steel-box concrete pier[J]. Journal of Architecture and Civil Engineering 2019;36(01):101–11.

Zhou XH, Liu YJ. Steel bridges[M]. Beijing: China Communications Press Co., Ltd; 2020.

Zhou XH, Liu YJ, Jiang L, et al. Review on mechanical behavior research of concrete filled rectangular hollow section tube stiffened with PBL[J]. China Journal of Highway and Transport 2017;30(11):45–62.

CHAPTER 3

Basic mechanical properties of concrete-filled rectangular steel tubular trusses

3.1 Overview

Concrete-filled rectangular steel tubular (CFRST) trusses serve as the primary load-bearing structure of composite CFRST truss-girder bridges. This chapter first analyzes the flexural behavior of concrete-filled steel tubular (CFST) trusses and establishes a calculation method for the flexural capacity of CFST trusses that accounts for joint bearing capacity. On the basis of summarizing the failure modes and bearing capacity calculation methods of rectangular steel tubular (RST) joints, the mechanism by which infilled concrete improves the performance of CFRST joints is analyzed. Finally, a series of truss tests are designed to investigate the effects of chord infilled concrete (i.e., CFRST trusses) on the failure modes and flexural capacity of trusses.

3.2 Mechanical behavior and flexural capacity of concrete-filled steel tubular trusses

3.2.1 Flexural failure modes and deformation characteristics of concrete-filled steel tubular trusses

Based on the differences in the configuration of CFST trusses and the types of loads they bear, their failure modes can be primarily categorized into two classes: flexural failure and shear failure. Flexural failure occurs when the chord members of the truss first reach the material's ultimate strength or undergo stability failure and lose their load-bearing capacity during the loading process. In the case of flexural failure, the upper and lower chords of the truss mainly bear axial forces. At the ultimate state, the material is fully utilized and the member deformation is fully developed, making this a ductile failure.

Shear failure refers to the scenario where the brace members or truss joints fail first before the chord members reach the material's ultimate strength during loading. In this case, the stress level of the chord members is relatively low and

Composite Truss Bridges With Concrete-filled Rectangular Steel Tube Members
DOI: https://doi.org/10.1016/B978-0-443-44011-3.00004-4

the overall deformation of the truss is small, classifying it as a brittle failure. Shear failure can be further divided into brace member failure and joint failure, depending on the variations in the truss brace members, chord members, and joint configurations. Specifically, brace member failure includes stability failure of compressed brace members and strength failure of tensioned brace members; joint failure, on the other hand, can be categorized into failure modes such as punching-shear failure and yield-line failure. Schematic diagrams of the failure modes of CFST trusses are presented in Fig. 3.1, while the schematic diagrams of the overall deformation of the truss under different failure modes are shown in Fig. 3.2.

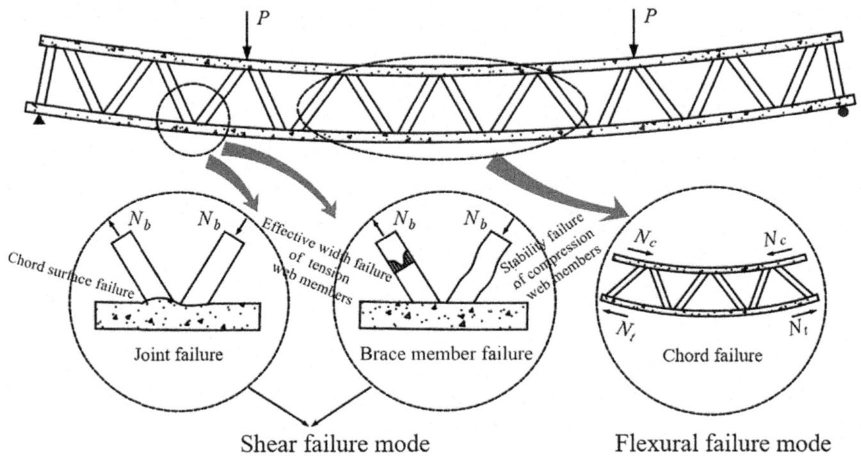

Figure 3.1 Schematic diagram of failure modes for concrete-filled steel tubular trusses.

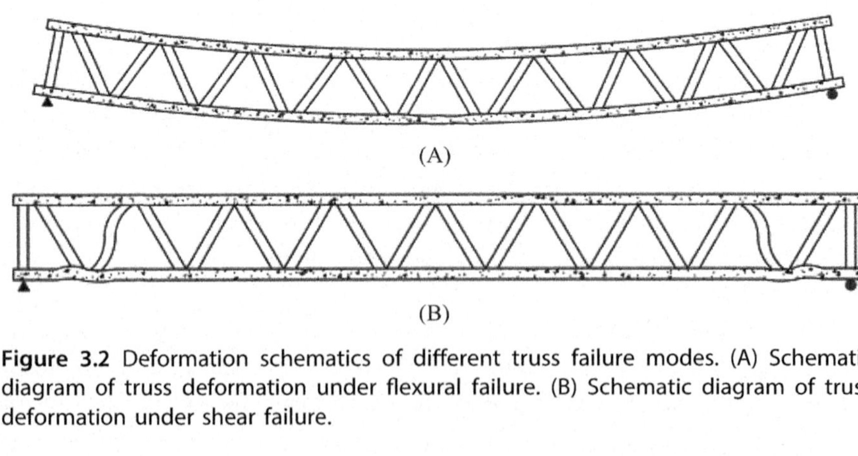

Figure 3.2 Deformation schematics of different truss failure modes. (A) Schematic diagram of truss deformation under flexural failure. (B) Schematic diagram of truss deformation under shear failure.

For steel tubular truss joints, the radial stiffness of the steel tubes is relatively low. Under the action of axial loads from brace members, noticeable deformation occurs on the surface of chord members at the joint location, which is referred to as joint deformation. When concrete is infilled into the truss chord members, the joint deformation caused by compressed brace members is significantly reduced; however, the joint deformation induced by tensioned brace members remains nonnegligible. Therefore, when calculating the flexural stiffness of CFST trusses under vertical loads, the influence of joint deformation should be taken into account.

For solid-brace concrete main girders with a small height-span ratio, the shear deformation generated under vertical loads is minimal, and their flexural stiffness can be calculated using the Bernoulli-Euler beam theory. For CFST trusses, however, joint deformation is nonnegligible. Consequently, even when the height-span ratio is small, the impact of joint deformation must be considered when calculating truss deformation.

Based on the above analysis, the deformation of CFST trusses under vertical loads can be divided into three components: chord member deformation, brace member deformation, and joint deformation, as illustrated in Fig. 3.3. Since trusses resist bending moments through the couple formed by the tension and compression forces in the upper and lower chord members, chord member deformation is defined as the bending deformation of the truss. The sum of brace member deformation (caused by axial forces in brace members) and joint deformation is defined as the shear deformation of the truss.

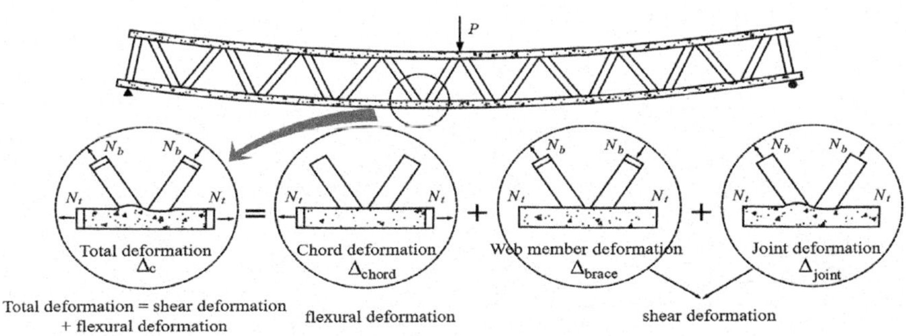

Figure 3.3 Schematic diagram for calculating the flexural stiffness of concrete-filled steel tubular trusses.

When the structural design of a steel tubular truss is reasonably optimized, the internal force distribution of its members under load can be calculated using the pin-jointed truss theory. Subsequently, the total deformation of the truss caused by the deformation of chord and brace members can be derived using structural mechanics principles. Joint deformation also exhibits a linear positive correlation with the axial force of brace members; thus the joint deformation induced by the axial load of brace members can be treated as part of brace member deformation, and the joint stiffness can be equivalent to a spring connected in series with the brace members. In this way, the deformation of the truss under vertical loads can be calculated, provided that the joint deformation under the axial load of brace members is accurately computed.

3.2.2 Flexural capacity of concrete-filled steel tubular trusses

In this section, a formula for calculating the flexural capacity of CFST trusses is proposed based on the calculation method for pin-jointed trusses. When a concentrated load P acts alone on a joint of the CFST truss, the simplified calculation model is illustrated in Fig. 3.4. Here, a denotes the panel length, N represents the number of panels in the composite truss girder, n is the number of panels between the load application point and the nearest support ($n \leq N/2$), H stands for the calculated height of the truss, and θ is the angle between the brace members and chord members. According to structural mechanics principles, the maximum internal forces N_c (for compressed chord members), N_t (for tensioned chord members), and N_b (for brace members) under joint loads can be calculated using Eqs. (3.1) to (3.3),

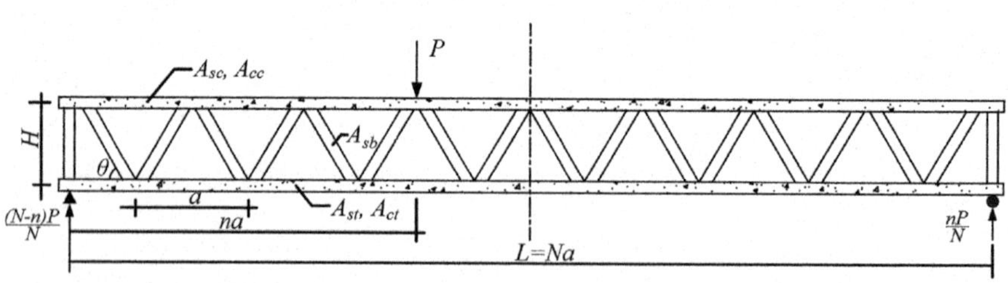

Figure 3.4 Simplified calculation model under joint load action.

respectively. The section resistances R_c (for compressed chord members), R_t (for tensioned chord members), and R_b (for brace members) of the members can be computed via Eqs. (3.4) to (3.6), respectively.

$$N_c = \frac{Pna}{H}\left(\frac{N - n - 0.5}{N}\right) \tag{3.1}$$

$$N_t = \frac{Pna}{H}\left(\frac{N - n}{N}\right) \tag{3.2}$$

$$N_b = \frac{(N - n)P}{N \sin \theta} \tag{3.3}$$

$$R_c = \chi(A_{sc}f_y + A_{cc}f_c) \tag{3.4}$$

$$R_t = A_{st}f_y \tag{3.5}$$

$$R_b = \eta_{\text{eff}} A_{sb}f_y \tag{3.6}$$

where A_{sc} and A_{st} are the cross-sectional areas of the compressed chord member and tensioned chord member, respectively; A_{sb} is the cross-sectional area of the brace member; A_{cc} is the cross-sectional area of concrete in the compressed chord member; χ is the stability coefficient of axially compressed members, whose value can be determined in accordance with relevant codes; f_y is the yield strength of steel; and η_{eff} is the effective reduction coefficient of brace member resistance considering brace member failure and joint failure. It shall be calculated using the coefficient corresponding to the maximum reduction degree, as shown in Eq. (3.7):

$$\eta_{\text{eff}} = \min\left\{\eta_1, \eta_2, \eta_{jc}\right\} \tag{3.7}$$

where η_1 is the stability reduction coefficient for the failure of compressed brace members; η_2 is the reduction coefficient for the failure of tensioned brace members; and η_{jc} is the joint bearing capacity efficiency coefficient. Since the joint bearing capacity is also calculated using the axial force of brace members as the expression, the ratio of the brace member axial force corresponding to the joint's ultimate state to the brace member's axial tension resistance ($A_{w}f_y$) can also be equivalent to a reduction in the effective cross-sectional area of the brace member. In this way, joint failure can be incorporated into the category of brace member failure.

Under load action, the efficiency coefficient S_i of each member in the CFST truss can be defined as the ratio of the internal force N_i of the member under load to its resistance R_i, that is,

$$S_i = \frac{N_i}{R_i} \tag{3.8}$$

where the subscript i can take the values of c, t, and b.

According to the definition of the efficiency coefficient, it represents the ratio of the internal force of a member under load to its resistance. The efficiency coefficient S_i ranges between 0 and 1, with a larger value indicating that the member is closer to failure. Substituting Eqs. (3.1) to (3.6) into Eq. (3.8) respectively yields:

$$S_c = \frac{Pna}{\chi(A_{sc}f_y + A_{cc}f_c)H}\left(\frac{N - n - 0.5}{N}\right) \tag{3.9}$$

$$S_t = \frac{Pna}{A_{st}f_y H}\left(\frac{N - n}{N}\right) \tag{3.10}$$

$$S_b = \frac{(N - n)P}{\eta_{\text{eff}} A_{sb}f_y N \sin\theta} \tag{3.11}$$

When the tensioned chord member of the CFST truss first reaches its ultimate state, its efficiency coefficient should be the first to reach 1. In this case, the efficiency coefficient of the tensioned chord member under load should be higher than those of the compressed chord member and brace members, that is, $S_t > S_c$ and $S_t > S_b$. Rearranging these inequalities yields Eqs. (3.12) and (3.13). These equations provide the ratio of member cross-sectional dimensions when the truss undergoes the corresponding failure mode. At this point, the flexural capacity M_u of the truss can be determined based on the ultimate bearing capacity of the tensioned chord member and the truss height, as shown in Eq. (3.14). Similarly, the relational expressions and corresponding bearing capacities for failures occurring in other truss members can be derived, which are summarized in Table 3.1. When the truss is subjected to complex loads, these loads can be simplified into a series of joint loads, and subsequent calculations can be performed in accordance with Table 3.1.

Table 3.1 Determination of truss loading failure modes and corresponding truss flexural bearing capacity.

Member cross-sectional dimension ratio	Failure mode	Flexural capacity
$\begin{cases} \frac{A_{sc}}{A_{st}} > \left(\frac{N-n-0.5}{N-n}\right)\frac{A_{sc}f_y}{\chi(A_{cc}f_c + A_{sc}f_y)} \\ \frac{\eta_{\text{eff}}A_{sb}}{A_{st}} > \frac{1}{\lambda \sin\theta} \end{cases}$	Failure of tension chord	$M_u = A_{st}f_y H$
$\begin{cases} \frac{\eta_{\text{eff}}A_{sb}}{A_{st}} < \frac{1}{\lambda \sin\theta} \\ \frac{\eta_{\text{eff}}A_{sb}f_y}{\chi(A_{cc}f_c + A_{sc}f_y)} < \frac{1}{\lambda \sin\theta}\frac{(N-n)}{(N-n-0.5)} \end{cases}$	Failure of brace member	$M_u = \eta_{\text{eff}}A_{sb}f_y na \sin\theta$
$\begin{cases} \frac{A_{sc}f_y}{\chi(A_{cc}f_c + A_{sc}f_y)} < \left(\frac{N-n-0.5}{N-n}\right) \\ \frac{\eta_{\text{eff}}A_{sb}f_y}{A_{cc}f_c + A_{sc}f_y} > \frac{1}{\lambda \sin\theta}\frac{(N-n)}{(N-n-0.5)} \end{cases}$	Failure of compression chord	$M_u = \chi(A_{sc}f_y + A_{cc}f_c)H\frac{(N-n)}{(N-n-0.5)}$

$$\frac{A_{sc}}{A_{st}} > \left(\frac{N-n-0.5}{N-n}\right)\frac{A_{sc}f_y}{\chi\left(A_{cc}f_c + A_{sc}f_y\right)} \tag{3.12}$$

$$\frac{\eta_{\text{eff}}A_{sb}}{A_{st}} > \frac{1}{\lambda \sin\theta} \tag{3.13}$$

where λ is the shear-span ratio, $\lambda = na/H$.

$$M_u = A_{st}f_y H \tag{3.14}$$

It can be derived from Fig. 3.1 that the stress state of truss joints is closely related to the axial force of brace members: when the joint bearing capacity is relatively low, the joints fail prior to the brace members; conversely, the brace members fail first. According to the definition in Eq. (3.7), the joint efficiency coefficient is defined as the ratio of the joint bearing capacity to the brace member bearing capacity at the onset of brace member strength failure, as expressed in Eq. (3.15):

$$\eta_{jc} = \frac{R_J}{A_{sb}f_y} \tag{3.15}$$

where η_{jc} is the joint bearing capacity efficiency coefficient, R_J is the joint bearing capacity under the action of brace member axial force, and A_{sb} is the cross-sectional area of the brace member.

It can be derived from the formula that a higher joint bearing capacity corresponds to a larger efficiency coefficient. This implies that when truss joints fail, the stress level of the brace members is also close to the ultimate state—indicating high material utilization and a reasonably designed truss joint. Conversely, a smaller joint bearing capacity efficiency coefficient makes trusses with such joints prone to joint failure; in this case, the stress level of other truss members remains relatively low, and the joints require further optimization. As a dimensionless parameter, the joint efficiency coefficient enables horizontal comparison of different types of truss joints and can thus serve as an evaluation index for the optimal design of trusses.

In essence, the joint bearing capacity efficiency coefficient quantifies the mechanical relationship between the joints and brace members of CFST trusses. It not only allows horizontal comparison of the mechanical performance of different joint types at the truss level but also serves as a universal evaluation index for assessing joint mechanical performance and optimizing design schemes.

As shown in Table 3.1, when the chord members of a CFST truss fail, the truss's flexural capacity is determined by the load-bearing capacity of the members. The calculation theories for the axial tension or compression capacity of hollow steel tubular members and CFST truss members are well established; the effective length and bearing capacity of truss members can be directly calculated with reference to current standards, such as Code for Design of Highway Steel Bridge Structures and Technical Specification for CFRST Structures. When a CFST truss experiences brace member or joint failure, the value of η_{eff} must first be calculated before substituting it into the formula for further computations. Combined with Eq. (3.15), to accurately determine the flexural capacity of a CFST truss under joint failure, the joint bearing capacity R_J must first be calculated.

For CFST trusses applied in composite truss-girder bridges, concrete can be infilled into both tensioned and compressed chord members or only into compressed chord members to form partially CFRST trusses. Correspondingly, the joint types include two categories: RST joints and CFRST joints. The basic mechanical performance of these joint types is investigated in Section 3.3 of this book.

3.3 Basic mechanical properties of shaped steel tubes and concrete-filled steel tubular truss joints

3.3.1 Classification of truss joint types

The main geometric forms of truss joints are K-type and N-type joints. The N-type joint is a special case of the K-type joint, where one brace member is perpendicular to the chord member; thus the K-type joint category includes N-type joints. When there is only one brace member, the joint becomes a Y-type or T-type joint. The T-type joint is a special case of the Y-type joint, with the brace member perpendicular to the chord member, so the Y-type joint category also encompasses T-type joints. When the gap between the brace members of a K-type joint is large, the K-type joint is equivalent to a pair of Y-type joints. X-type joints are located at supports or concentrated load application points and also appear in crossed brace member systems and bracing systems. T-type, Y-type, and X-type joints are the most fundamental joint forms in trusses and serve as the basis for studying other joint configurations.

Currently, most relevant domestic and international codes and specifications—including China's Code for Design of Steel Structures (GB 50017-2017)—classify joint types based on the aforementioned geometric forms. However, some classifications are based on the internal force equilibrium mode of the members in the joint. For example, the American Institute of Steel Construction (AISC) Specification for Hollow Structural Sections proposes that joint types should be classified according to the equilibrium mode of forces in the members connected to the joint.

A joint is classified as a T-type or Y-type joint when the vertical components of the brace member axial forces (perpendicular to the chord member) are fully balanced by the shear force in the chord member's cross section, as shown in Fig. 3.5. A joint is classified as an X-type joint when the axial force of a brace member is balanced by the opposite brace member

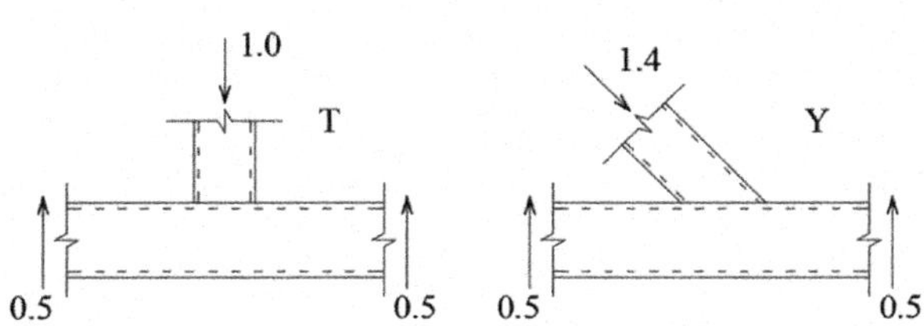

Figure 3.5 T-type and Y-type joints.

via the chord member, as illustrated in Fig. 3.6. A joint is classified as a K-type or N-type joint when the axial forces of brace members on the same side of the chord member can self-balance in the direction perpendicular to the chord length, as depicted in Fig. 3.7. For the joint with a K-type geometric shape shown in Fig. 3.8, according to the principle of classifying joint types based on the equilibrium mode of the vertical components of brace member axial forces, 50% of the axial force of the inclined brace member is balanced by the vertical brace member, and the remaining 50% is balanced by the shear force in the chord member's cross section. Therefore this joint is classified as a partial K-type and partial Y-type joint.

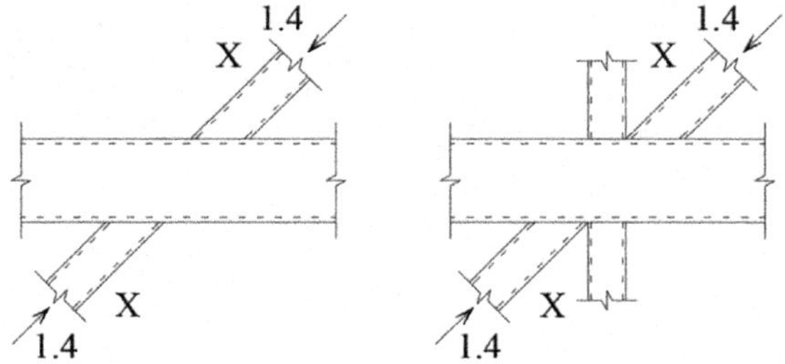

Figure 3.6 Y-type joints.

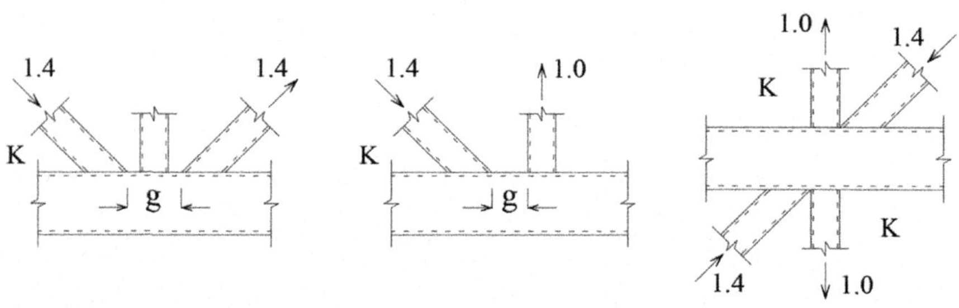

Figure 3.7 K-type joints.

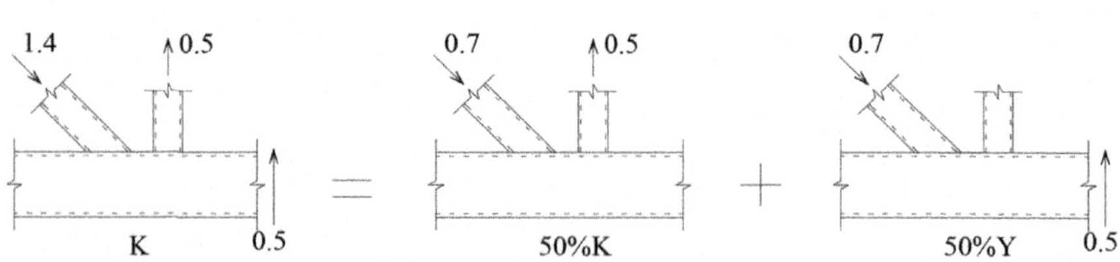

Figure 3.8 Partially K-type and Y-type joints.

In the examples shown in Figs. 3.5 to 3.8, the angle between the inclined brace members and chord members is 45 degrees, and the internal force values in the figure are relative values. The aforementioned joint classification method in the AISC Specification for Hollow Structural Sections reflects the essence of the joint's force transmission mechanism but is less convenient for engineering applications.

Considering common practices in engineering applications and to align with China's Code for Design of Steel Structures (GB 50017-2017), the joint types in this paper are still classified based on their geometric forms. Meanwhile, according to the load types of brace members, the joint regions where axially compressed, axially tensioned, and flexural brace members connect to chord members are respectively referred to as compressed joints, tensioned joints, and flexural joints.

3.3.2 Joint configuration

Both the chord members and brace members of RST joints consist of rectangular cross sections. The specifications and definitions of the joint symbols are illustrated in Fig. 3.9. Meanwhile, domestic and international scholars have conducted extensive experimental studies and finite element analyses on RST joints, leading to the development of corresponding design codes for tubular joints and the establishment of requirements for the applicable range of joint parameters. Table 3.2 presents the structural requirements for joints specified in the Code for Design of Steel Structures (GB 50017-2017).

Through extensive experimental research and theoretical analysis on RST joints conducted both domestically and internationally, the following structural requirements for the geometric dimensions of RST joints have been proposed: (1) The angle θ_i between the axes of the branch member

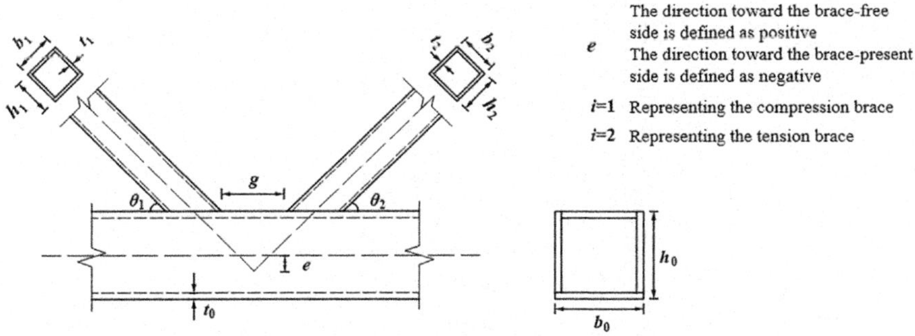

Figure 3.9 Joint notation convention.

Table 3.2 Applicable range of geometric parameters for rectangular steel tubular joints.

Joint type	b_i/b_0&h_i/b_0	b_i/t_i&h_i/t_i		h_i/b_i	b_0/t_0 h_0/t_0	g
		Compression	Tension			
T-type joint, Y-type joint, X-type joint	≥0.25	≤37ε_{ki} and ≤35	≤35	0.5–2	≤35	T-type joint, Y-type joint, X-type joint
K-type and N-type gap joints	≥0.1+0.01 b_0/t_0 β≥0.35					K-type and N-type gap joints

Notes: (1) $\varepsilon_{ki} = (235/f_y)^{0.5}$. (2) Both the branch members and chord members have rectangular cross sections; (3) When $g/b_0 > 1.5(1-\beta)$, the joint shall be calculated as a T-type or Y-type joint.

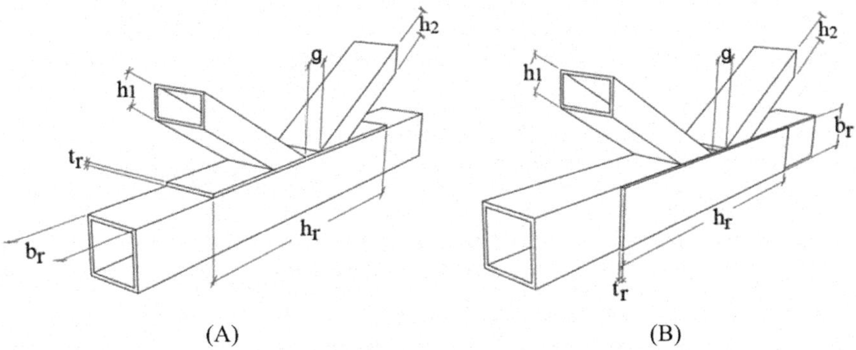

Figure 3.10 Strengthening of rectangular steel tubular joints using stiffener plates. (A) Flange plate strengthening. (B) Sidewall strengthening of the chord.

and chord member shall be $\theta_i \geq 30^0$. (2) The width-thickness ratio of the chord member shall satisfy $b_0/t_0 \leq 35$. (3) The width-thickness ratio of the branch member shall satisfy $b_i/t_i \leq 35$, and its height-thickness ratio shall satisfy $h_i/t_i \leq 35$. (4) The height-width ratio of all members shall be within the range of 0.5 to 2.0, that is, $0.5 \leq h_i/b_i \leq 2$.

For T-type, Y-type, and X-type joints, the ratios of the branch member's width or height to the chord member's width shall satisfy the following constraints: $b_i/b_0 \geq 0.25$ and $h_i/b_0 \geq 0.25$.

For K-type gapped joints, the following requirements shall be satisfied: (1) The ratios of the branch member's width or height to the chord member's width, that is, b_i/b_0 & $h_i/b_0 \geq 0.1 + 0.01b_0/t_0$. (2) The ratio of the average value of the two branch members' width and height to the chord member's width (denoted as β) shall be ≥ 0.35. (3) The gap g between the two branch members shall meet: $0.5(1 - \beta) \leq \frac{g}{b_0} \leq 1.5(1 - \beta)$ and $g \geq t_1 + t_2$. (4) The eccentricity e of the axes of the intersecting members shall satisfy: $0.55 \leq e/h_0 \leq 0.25$.

Typically, the cross-sectional dimensions of truss members are selected based on the most unfavorable internal forces. However, due to the relatively high flexibility of the chord member walls in RST trusses, the joint bearing capacity may fail to meet the requirements. Therefore it is necessary to strengthen the joints. Fig. 3.10 presents a method for joint strengthening using stiffening plates, where the dimensions of the stiffening plates shall satisfy the following requirements:

(1) Width of the stiffening plate: $b_r \geq h_0 - 2t_0$.
(2) Length of the stiffening plate: $h_r \geq 1.5(h_1/\sin\theta_1 + g + h_2/\sin\theta_2)$, for K-type joints; $h_r \geq 1.5h_1/\sin\theta_1$ (for top plate stiffeners of T-type,

Y-type, and X-type joints); $h_r \geq 1.5h_1/\sin\theta_1$ or $h_r = h_1/\sin\theta_1 + \sqrt{b_r(b_r - b_1)}$ (for sidewall stiffeners of T-type, Y-type, and X-type joints).

(3) When the thickness of the stiffening plate (t_r) meets $t_r \geq 2t_i$ for K-type joints, or $t_r \geq 4t_1 - t_0$ for T-type, Y-type, and X-type joints, it is not necessary to check the joint bearing capacity.

Strengthening joints with stiffening plates is relatively labor intensive and time consuming, which reduces fatigue performance and impairs the aesthetic appearance of the structure. By infilling concrete into the chord members to form CFRST truss joints, the structural appearance can be preserved while still achieving the goal of joint strengthening; additionally, this method can enhance the bearing capacity of the chord members. To investigate the effect of infilled concrete on the bearing capacity of RST joints, it is first necessary to understand the failure modes of RST joints and their corresponding bearing capacity calculation methods.

3.3.3 Failure modes and bearing capacity calculation of rectangular steel tubular joints

3.3.3.1 Rectangular steel tubular joints with axial force in branch members

Different joint types, load conditions, and joint geometric parameters may lead to distinct failure modes. Based on the experimental results of RST joints with axial force in branch members (conducted both domestically and internationally), the potential failure modes of such joints have been summarized, as illustrated in Fig. 3.11:

Mode A: Plastic failure of the chord member surface (one branch member indents the chord surface while the other pulls it outward)

Mode B: Punching-shear failure of the chord surface around the branch member (under tension or compression in the branch member)

Mode C: Fracture of the tensioned branch member or its weld

Mode D: Local buckling of the steel tube in the compressed branch member

Mode E: Shear failure of the chord member at the gap

Mode F: Crushing or local buckling of the chord sidewall under the action of the compressed branch member

Mode G: Local buckling of the chord top plate on the opposite side of the tensioned branch member

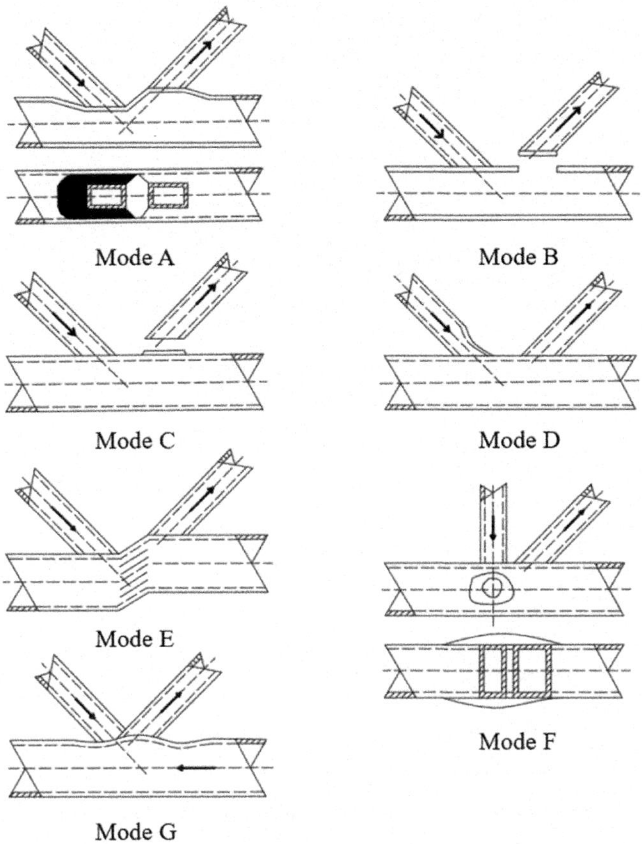

Figure 3.11 Failure modes of rectangular hollow section joints.

The failure mode of an actual joint may be a combination of the above modes. However, for ease of application, a single bearing capacity calculation typically corresponds to one failure mode of the joint.

Mode A typically occurs in joints where the width ratio of branch-to-chord members is relatively small. The Comité International pour le Développement et l'Étude de la Construction Tubulaire (CIDECT) Design Guide specifies that its bearing capacity can be calculated using the yield-line model, as illustrated in Fig. 3.12:

T-type, Y-type, and X-type joints (where $\beta \leq 0.85$):

$$N_{\mathrm{iu}} = \frac{f_{\mathrm{y0}} t_0^2}{(1-\beta)\sin\theta_{\mathrm{i}}}\left[\frac{2h_{\mathrm{i}}}{b_0 \sin\theta_{\mathrm{i}}} + 4(1-\beta)^{0.5}\right] f(n) \tag{3.16}$$

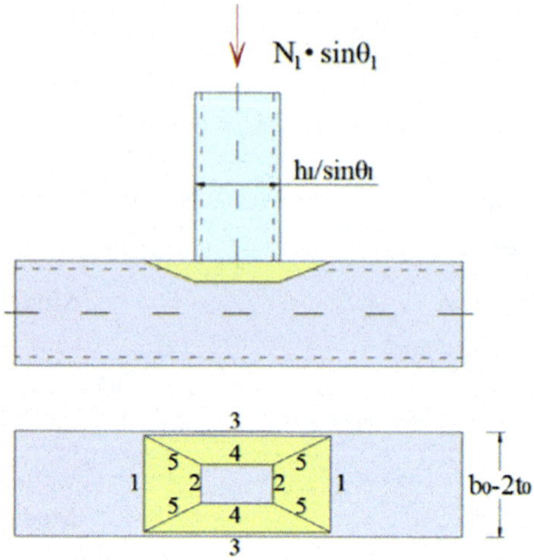

Figure 3.12 Yield-line model.

K-type joints:

$$N_{iu} = 8.9\frac{f_{y0}\,t_0^2}{\sin\theta_i}\left(\frac{b_1 + b_2 + h_1 + h_2}{4b_0}\right)\gamma^{0.5}f\,(n) \tag{3.17}$$

In Eqs. (3.16) and (3.17), $f\,(n)$ denotes the influence coefficient of the chord member's internal forces on the joint bearing capacity, where $n = \frac{N_0}{A_0 f_{yo}} + \frac{M_0}{W_0 f_{y0}}$, for $n \geq 0$ (chord member under compression), and $f\,(n) = 1.3 + \frac{0.4}{\beta}n$, with the constraint that $f\,(n) \leq 1.0$. Here, N_0 and M_0 represent the axial force and bending moment of the chord member, respectively.

When $0.85 \leq \beta \leq (1 - 1/\gamma)$, the joint may undergo Failure Mode B. The corresponding joint bearing capacity can be calculated using the punching-shear model (as shown in Fig. 3.13).

T-type, Y-type, and X-type joints:

$$N_{iu} = \frac{f_{y0}\,t_0}{\sqrt{3}\,\sin\theta_i}\left(\frac{2h_i}{\sin\theta_i} + 2b_{ep}\right) \tag{3.18}$$

K-type joints:

$$N_{iu} = \frac{f_{y0}\,t_0}{\sqrt{3}\,\sin\theta_i}\left(\frac{2h_i}{\sin\theta_i} + b_i + b_{ep}\right) \tag{3.19}$$

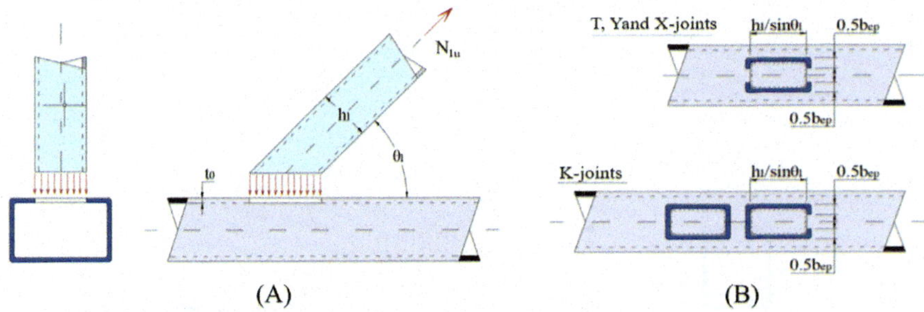

Figure 3.13 Punching-shear model. (A) Punching-shear model. (B) Effective perimeter length for punching shear in the chord wall.

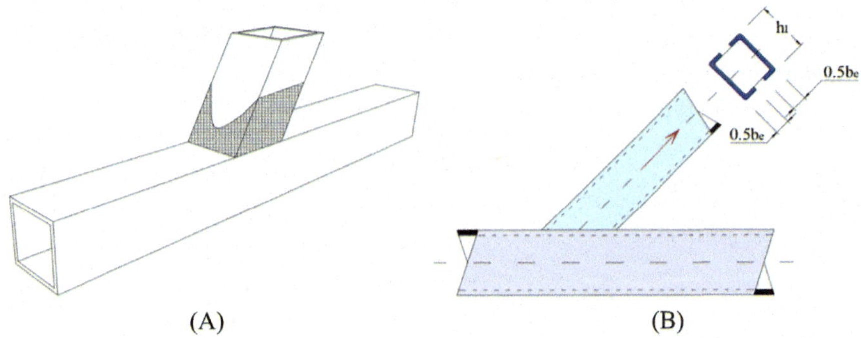

Figure 3.14 Effective width failure. (A) Stress distribution at the brace end. (B) Effective width.

In Eqs. (3.18) and (3.19), the content inside the parentheses represents the effective side length of the punching-shear edge on the chord pipe wall, as illustrated in Fig. 3.13B. The effective width of the punching-shear edge on the chord pipe wall is given by $b_{ep} = \frac{10}{b_0 / t_0} b_i \le b_i$.

Both Failure Mode C and Mode D are branch pipe failures, which are related to the stress distribution at the end of the branch pipe and classified as member failures. When the width of the branch pipe is close to that of the chord pipe, the stress on the two sidewalls of the branch pipe (parallel to the sidewalls of the chord pipe) at the joint is relatively high, while the stress distribution on the other two sidewalls is uneven (with lower stress in the middle), as illustrated in Fig. 3.14. Consequently, the sidewalls with high stress are more prone to yielding. In the calculation of bearing capacity, this phenomenon is treated as a

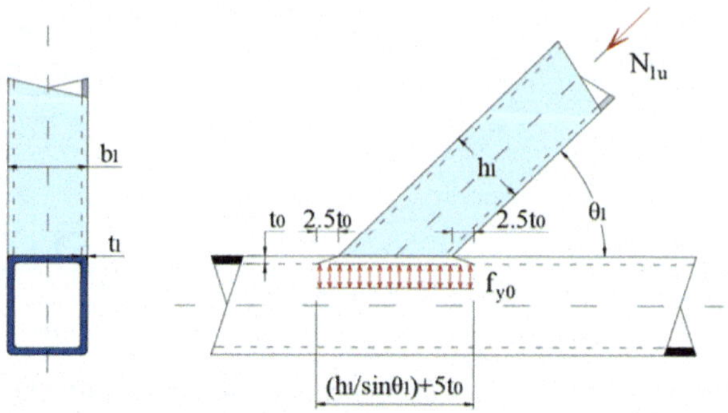

Figure 3.15 Width of the equivalent column.

specific failure mode, namely, "effective width failure," and the corresponding bearing capacity calculation formula is as follows:

T-type, Y-type, and X-type joints ($\beta > 0.85$):

$$N_{iu} = f_{yi}\, t_i\,(2h_i - 4t_i + 2b_e) \tag{3.20}$$

K-type joints:

$$N_{iu} = f_{yi}\, t_i\,(2h_i - 4t_i + b_i + b_e) \tag{3.21}$$

In Eqs. (3.20) and (3.21), b_e denotes the effective width of the branch pipe flange. Its calculation formula is given by: $b_e = \frac{10}{b_0 / t_0} \frac{f_{y0}\, t_0}{f_{yi}\, t_i} b_i$, with the constraint that $b_e \leq b_i$.

Failure Mode E refers to the shear failure of the entire cross section of the chord pipe in K-type joints, which is categorized as a member-level failure.

For T-type, Y-type, and X-type joints, Failure Mode F may occur when the branch member is in compression and its width is close to or equal to that of the chord member. The joint's bearing capacity can be calculated using the equivalent column model:

T-type, Y-type, and X-type joints (where β=1.0):

$$N_{1u} = \frac{f_{k0}\, t_0}{\sin\theta_1}\left(\frac{2h_1}{\sin\theta_1} + 10t_0\right) \tag{3.22}$$

In Eq. (3.22), the content inside the parentheses represents the effective width of the equivalent column, as illustrated in Fig. 3.15. Here, f_{k0} denotes the buckling stress: for T-type and Y-type joints, $f_{k0}=\phi f_{y0}$, and for X-type joints, $f_{k0}=0.8\sin\theta_1(\phi f_{y0})$. Φ is the equivalent column buckling coefficient, which is determined based on the column slenderness ratio $\lambda = 3.46\left(\frac{h_0}{t_0} - 2\right)\left(\frac{1}{\sin\theta_1}\right)^{0.5}$.

Since the in-plane stiffness and strength of the chord wall steel plate are far greater than its out-of-plane stiffness and strength, Failure Mode G may occur under the following two scenarios: (1) When the angle between the branch and chord members is very small, the horizontal component of the branch member force is much larger than the vertical component, and the connected chord wall is mainly subjected to in-plane loads. (2) For K-type lapped joints, the vertical components of the branch member forces are basically self-balancing, so the out-of-plane load borne by the chord wall steel plate is very small. In other words, the RST joints that meet the construction requirements will not experience Failure Mode G.

3.3.3.2 Rectangular steel tubular joints with flexure in branch members

Due to the presence of joint rigidity and joint eccentricity, truss members are also affected by secondary bending moments; for truss joints, this manifests as flexure in the branch members. In the case of vierendeel trusses, the branch members also sustain significant bending moments. Fig. 3.16 illustrates the primary failure modes of RST joints under bending loads. Since bending is a combination of tension and compression, the calculation formulas for the load-carrying capacity of RST joints under axial branch loading can be referenced. For joints with $\beta \leq 0.85$, the yield-

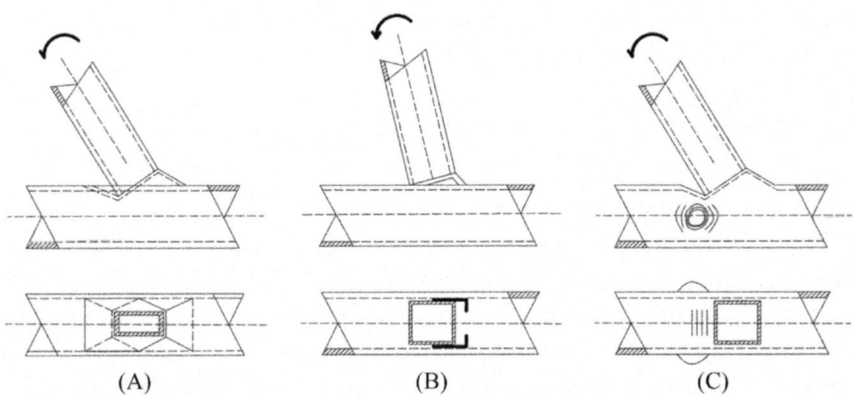

Figure 3.16 Failure modes of rectangular steel tubular joints under brace bending. (A) Yield-line failure mode of the chord face. (B) Effective-width failure mode of the branch member. (C) Chord sidewall buckling failure mode.

line failure mode (as shown in Fig. 3.16A) may occur, and the corresponding flexural capacity M_u is given by:

$$M_u = f_{y0}\, t_0^2 h_1 \left[\frac{1}{2h_1/b_0} + \frac{2}{\sqrt{1-\beta}} + \frac{h_1/b_0}{(1-\beta)} \right] \tag{3.23}$$

For Mode (b), the flexural capacity of the branch member is reduced using the effective width method, and the joint flexural capacity is taken as equal to the reduced flexural capacity of the branch member:

$$M_u = f_{y1}\left[Z_1 - (1 - b_e/b_1)\, b_1 t_1 (h_1 - t_1)\right] \tag{3.24}$$

In Eq. (3.24), Z_1 denotes the plastic section modulus of the branch member, and b_e represents the effective width of the branch member's flange plate. The effective width is calculated as: $b_e = \frac{10}{b_0/t_0}\frac{f_{y0}\,t_0}{f_{yi}\,t_i} b_t \le b_i$.

Mode (c) applies to joints with β=1.0. The bending moment is converted into a couple of forces with a lever arm equal to $h_1/2+2.5t_0$, and the flexural capacity depends on the buckling load of the chord member's sidewall:

$$M_u = 0.5 f_{k0}\, t_0 (h_1 + 5t_0)^2 \tag{3.25}$$

When the joint parameters meet the construction requirements, the buckling stress f_{k0} in Eq. (3.25) is determined as follows: $f_{k0}=f_{y0}$ for T-type and Y-type joints and $f_{k0}=0.8f_{y0}$ for X-type joints.

3.3.4 Combined punching-shear and yield-line failure model for rectangular tubular joints

Section 3.3.3 of this chapter introduces the joint bearing capacity specified in the CIDECT design guide. For RST joints: when $\beta \le 0.85$, the joint bearing capacity is calculated using the yield-line model. When $0.85 \le \beta \le 1-1/\gamma$, it is calculated using the punching-shear model. Taking RST T-type joints as an example, this section compares the joint bearing capacities under the two failure modes, with the results shown in Fig. 3.17. Here P_Y and P_Q respectively correspond to the joint bearing capacities under yield-line failure and punching-shear failure modes, which align with the joint bearing capacity N_{ui} defined in Section 3.3.3; P_{CIDECT} denotes the calculated value of joint bearing capacity provided in Section 3.3.3.

As shown in Fig. 3.17: when the β-value is relatively small, the yield-line capacity is lower than the punching-shear capacity, making the yield-line mechanism the governing factor for joint capacity. When the β-value is

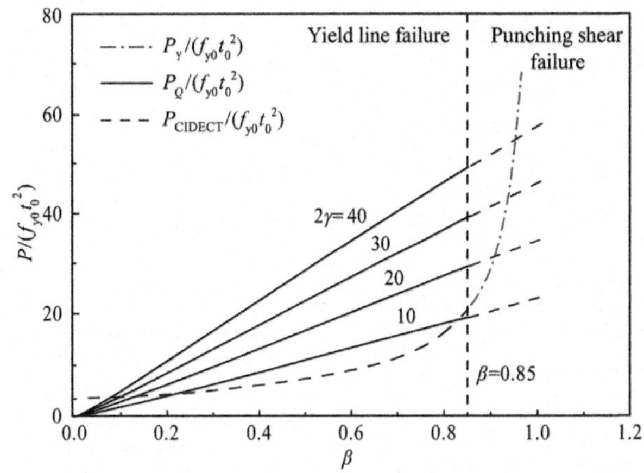

Figure 3.17 Chord failure capacity of rectangular steel tubular joints according to Comité International pour le Développement et l'Étude de la Construction Tubulaire Code.

relatively large, the punching-shear capacity becomes the governing factor. With β=0.85 as the dividing threshold, calculations are performed using the yield-line model and punching-shear model respectively. However, there is a significant discontinuity in the calculated results around β=0.85, and two distinct calculated values are obtained when directly substituting β=0.85 into the two models. This indicates that the calculation results of the current CIDECT design guide method are not continuous with respect to the variation of the β-parameter. In reality, the joint bearing capacity should not exhibit such a "breakpoint." Therefore, based on the yield-line model and punching-shear model, this section proposes a combined yield-line-and-punching-shear failure model. This model not only achieves continuity of bearing capacity results across the full range of β-values but also more reasonably reveals the failure mechanism of RST joints from a mechanical perspective.

3.3.4.1 Combined yield-line and punching-shear failure model

In the initial loading stage of the joint, the surface of the chord member deforms and reaches the yield stress. Subsequently, punching-shear cracks initiate at locations on the branch member where the stress level is high (and stiffness is large). Based on the stress distribution law of the branch member in RST joints, it is assumed that punching-shear failure occurs along the full length of the branch member in the height direction and within the effective width range of the branch member in the width direction, while yield-line failure occurs at all other locations. The combined yield-line and

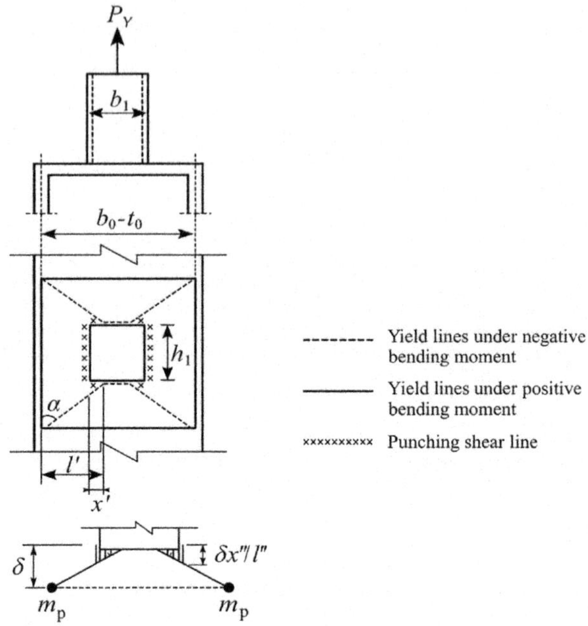

Figure 3.18 Combined yield-line and punching-shear model for rectangular hollow section joints.

punching-shear failure model is illustrated in Fig. 3.18. This model fully accounts for the variation in the stiffness of the branch member, as well as the yielding and shear behaviors of the chord member surface.

The position of the yield line in Fig. 3.18 can be described using x' and l', where x' denotes the length of the punching-shear region along the width direction of the branch tube and l' represents the distance from the rightmost end of the punching-shear line (along the width direction of the branch tube) to the positive moment yield line. Additionally, δ is the virtual displacement, and α is the angle of the yield line.

$$\frac{x'}{l'} = 1 - \left(\frac{b_0}{2l'}\right)(1 - \beta) > 0 \tag{3.26}$$

The work done by the yield line, denoted as E_Y, is expressed as follows:

$$E_Y = 2\left\{2b_0 m_p \frac{\delta}{l' \cot\alpha} + (4l' \cot\alpha + h_1) m_p \frac{\delta}{l'}\right\} \tag{3.27}$$

The work done by the punching-shear line, denoted as E_Q, is expressed as follows:

$$E_\varrho = 2\left\{\frac{f_{y0}}{\sqrt{3}} t_0 h_1 \delta \frac{x'}{l'} + 2\frac{f_{y0}}{\sqrt{3}} t_0 x' \delta \frac{x'}{2l'}\right\} \tag{3.28}$$

In accordance with the principle of virtual work:

$$\begin{aligned} P_{YQ}\delta = 4m_p\Bigg\{&\frac{b_0}{l'}\tan\alpha + \left(2\cot\alpha + \frac{h_1}{2l'}\right) \\ &+ \frac{2}{\sqrt{3}}\left(\frac{h_1}{t_0}\frac{x'}{l'} + \frac{x'}{t_0}\frac{x'}{l'}\right)\Bigg\}\delta \end{aligned} \tag{3.29}$$

In the formula, P_{YQ} represents the comprehensive failure load corresponding to the combined failure of the yield line and punching shear; m_p denotes the plastic yield moment of the main tube wall, with $m_p = f_y t_0^2/4$. Assuming $\gamma' = \frac{2l'}{b_0}$, the following expression can be derived:

$$\begin{aligned} \frac{P_{12}}{m_p} = &\frac{8}{\gamma'}\tan\alpha + 4\left(2\cot\alpha + \frac{h_1}{\gamma' b_0}\right) \\ &+ \frac{8}{\sqrt{3}}\left\{\frac{h_1}{t_0}\left[1 - \frac{1-\beta}{\gamma'}\right] + \frac{1}{2}\gamma'\frac{b_0}{t_0}\left[1 - \frac{1}{\gamma'}(1-\beta)\right]^2\right\} \end{aligned} \tag{3.30}$$

Different values of α correspond to different yield-line modes and thus result in different calculated values of P_{YQ}. Therefore the minimum value of P_{YQ} should be adopted. By setting the derivativc $\frac{dP_{YQ}}{d\alpha} = 8\left(\frac{1}{\gamma'}\sec^2\alpha - \csc^2\alpha\right) = 0$, the following is obtained:

$$\tan\alpha = \sqrt{\frac{2l'}{b_0}} = \sqrt{\gamma'} \tag{3.31}$$

By substituting Eq. (3.31) into Eq. (3.30) and assuming the branch tube is a square tube, the following expression can be obtained:

$$\frac{P_{YQ\,\min}}{f_{y0}\,t_0^2} = \frac{4}{\sqrt{\gamma'}} + \frac{\beta}{\gamma'} + \frac{2}{\sqrt{3}}\left\{\beta\cdot(2\gamma)\cdot\left[1 - \frac{1-\beta}{\gamma'}\right] + \gamma' \cdot\gamma\cdot\left(1 - \frac{1-\beta}{\gamma'}\right)^2\right\} \tag{3.32}$$

From the effective width of the branch tube specified in the CIDECT Code, the following can be obtained:

$$\gamma' = \frac{2l'}{b_0} = \frac{b_0 - b_1 + 10b_1/(b_0/t_0)}{b_0} = 1 - \beta + \frac{10\beta}{2\gamma} \tag{3.33}$$

3.3.4.2 Formula verification and comparison

The calculation formula for the combined failure model of yield line and punching shear was verified. A total of 42 sets of test data on the axial force of branch tubes in rectangular steel tube joints with a β range of 0.19–0.91 were selected, as presented in Table 3.3. The joint load-bearing capacity was calculated using Eq. (3.32) and the method specified in the CIDECT Code, respectively. These calculated results were then compared with the test values, as illustrated in Fig. 3.19. Herein, N_c denotes the calculated value, and N_e represents the test value.

As can be seen from Fig. 3.19, for the combined failure model of yield line and punching shear, the mean value (μ) of the ratio between the calculated values and the test values is 1.085, the standard deviation (σ) is 0.209, and the coefficient of variation (σ/μ) is 0.192. For the CIDECT Code method, the mean value (μ) of the ratio between the calculated values and the test values is 0.817, the standard deviation (σ) is 0.215, and the coefficient of variation (σ/μ) is 0.264. A comparison of the mean values shows that the calculation results from the CIDECT Code are conservative. In contrast, the results obtained by the proposed method in this study are basically consistent with the test results, though the calculated values are slightly larger. Meanwhile, a comparison of the coefficients of variation indicates that the proposed method in this study exhibits better agreement with the test values. Therefore it is recommended that a certain safety factor be applied when using the calculation formula proposed in this study.

The load-bearing capacities obtained from the formula proposed in this study and the CIDECT Code are plotted in Fig. 3.20 for comparative analysis.

As can be seen from Fig. 3.20, within the range of $\beta < 0.85$, the calculation results of the formula proposed in this study are basically larger than those of the

Table 3.3 Test data for rectangular hollow section joints under brace uniaxial loading.

Node number	Chord			Brace			β	η	$2\gamma\ h_0$ (mm)	$f_{y0}\ b_0$ (mm)	Test value t_0 (mm)
	h_0 (mm)	b_0 (mm)	t_0 (mm)	h_1 (mm)	b_1 (mm)	t_1 (mm)					
1	102	102	9.5	51	51	4.9	0.5	0.5	10.7	421	410
2	102	102	6.3	51	51	4.9	0.5	0.5	16.2	412	137
3	102	102	4	51	51	4.9	0.5	0.5	25.5	417	63
4	127	127	7.9	102	102	6.4	0.8	0.8	6.1	404	563
5	127	127	7.9	102	102	6.4	0.8	0.8	6.1	342	359
6	150	150	6	125	125	6	0.83	0.83	5	366	353
7	150	150	6	125	125	6	0.83	0.83	5	328	332
8	200	200	6	178	178	12.7	0.89	0.89	33.3	368	514
9	127	127	3	102	102	6.4	0.8	0.8	42.3	382	88
10	127	127	3	51	102	6.4	0.8	0.4	42.3	382	76
11	127	127	3	152	102	6.4	0.8	1.2	42.3	382	111
12	127	127	3	203	102	6.4	0.8	1.6	42.3	382	148
13	203	203	4.8	178	178	12.7	0.88	0.88	42.3	348	270
14	150	150	6	100	100	6	0.67	0.67	25	366	195
15	200	200	6	150	150	6	0.75	0.75	33.3	368	255
16	250	250	6	200	200	9	0.8	0.8	41.7	400	297
17	150	250	9	178	178	12.7	0.71	0.71	27.8	387	569
18	150	150	6	75	75	3.2	0.5	0.5	25	366	112
19	200	200	6	125	125	6	0.63	0.63	33.3	368	158
20	178	229	4.6	102	102	6.4	0.45	0.45	49.8	375	59
21	254	254	9.5	127	127	6.4	0.5	0.5	26.7	380	305
22	127	127	7.9	102	102	4.8	0.8	0.8	16.1	404	544
23	150	150	6	127	127	6.4	0.85	0.85	25	366	430
24	250	250	6	75	75	2.3	0.3	0.3	41.7	400	83
25	150	150	6	75	75	2.3	0.5	0.5	25	366	97

(*Continued*)

Table 3.3 (Continued)

Node number	Chord			Brace			β	η	2γ h_0 (mm)	f_{y0} b_0 (mm)	Test value t_0 (mm)
	h_0 (mm)	b_0 (mm)	t_0 (mm)	h_1 (mm)	b_1 (mm)	t_1 (mm)					
26	150	150	6	102	102	3.2	0.68	0.68	25	366	193
27	350	350	12	102	102	2.4	0.3	0.3	29.2	264	216
28	254	254	9.5	127	127	3	0.5	0.5	26.7	380	264
29	150	150	6	127	127	3	0.85	0.85	25	366	316
30	100	199.1	7.85	40.3	39.8	1.99	0.19	0.2	25.4	288	111
31	100	199.1	7.84	39.7	60.1	1.97	0.3	0.2	25.4	288	119
32	100	199	7.84	79.4	99.8	1.98	0.5	0.4	25.4	288	113
33	100	199.1	7.72	80.8	120.6	1.99	0.61	0.41	25.8	288	224
34	100	199.3	7.72	39.2	118.7	3.88	0.6	0.2	25.8	288	174
35	100	199.3	7.69	80.4	159.2	3.9	0.8	0.4	25.9	288	363
36	100	199.2	7.8	79.8	181	3.9	0.91	0.4	25.5	288	505
37	100	199.2	7.86	122	159.4	3.9	0.8	0.6	25.3	288	448
38	100	199.2	7.69	122.2	180.5	3.9	0.91	0.6	25.9	288	642
39	100	199	7.76	159.3	159.6	3.96	0.8	0.8	25.6	288	491
40	100	199	7.8	81	158.8	5.99	0.8	0.4	25.5	288	392
41	100	198.8	7.8	79.5	178.7	6.3	0.9	0.4	25.5	288	630
42	100	199.2	7.84	119.7	179.7	6.02	0.9	0.6	25.4	288	776

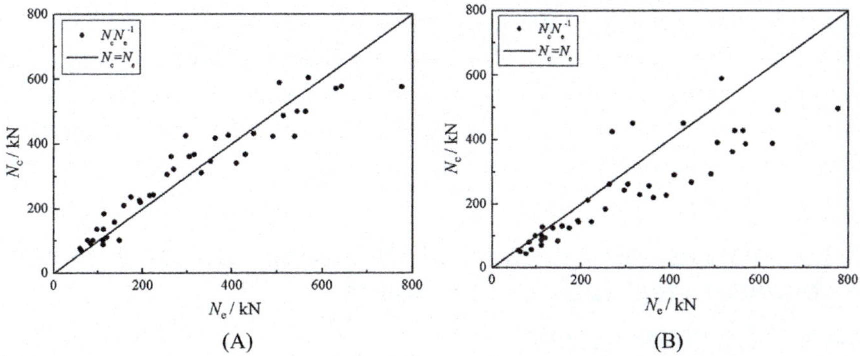

Figure 3.19 Comparison between the calculated results and test results. (A) Comparison between the calculated results from the proposed formula and test results. (B) Comparison between the calculated results from the Comité International pour le Développement et l'Étude de la Construction Tubulaire specification and test results.

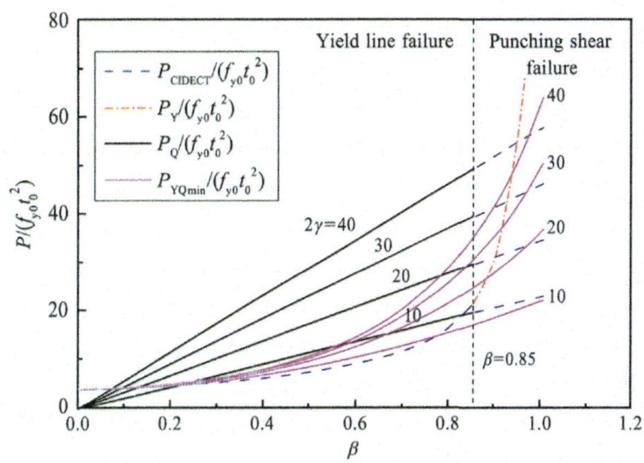

Figure 3.20 Comparison of bearing capacity between the formula proposed in this book and the Comité International pour le Développement et l'Étude de la Construction Tubulaire Code formula.

yield-line model in the CIDECT Code. This indicates that the CIDECT Code, which only considers the yield deformation of the main tube, underestimates the joint load-bearing capacity. Within the range of $0.85 < \beta < 1-1/\gamma$, the calculation results of the proposed formula are basically smaller than those of the punching-shear model in the CIDECT Code. It can be observed that the CIDECT Code, which only accounts for the shear failure of the main tube, overestimates the joint load-bearing capacity. Meanwhile, the calculation results of the formula proposed in this study are continuous within the range of the β parameter. This continuity simplifies the determination of β when checking the joint load-bearing capacity, making it easier for engineering practitioners to use. Furthermore, the proposed formula better reveals the failure mechanism of the main tube in rectangular steel tube joints.

3.3.5 Analysis on mechanical behavior of concrete-filled rectangular steel tubular truss joints

3.3.5.1 Compression joints

For T-type compression joints, the transverse load exerted by the compressed branch tube on the main tube is balanced by the shear force of the main tube's cross section. This scenario is analogous to a simply supported beam with a short span subjected to transverse concentrated loads, as illustrated in Fig. 3.21A.

In the case of rectangular steel tube joints, the wall of the main tube is relatively flexible, leading to deformation that is primarily concentrated in the joint region. This deformation is much smaller than the deflection of a simply supported beam, and the joint exhibits a relatively low load-bearing capacity. Consequently, the simply supported beam model is generally employed directly for model test research on rectangular steel tube joints. To mitigate the influence of end effects at the supports, the span of the simply supported beam model should not be excessively small; it is typically recommended to be 5 to 10 times the cross-sectional dimension of the steel tube.

For CFRST truss joints, the concrete infilled in the main tube enhances both the transverse compressive stiffness and compressive load-bearing capacity of the chord member. This enhancement prevents excessive flexural

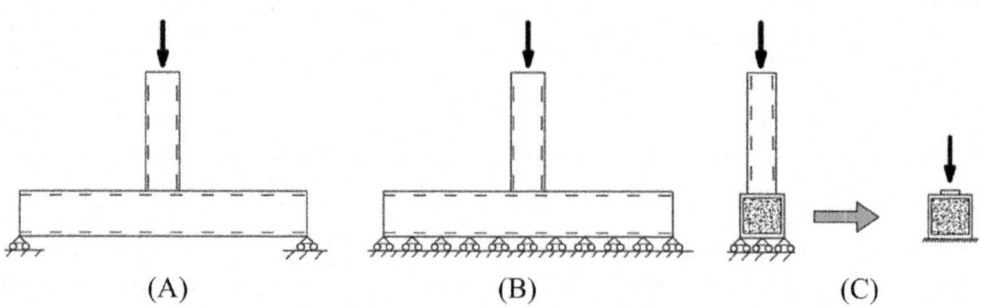

Figure 3.21 Load transfer mechanism in concrete-filled rectangular steel tubular T-type joints. (A) Simply supported beam model. (B) Equivalent model with continuous support reaction on the opposite side of the chord. (C) CFRST transverse local compression model.

deformation of the steel plate at the connection surface of the main tube in CFRST compression joints, resulting in a higher load-bearing capacity of the joint. If the simply supported beam model were adopted for such joints, it would likely lead to the failure of the beam prior to the failure of the joint.

To achieve joint failure, the shear force acting on the cross section of the chord member can be equivalent to a continuous support reaction force on the opposite side of the chord member, as shown in Fig. 3.21B. However, for the model illustrated in Fig. 3.21B, there is a high probability that the joint's load-bearing capacity will depend on the compressive capacity of the branch tube. In such cases, joint failure may still not be achieved. To avoid the occurrence of the branch tube failure mode and reveal the mechanical behavior of CFRST compression joints, the model depicted in Fig. 3.21B can be further equivalent to the CFRST transverse local compression model shown in Fig. 3.21C.

In an X-type compression joint, the compressive force of one branch is balanced by another branch through the chord, resulting in transverse local compression of the rectangular CFST chord member and local compression of the concrete inside the chord. To prevent branch failure, this condition can also be equivalently modeled as a transverse local compression model, as shown in Fig. 3.22, where the transverse local bearing area is given by $A_b=b_1h_1/\sin\theta_1$. The behavior of a rectangular CFST Y-type compression joint is similar to that of a T-type joint, where the vertical component of the branch force also induces transverse local compression in the rectangular CFST chord member.

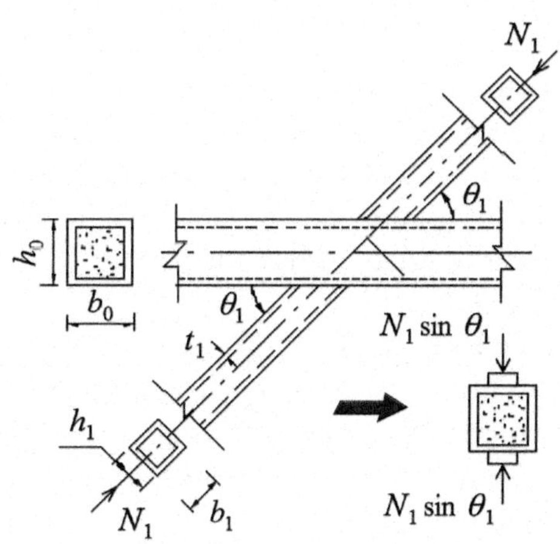

Figure 3.22 Force characteristics of X-type joints.

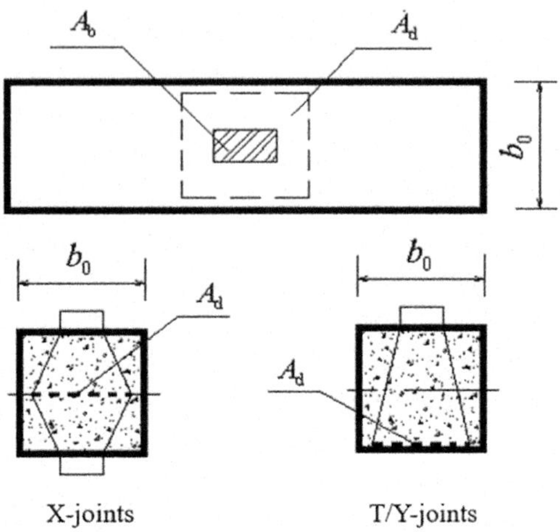

Figure 3.23 Local compression of infill concrete.

To further analyze the characteristics of local compression of the concrete inside the chord at the junction between a compressed branch and the chord, Fig. 3.23 illustrates the schematic diagram of the load dispersion of the locally compressed concrete at a certain angle. For an X-type joint, the local compression load disperses to half the section height of the concrete, that is, the dispersion area or bearing area Ad is located at half the section height. In contrast, for T-type and Y-type joints, the load disperses throughout the entire section height of the chord concrete, as shown in Fig. 3.23. According to the provisions of the Chinese Code for Design of Concrete Structures, the effective bearing area for local compression should be determined based on the principle of "concentric symmetry," meaning that the calculated bearing area depends only on the size and shape of the local compression area. Therefore, to remain consistent with the Chinese Code, the calculated bearing area Ad of rectangular CFST T-type and Y-type compression joints is taken as the same as that of X-type joints. This provides the basis for adopting similar calculation methods for the bearing capacity of rectangular CFST T- type, Y- type, and X-type compression joints.

For rectangular CFST K-type gap joints, due to the presence of concrete inside the chord, the area of the joint subjected to the branch compression also experiences transverse local compression of the rectangular CFST chord member, with stress characteristics similar to those of T-type and Y-type compression joints.

In summary, whether it is an X-type joint or a T-type, Y-type, or K-type joint, the vertical component of the branch force in rectangular CFST compression joints always induces transverse local compression of the rectangular CFST chord member and local compression of the concrete inside the chord. Therefore transverse local compression of the rectangular CFST chord is the primary characteristic of rectangular CFST compression joints.

3.3.5.2 Tension joints

Under the action of the vertical component force of the branch tube, the top plate of the main tube in CFRST X-type, T-type, and Y-type tension joints undergoes significant flexural deformation. This deformation behavior is quite similar to that of hollow rectangular steel tube joints, and the load-bearing capacity of such joints is controlled by the vertical component force of the branch tube. However, the concrete infilled inside the main tube restrains the inward concave deformation of the main tube's side plates (i.e., restrains the rotation at the corner of the main tube), which in turn induces significant membrane tension in the top plate of the main tube. Consequently, the stiffness of CFRST tension joints is higher than that of rectangular hollow steel tube tension joints.

Fig. 3.24 simplifies the CFRST tension joint into a fixed-end beam model and the rectangular steel tube tension joint into a frame model, so as to compare the deformation characteristics of the CFRST and rectangular steel tube X-type tension joints. In Fig. 3.24, the cross-sectional width of both the fixed-end beam and frame members is taken as a unit width, while the cross-sectional height is equal to the wall thickness of the main tube. The load on the branch tube is equivalent to concentrated loads acting on the two side plates of the branch tube. In accordance with the principles of structural mechanics, the concentrated force required for unit deformation in the model, that is, the stiffness of the model, can be calculated. Through derivation, the ratio of the elastic stiffness of the simplified model for the

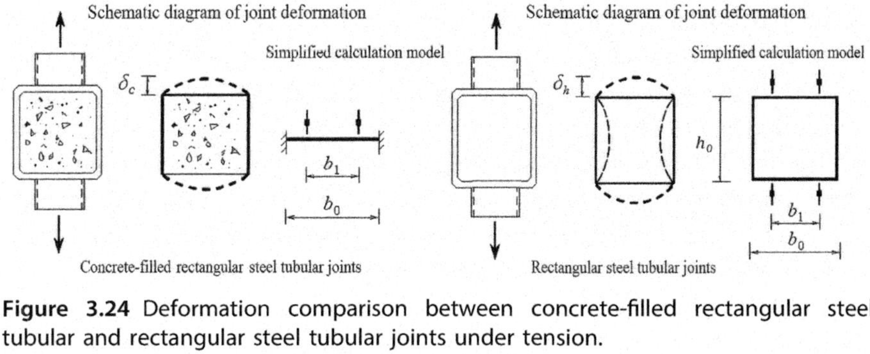

Figure 3.24 Deformation comparison between concrete-filled rectangular steel tubular and rectangular steel tubular joints under tension.

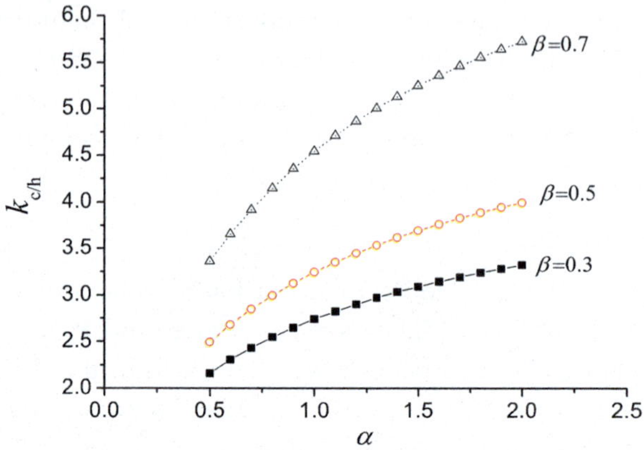

Figure 3.25 Relationship between $k_{c/h}$ and α.

CFRST tension joint to that of the rectangular steel tube tension joint (denoted as $k_{c/h}$) in Fig. 3.24 is obtained as follows:

$$k_{c/h} = 1 + \frac{3\alpha(1+\beta)}{(1+\alpha)(1-\beta)(1+2\beta)} \tag{3.34}$$

In the formula, $\beta = b_1/b_0 < 1$ and $\alpha = h_0/b_0$. The relationship curve between $k_{c/h}$ and $\alpha$was plotted based on Eq. (3.34), as shown in Fig. 3.25. It can be observed from Fig. 3.25 that, in the simplified model, the elastic stiffness of the rectangular CFST tension joint is more than twice that of the RST tension joint. Additionally, this stiffness increases with the increase in both the chord height-to-width ratio α and the branch-to-chord width ratio β.

Test results of RST T- type joints show that when the branch-to-chord width ratio β ranges between 0.3 and 0.6, the load-carrying capacity of the joint is controlled by joint deformation. In this case, the load corresponding to a 1% deformation over a length of 1.5 times the chord width can be taken as the joint load-carrying capacity. Therefore, for CFRST tension joints of X- type, T- type, and Y-types: when the branch-to-chord width ratio β is small (and the joint load-carrying capacity is controlled by joint deformation), their load-carrying capacity is higher than that of the corresponding RST joints. When β is large (and the joint load-carrying capacity is governed by the punching-shear strength of the chord flange plate or the yield strength of the chord sidewall), the load-carrying capacities of these two types of joints (CFST joints and steel tubular joints) are comparable.

For CFRST K-type joints, the joint region at the compressive branch member has high bearing strength; thus the load-carrying capacity of the K-

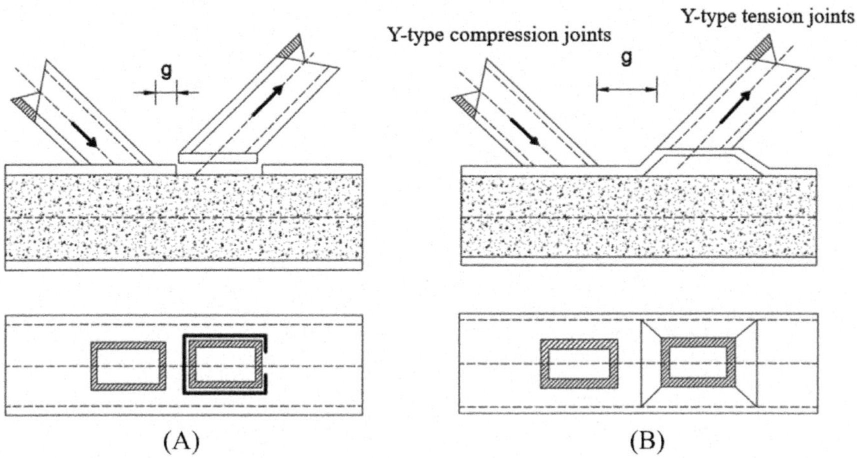

Figure 3.26 Analysis of failure modes in concrete-filled rectangular steel tubular joints under tension. (A) Punching-shear failure. (B) Yield-line failure.

type joint depends on the connection strength at the tensile branch member. The failure mode of the K-type joint is related to the gap g between the two branch members, as illustrated in Fig. 3.26. When g is small, the flexural deformation of the chord's top plate at the gap is minimal, and the joint tends to undergo punching-shear failure at the position where the tensile branch member connects to the chord, as shown in Fig. 3.26A. When g is large, significant flexural deformation occurs in the chord's top plate at the gap, and yield lines are likely to form at the connection between the tensile branch member and the chord. As a result, the CFRST K-type joint functions as a pair of Y-type joints, as presented in Fig. 3.26B.

3.3.5.3 Bending joint

The bending moment acting on the branch member of a CFRST bending joint can be equivalent to a combined tension-compression effect. Similar to CFRST compression joints, the downward concave deformation of the chord's top plate (on the side of the branch's compression flange) is restricted by the concrete infilled in the chord. On the tension side of the branch member, the behavior is analogous to that of CFRST tension joints: the concrete infill prevents the inward concave deformation of the chord's sidewalls, resulting in smaller flexural deformation of the top plate compared to the corresponding RST joints. Consequently, both the load-carrying capacity and flexural stiffness of CFRST bending joints are higher than those of their RST counterparts.

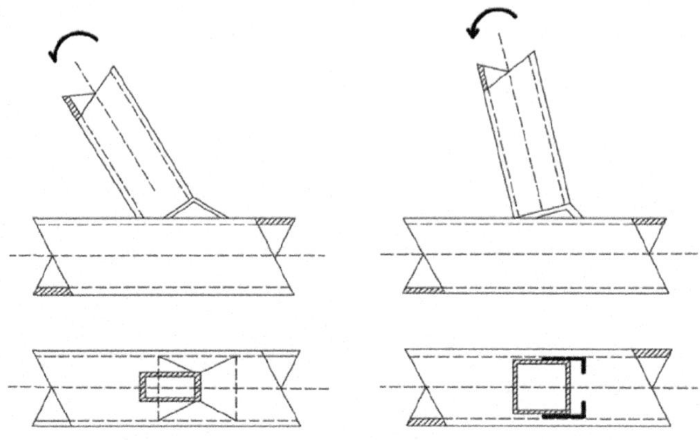

Figure 3.27 Analysis of failure modes in concrete-filled rectangular hollow section joints under bending. (A) Yield-line failure mode. (B) Punching-shear failure of the chord top plate or effective-width failure of the tension flange of the branch member.

Fig. 3.27 illustrates the potential failure modes of CFRST bending joints: Fig. 3.27A depicts the yield-line failure mode (occurring when the width ratio β is small). Under this condition, significant flexural deformation develops in the top plate of the chord member on the tension side of the branch member. Fig. 3.27B illustrates the failure modes observed when the width ratio β is large: these include either the punching-shear failure mode of the chord's top plate (on the tension side of the branch member) or the effective width failure mode of the tension flange of the branch member.

3.3.6 Analysis on influencing factors of mechanical performance of concrete-filled rectangular steel tubular truss joints

The above analysis of the basic working performance of CFRST truss joints reveals the following: filling the chord members with concrete improves the joint's mechanical performance. Compared with RST joints, CFRST joints exhibit changes in both their working mechanism and failure morphology. Additionally, the influence of joint geometric parameters and physical parameters on the mechanical performance of CFRST truss joints varies depending on the joint type and the loading mode of the branch members.

Filling the chord members with concrete has the most significant impact on the mechanical performance of CFRST compression joints. For these compression joints, the vertical component of the branch member force induces local transverse compression in the chord member. Consequently, the key factors influencing the performance of CFRST compression joints include the strength of the infilled concrete, the filling length of the concrete, the local bearing area of the joint, the branch-to-chord width ratio (β), and the angle between the branch and chord members. In contrast, the influence of infilled concrete on the performance of CFRST tension joints is primarily reflected in its effect on joint stiffness; the strength of the infilled concrete has no significant impact on tension joint performance. Therefore, similar to RST joints, the main factors affecting the performance of CFRST tension joints are the branch-to-chord width ratio (β), the chord width-thickness ratio, and the yield strength of the chord steel. The load-carrying capacity of CFRST bending joints is governed by the tension side of the branch members. As a result, the key factors influencing the performance of CFRST bending joints are identical to those of tension joints.

In a CFRST K-type joint, one branch member is in compression while the other is in tension. As the gap between the two brace members increases, the K-type joint transforms into a pair of Y-type joints. Therefore, in addition to the primary influencing factors mentioned above, joint gap is another critical factor affecting the performance of CFRST K-type joints.

When calculating the load-carrying capacity of truss joints, the influence of the axial force in the joint chord member (i.e., the truss chord member) should also be considered. For RST joints, axial compression in the chord member reduces their load-carrying capacity, whereas axial tension has little effect on their load-carrying capacity. For CFRST truss joints, the concrete infilled in the chord member can share the axial force of the chord and increase the local buckling stress of the chord's sidewalls. Consequently, the reduction in the joint's compressive load-carrying capacity caused by the chord's axial force is less pronounced for CFRST joints than for the RST joints.

3.4 Rectangular concrete-filled steel tubular truss tests

When a CFST truss fails due to member damage, its flexural capacity can be calculated based on the load-carrying capacity of the failed members. In contrast, when truss failure is caused by joint damage, the truss's flexural capacity is closely related to the load-carrying capacity of the joints. To analyze the correlation between the joints and the overall load-carrying capacity of CFRST trusses, a series of CFRST trusses with relatively large height-span ratios were designed. This study focused on investigating the failure modes and load-carrying capacities of CFRST trusses when failure initiates at the joints.

3.4.1 Test overview

A total of 13 CFST truss specimens were tested, categorized as follows for comparative analysis: 2 empty steel tubular truss specimens (R0, C0, A0), 2 circular CFST truss specimens (C1, C2), 5 rectangular CFST truss specimens (R1–R2, A1–A2, reinforced concrete-filled steel tubular), and 3 rectangular CFST truss specimens with Perfobond Leister ribs (PBL) stiffeners (PBL-1 to PBL-3). The truss parameters and test results are presented in Table 3.4. All trusses were loaded via a single-point loading method applied at the mid-span, and the schematic diagram of the test loading setup is shown in Fig. 3.28.

3.4.2 Test results of specimens

Specimen A0 is an RST truss with RST joints. The failure of the specimen occurred at the compression joints of the top chord. When vertical loads were applied at the mid-span joints of the truss top chord, the top chord members were subjected to compression. For truss joints where the chord member is in compression, their load-carrying capacity further decreases as the compressive stress level of the chord member increases. Additionally, under the action of compressive branch members, empty tubular joints (i.e., nonstiffened tubular joints) are prone to two phenomena: surface depression of the chord member and sidewall buckling of the chord member. The failure mode of the joints in Specimen A0 is schematically illustrated in Fig. 3.29, which corresponds to Failure Mode F depicted in Fig. 3.11.

Specimen A1 has identical structural dimensions to Specimen A0, with the key difference being that the compressive top chord members of Specimen A1 were filled with concrete. Due to the presence of the concrete infill, the load-carrying capacity of the top chord joints (under the action of compressive brace members) was significantly enhanced. However, the bottom chord members of Specimen A1 remained unfilled steel tubes; consequently, the load-carrying capacity of the bottom chord joints (under the action of compressive brace members) was relatively low. For this reason, the failure of Specimen A1 occurred at the compressive joints of the truss bottom chord, as schematically illustrated in Fig. 3.30. The joint failure mode—characterized by chord member surface depression and sidewall buckling—remained consistent with Failure Mode F depicted in Fig. 3.31.

Specimen A2 has identical structural dimensions to Specimens A0 and A1, with the distinction that both its top and bottom chord members were

Table 3.4 Key parameters of specimens and test results.

Specimen number	H/mm	L/mm	H/L	β	θ/°	Chord dimension/mm	Brace member dimension/mm	Truss type	Failure mode of trusses	Corresponding to the joint failure modes in Section 1.13.3	Flexural capacity (kN·m)
A0	1000	4800	1/4.8	0.8	55	100*100*4	80*80*4	I-0	Failure of top chord hollow section joint in compression	Mode F	369.36
A1	1000	4800	1/4.8	0.8	55	100*100*4	80*80*4	I-1	Failure of bottom chord hollow section joint in compression	Mode F	391.53
A2	1000	4800	1/4.8	0.8	55	100*100*4	80*80*4	I-2	Failure of bottom chord CFST joint in tension	Mode B	424.78
R0	343	525	1/6.5	0.6	51	100*80*3	60*40*3	I-0	Failure of top chord hollow section joint in compression	Mode F	154.44
R1	343	525	1/6.5	0.6	51	100*80*3	60*40*3	I-1	Failure of top chord hollow section joint in tension	Mode B	158.73
R2	343	525	1/6.5	0.6	51	100*80*3	60*40*3	I-2	Failure of bottom chord CFST joint in tension	Mode C	177.61
C0	3461	525	1/6.6	0.64	52	φ89*4	φ57*3.5	II-0	Failure of top chord hollow section joint in compression	–	180.02

(Continued)

Table 3.4 (Continued)

Specimen number	H/mm	L/mm	H/L	β	$\theta/^{\circ}$	Chord dimension/mm	Brace member dimension/mm	Truss type	Failure mode of trusses	Corresponding to the joint failure modes in Section 1.13.3	Flexural capacity (kN·m)
C1	3461	525	1/6.6	0.64	52	φ89*4	φ57*3.5	II-1	Failure of bottom chord hollow section joint in compression	–	199.07
C2	3461	525	1/6.6	0.64	52	φ89*4	φ57*3.5	II-2	Failure of top chord hollow section joint in tension	–	268.31
RCFST	4000	880	1/4.5	0.75	60	160*160*3	120*80*3	I-2	Failure of bottom chord PBL-strengthened CFST joint in tension	Mode B	357.05
PBL-1	4000	880	1/4.5	0.5	60	160*160*3	80*80*3	III-2	Failure of bottom chord PBL-strengthened CFST joint in tension	Mode B	453.55
PBL-2	4000	880	1/4.5	0.75	60	160*160*3	120*80*3	III-2	Failure of bottom chord PBL-strengthened CFST joint in tension	Mode B	525.93
PBL-3	4000	880	1/4.5	0.875	60	160*160*3	140*80*3	III-2	Failure of bottom chord PBL-strengthened CFST joint in tension	Mode B	564.53

Notes: Truss type I represents rectangular truss, II represents circular truss, III represents Perfobond Leister ribs *(PBL)*-stiffened rectangular truss, 0 represents empty pipe truss, 1 represents truss with only the compression chord poured with concrete, and 2 represents truss with the both upper and lower chord poured with concrete. CFST, Concrete-filled steel tubular; RCFST, reinforced concrete-filled steel tubular.

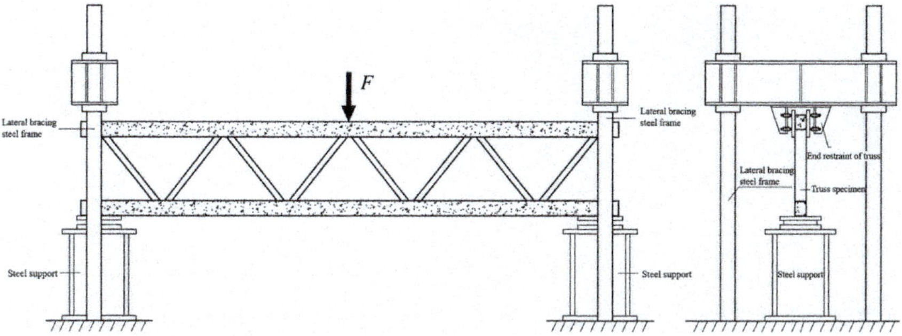

Figure 3.28 Schematic diagram of the test loading setup.

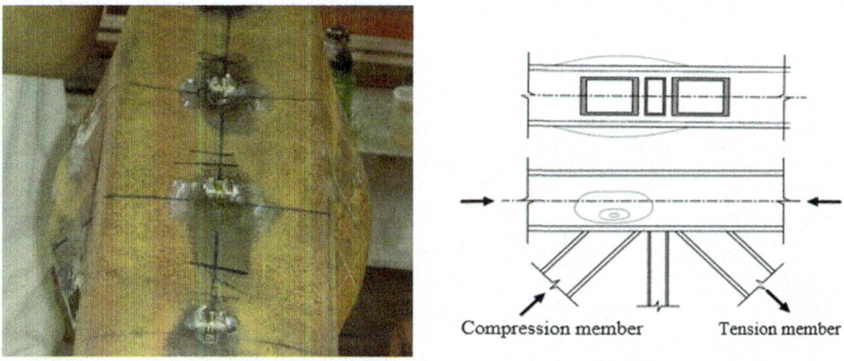

(A) (B)

Figure 3.29 Failure of the upper chord hollow section joint in compression for Specimen A0. (A) Test joint failure diagram. (B) Schematic diagram of joint failure.

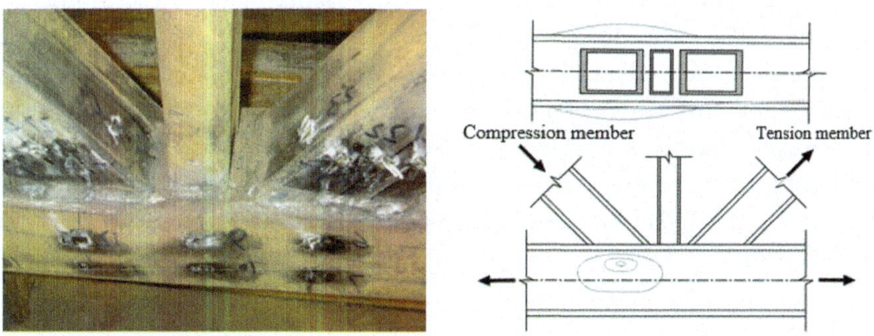

(A) (B)

Figure 3.30 Failure of the upper chord hollow section joint in compression for specimen A1. (A) Test joint failure diagram. (B) Schematic diagram of joint failure.

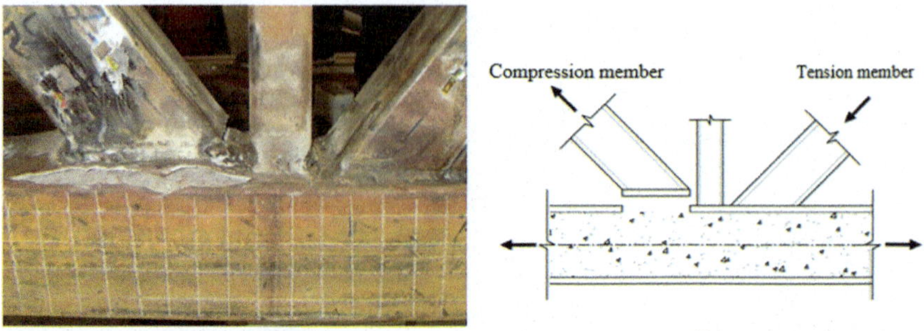

(A) (B)

Figure 3.31 Punching-shear failure of the lower chord concrete-filled rectangular steel tubular joint in tension for specimen A2. (A) Test joint failure diagram. (B) Schematic diagram of joint failure.

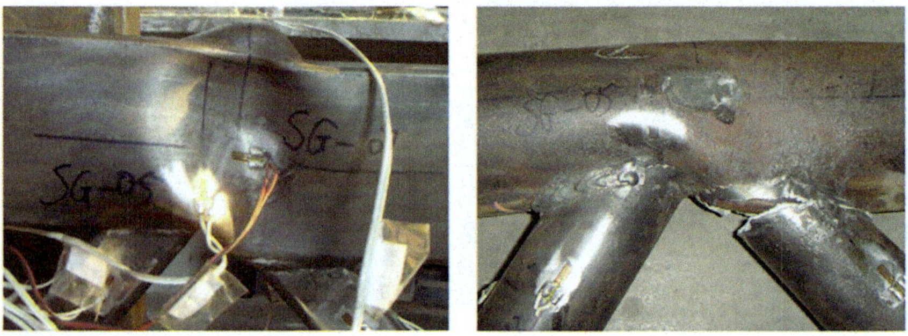

(A) (B)

Figure 3.32 Comparison of failure modes between hollow steel tubular truss Specimens R0 and C0. (A) Specimen R0. (B) Specimen C0.

filled with concrete—resulting in all truss joints being CFST joints. In Specimen A2, the load-carrying capacity of all CFST joints under compression was effectively enhanced. At this point, the load-carrying capacity of the truss's tension joints became lower than that of its compression joints. Consequently, the failure mode of Specimen A2 was punching-shear failure of the tension joints, as schematically illustrated in Fig. 3.31. This failure mode corresponds to Failure Mode B of joints depicted in Fig. 3.11.

3.4.3 Test results of specimens R0–R2 and C0–C2

Fig. 3.32 shows the failure morphologies of the empty steel tubular truss specimens. Since no concrete was filled in the chord members, all truss

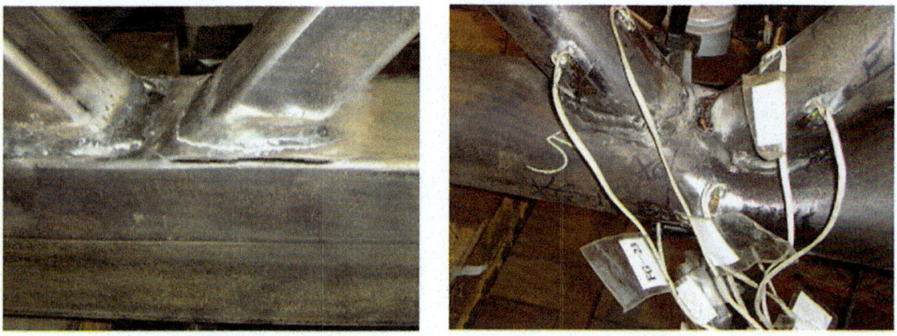

(A) (B)

Figure 3.33 Comparison of failure modes for truss specimens with concrete filling only in the compression chord. (A) Specimen R1. (B) Specimen C1.

joints were empty tubular joints. For Specimen R0, failure occurred at the compression chord joints: the brace plates on both sides of the chord member (connected to the compressive branch members) bulged outward, as illustrated in Fig. 3.32A. For Specimen C0, failure also occurred at the compression chord joints: the chord member bulged outward on both sides at the connection with the compressive branch members, accompanied by surface depression, leading to yield failure of the chord member, as illustrated in Fig. 3.32B.

Fig. 3.33 presents the failure morphologies of truss Specimens R1 and C1, where only the compressive top chord members were filled with concrete. For Specimen R1, failure occurred at the empty steel tubular joints: the top plate of the chord member was torn under tension, resulting in punching-shear failure (as shown in Fig. 3.33A). Meanwhile, the compression joints of the bottom chord also exhibited top plate depression and sidewall bulging. Similar to Specimen R1, Specimen C1 also failed at the empty steel tubular joints of the bottom chord. Specifically, at the connection between the bottom chord (empty tube) and the compressive branch members, the chord member bulged outward on both sides, leading to yield failure of the chord member (as shown in Fig. 3.33B).

Fig. 3.34 illustrates the failure morphologies of two specimens: Specimen R2 (a rectangular CFST truss) and Specimen C2 (a circular CFST truss). All joints of both specimens are CFST joints, and their failures are characterized by tension joint failure, as shown in Fig. 3.34.

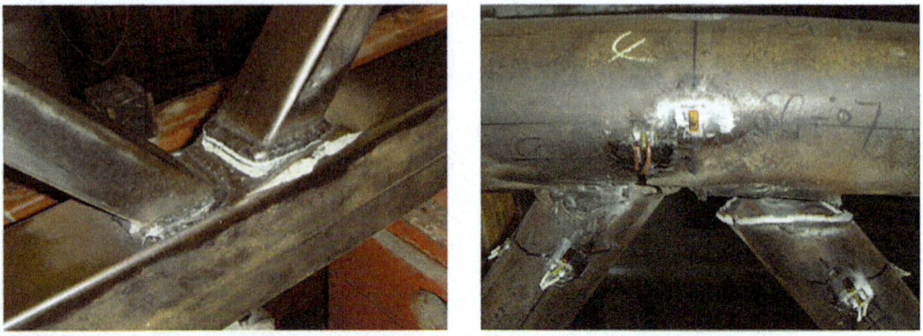

(A) (B)

Figure 3.34 Comparison of failure modes for truss specimens with concrete filling in both tension and compression chords. (A) Specimen R2. (B) Specimen C2.

3.4.4 Test results of specimens reinforced concrete-filled steel tubular and perforated plate shear 1 to perforated plate shear 3

From the experimental analysis of trusses with concrete-filled chord members, the following conclusions can be drawn: filling the chord members of joints with concrete prevents premature failure of compression joints, thereby shifting the truss's failure mode to tension joints. Additionally, the load-carrying capacity of tension joints can be effectively enhanced by installing PBL stiffeners. This section presents an experimental study on CFRST truss specimens equipped with PBL stiffeners.

Specimens PBL-1 to PBL-3 had their chord members filled with concrete and were equipped with PBL stiffeners. The vertical shear resistance provided by the concrete dowels formed in the holes of the PBL stiffeners enhanced the load-carrying capacity of the tension joints and restricted the deformation on the chord member surface. During the loading process, the surface deformation of the chord members (at the tension joint locations under the action of tensile brace members) was smaller for Specimens PBL-1 to PBL-3 compared to the CFRST specimens. As the load continued to increase, minor cracks first initiated in the chord members of all specimens at the connections with the tensile branch members. Subsequently, additional cracks propagated, and further cracking occurred at the chord member connections to the tensile branch members within the joint region. Finally, all specimens experienced punching-shear failure at the chord member surface where the tensile branch members connected to the joints; concurrently, the bottom plates of the chord members at the joints were torn by the tensile branch members. For the CFRST specimens (without

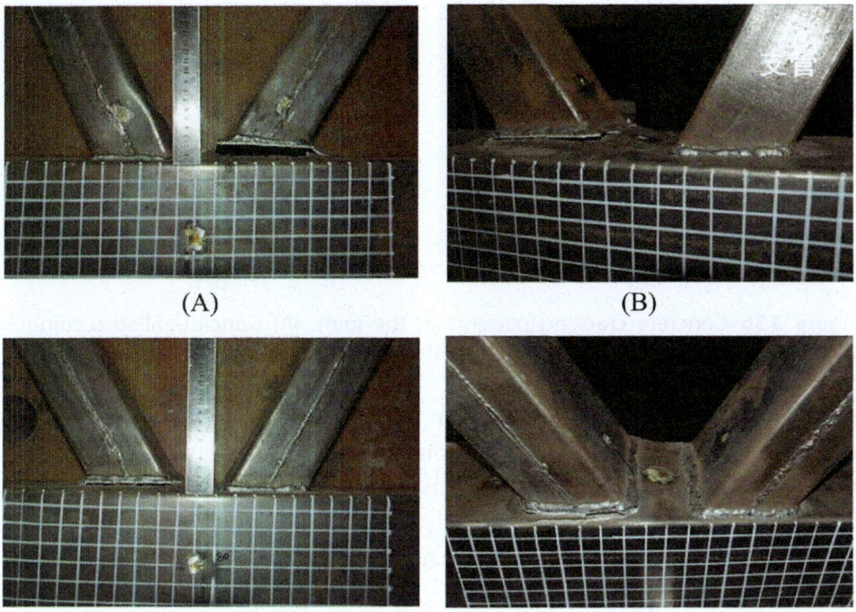

(A) (B) (C) (D)

Figure 3.35 Specimen failure modes. (A) Specimen concrete-filled rectangular steel tubular. (B) Specimen Perfobond Leister ribs (PBL)-1. (C) Specimen PBL-2. (D) Specimen PBL-3.

PBL stiffeners), the compressive branch members also exhibited buckling deformation. All four specimens ultimately failed due to joint failure, with their failure modes illustrated in Fig. 3.35. Compared to empty tubular joints, the CFRST joints with PBL stiffeners showed significant improvements in both tensile and compressive performance. As can be seen from Table 3.4, the load-carrying capacities of Specimens PBL-1 to PBL-3 were substantially higher than that of the CFRST specimen without PBL stiffeners.

After cutting the steel plates at the tension chord joints, the cracking patterns of the concrete within the joints were observed, as shown in Fig. 3.36. The observations reveal distinct differences: for the CFRST specimen, the concrete cracking pattern at the joint manifested as vertical through-cracks extending from the bottom upward, with the width of the cracks on the lower side reaching 0.4 mm. For the PBL-2 specimen, the concrete cracking pattern at the joint was characterized by vertical cracks on the lower side of the chord member and dense localized vertical cracks near the tensile branch members, with all crack widths less than 0.2 mm. These observations indicate the following: in CFRST truss tension chord

(A) (B)

Figure 3.36 Concrete cracking patterns at the joint. (A) Concrete-filled rectangular steel tubular specimen. (B) Perforated plate shear-2 specimen.

joints without PBL stiffeners, the infilled concrete undergoes global cracking under tensile stress. In contrast, for joints with PBL stiffeners, the longitudinal shear resistance of the PBL stiffeners enables the steel tube and concrete to work synergistically. This synergy not only restricts the propagation of concrete cracks and promotes more uniform crack distribution, but also, at the joint regions connected to the tensile branch members, the vertical shear resistance of the PBL stiffeners effectively limits surface deformation of the chord member. Consequently, dense localized vertical cracks are observed near the tensile branch members.

3.4.5 Analysis on failure mechanism of concrete-filled rectangular steel tubular trusses

To investigate the correlation between joint load-carrying capacity and truss flexural capacity, all trusses were designed with a relatively large height-span ratio. This design ensured that truss joints tended to fail first, providing robust experimental support for the analysis of truss failure mechanisms in this section.

According to structural mechanics principles, for a simply supported truss subjected to a concentrated load at mid-span, the axial forces in all diagonal brace members are identical. Therefore it can be assumed that, under the same loading step, the joint loads within a single truss are uniform. When considering the influence of the stress level in the joint chord members (i.e., truss chord members), joints with higher compressive stress levels in the chord members exhibit lower load-carrying capacity and are more prone to failure.

From the mechanism analysis of RST joints and CFST joints in Section 1.13 of this chapter, it is concluded that filling the chord members of joints with concrete significantly enhances the load-carrying capacity of compression joints. This conclusion is illustrated through the analysis of truss Specimens R0 to R2: in the empty steel tubular truss Specimen R0, the actual failure mode was joint failure caused by compressive brace members at the top chord joints (Fig. 3.37A), which is consistent with Failure Mode F in Fig. 3.11. This further confirms the low compressive load-carrying capacity of empty tubular joints. After filling the top chord members of truss Specimen R1 with concrete, the load-carrying capacity of the top chord compression joints was significantly improved. However, the bottom chord members remained empty steel tubes, so the failure mode shifted to the compression joints of the empty bottom chord (Fig. 3.37B). The observed joint failure mode was identical to that in Specimen R0. When both the top and bottom chord members of truss Specimen R2 were filled with concrete, the load-carrying capacity of all truss compression joints was markedly enhanced. Once the compressive load-carrying capacity of the CFRST joints exceeded that of the tension joints, the truss failure mode transitioned to tension joint failure (Fig. 3.37C). In the experiment of CFRST truss Specimen R2, the actual observed failure mode was punching-shear failure of the bottom chord tension joints, which aligns with the above analysis. For CFST trusses, although filling the tensile chord members with concrete does not significantly improve the axial tensile capacity of the chords themselves, the mechanical characteristics of trusses mean that tensile chords are also subjected to forces from compressive brace members. In this case, the infilled concrete can substantially enhance the joint load-carrying capacity at the locations of compressive brace members, thereby improving the truss's flexural capacity. Thus filling both compressive and tensile chord members with concrete effectively prevents premature joint failure and further enhances the truss's flexural capacity. After filling the chord members with concrete, the compressive load-carrying capacity of CFRST truss joints is significantly improved, and truss failures predominantly occur at tension joints. Therefore further enhancing the load-carrying capacity of CFRST trusses requires improving the capacity of CFRST tension joints. This can be achieved through two approaches:

Installing longitudinal PBL stiffeners in the chord members: The infilled concrete forms concrete dowels in the holes of the stiffeners, which enhances the tensile performance of the joints. A comparison of test results between truss Specimen PBL-2 and the CFRST specimen shows that both the tensile and compressive joint capacities were significantly improved, leading to a corresponding increase in truss capacity.

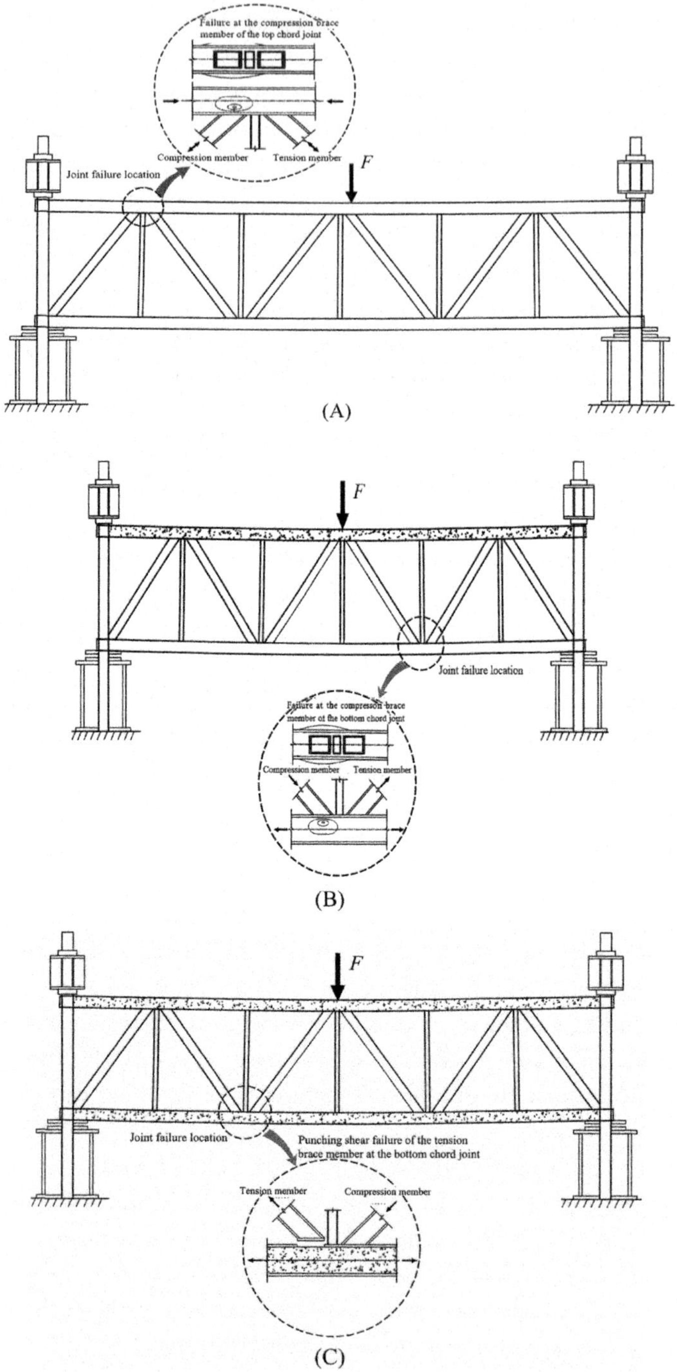

Figure 3.37 Analysis of failure mechanisms in concrete-filled steel tubular (CFST) trusses. (A) Hollow steel tubular truss. (B) Truss with CFST chords (top chords) under compression only. (C) CFST truss.

Leveraging the characteristics of RST joints by adopting CFRST equal-width joints with $\beta = 1$: When subjected to tension, the load from the branch members of this joint is directly transferred to the side plates of the chord member, reducing the risk of failure in the chord's top plate. This design significantly improves joint load-carrying capacity.

A more detailed investigation into the tensile and compressive capacities of CFRST joints is presented in Chapter 4 of this book.

3.5 Summary

(1) The mechanical characteristics of CFST trusses were analyzed. Specifically, methods for determining the failure modes and calculating the load-carrying capacity of CFST trusses were proposed, the concept of a "joint efficiency factor" was introduced, and the correlation between joint load-carrying capacity and truss flexural capacity was established.

(2) The basic mechanical properties of RST truss joints and CFRST truss joints were analyzed, laying a foundation for subsequent research on the mechanical properties of CFRST trusses.

(3) Experimental studies on CFRST trusses were conducted. The results indicate that joints of empty steel tubular trusses are prone to failure under the action of compressive branch members. Filling concrete into both the tensile and compressive chord members of steel trusses can enhance the load-carrying capacity of compression joints; this shifts the truss joint failure location to tension joints and concurrently improves the overall load-carrying capacity of the trusses.

Extended reading

CIDECT Design Guide No.3, Design Guide for Rectangular Hollow Section (RHS) Joints Under Predominantly Static Loading. Cologne: Verlag Tuv Rheinland Press, 2009.

Gao YM, Liu YJ, Jiang L, et al. Experiment on flexural behaviour of rectangular concrete filled steel tubular truss stiffened with PBL. Journal of Architecture and Civil Engineering 2017;34(5):171–80.

Liu JP, Liu YJ. Research on influence of chord concrete-inside on overall stiffness of rectangular hollow section steel tube truss girder. Journal of Building Structures 2009;30(S2):202–7.

Liu JP, Liu YJ. Effect of concrete-filled in chord tube on the mechanical behavior of RHS steel tube joints. Journal of Xi'an University of Architecture & Technology(Natural Science Edition) 2011;43(01):18–24.

Liu JP, Liu YJ, Huang WJ. Effect of concrete-filled chord tubes on the mechanism of RHS steel tubular trusses. Journal of Guangxi University(Natural Science Edition) 2012;37(01):81–7.

Liu YJ. Experiments on ultimate bearing capacity and research on design method of joints of concrete-filled rectangular steel tube truss. Changsha: Hunan University; 2003.

Liu YJ, Jiang L, Xiong ZH. Comprehensive failure model of yield line and punching shear for rectangular steel tube joint. Journal of Architecture and Civil Engineering 2016;33(06):6–13.

Liu YJ, Li YX, Liu JP, Yang GJ. Static behavior analysis of RHS trusses with concrete-filled compression chord member. Journal of Architecture and Civil Engineering 2008;25(04):65–72.

Liu YJ, Liu JP, Yang GJ, Deng YB. Experimental research on mechanical behavior of RHS trusses with concrete-filled in chord. Journal of Building Structures 2009;30(06):107–12.

Liu YJ, Liu JP, Zhang JG. Experimental research on RHS and CHS truss with concrete-filled chord. Journal of Building Structures 2010;31(04):86–93.

Ma YP, Liu YJ, Ma TY, et al. Flexural stiffness of rectangular hollow section (RHS) trusses. Engineering Structures 2021;239:112336.

Ma YP, Liu YJ, Long X, et al. Simplified calculation method for flexural strength of CFST composite truss girder. Journal of Building Structures 2020;41(05):76–84.

CHAPTER 4

Static performance of concrete-filled rectangular steel tubular truss joints

4.1 Overview

Joints are the weak parts of steel tubular trusses and are prone to failure before the truss members reach their ultimate state. At this point, the material utilization of the truss is relatively low, and the corresponding load-bearing capacity is also limited, which should be avoided. After the main chord is filled with concrete, the mechanical performance of the joints is improved to some extent, and the joints exhibit different stress characteristics under the tensile and compressive loads of the branch members. This chapter presents a series of experimental studies on concrete-filled rectangular steel tubular (CFRST), analyzing the mechanical behavior of the joints when the main chord is filled with concrete and the branch members are subjected to tension, compression, and bending.

4.2 Experimental study on the compressive performance of concrete-filled rectangular steel tubular truss joints

4.2.1 Specimen design and fabrication

The steel tubes used in Packer's test had a rectangular section of 127.5 × 177.8 × 4.75 mm. To reduce the experimental scale and make effective use of Packer's test results, the specimens in this study adopted a different square steel tube section, with dimensions of 80 × 80 × 3 mm, a length of 750 mm, and made of high-frequency welded tubes. The mechanical properties of the steel tube material were obtained by tensile tests in accordance with the method specified in Test Methods for Metallic Materials in Tension (GB228-87). The measured ultimate tensile strength was $f_u = 580$ MPa, and the yield strength was $f_y = 325$ MPa.

The concrete inside the tubes was cast manually. To analyze the influence of the infilled concrete on the transverse local bearing

Composite Truss Bridges With Concrete-filled Rectangular Steel Tube Members
DOI: https://doi.org/10.1016/B978-0-443-44011-3.00005-6

performance of CFRST members, the filling length of the concrete inside the tube, Lc, was varied. A total of 12 specimens were fabricated, including 9 concrete-filled steel tubular (CFST) specimens and 3 hollow steel tube specimens. Before casting the concrete, a wooden formwork was placed along one side inside the steel tube. During casting, strict control was maintained to ensure equal concrete length on both sides of the specimen's center, with the concrete length determined by measuring the unfilled portion of the tube. The cube compressive strength fcu of the concrete was determined in accordance with the Standard for Test Method of Mechanical Properties of Ordinary Concrete (GBJ81-85). The measured compressive strength of 150 mm cube specimens was f_{cu}=31.1 MPa.

Following J.A. Packer's experimental method, a steel bearing plate was used in place of the compressed branch member. The local bearing steel plate had a thickness of 12 mm and was welded at the center of the upper and lower flanges of the steel tube, with a weld thickness of 5 mm. The construction details of the specimens are shown in Fig. 4.1.

The areas of the steel bearing plates were 30 × 30 mm, 40 × 40 mm, 60 × 60 mm, and 80 × 80 mm, corresponding to branch-to-chord width ratios ($\beta = b_1/b_0$) of 0.375, 0.5, 0.75, and 1.0 for the rectangular concrete-filled steel tubular X-type compression joints. The width-to-thickness ratio of the steel tube was 28.3, with $\gamma = b_0/2t_0 = 14.15$. The geometric parameters of the specimens are listed in Table 4.1, which also includes the parameters of Packer's test specimens.

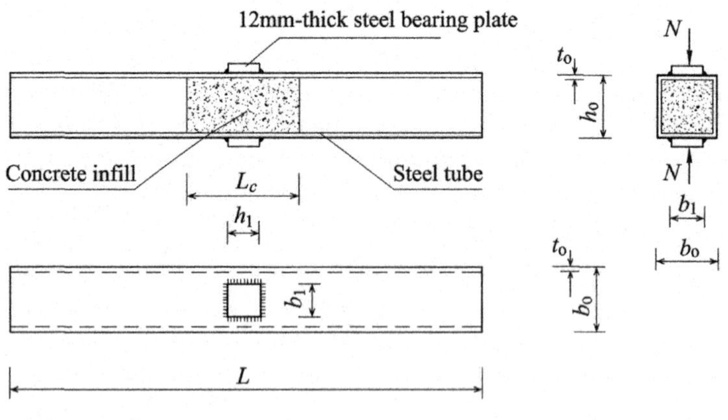

Figure 4.1 Test configuration.

Table 4.1 Specimen parameters.

Specimen number			Chord					Bearing plate		Note
			b_0 (mm)	h_0 (mm)	t_0 (mm)	L (mm)	L_c (mm)	b_1 (mm)	h_1 (mm)	
Test specimen	Xc-30-150		80.0	80.0	2.83	750	165	30	30	Specimen number
	Xc-30-350						357			(Xc-××-×××):
	Xc-30-750						750			Xc—Corresponding to X-type
	Xc-40-000						0	40	40	compression joints
	Xc-40-150						160			××—Width of steel bearing
	Xc-40-750						750			plate (mm)
	Xc-60-000						0	60	60	×××—Length of concrete
	Xc-60-350						362			infill (mm)
	Xc-60-750						750			Yield strength of steel tube:
	Xc-80-000						0	80	80	f_y =325 MPa
	Xc-80-150						160			Cube compressive strength of
	Xc-80-750						750			concrete: f_{cu}=31.1 MPa
Test specimen with Packer	A	1	127.5	177.8	4.75	750	0	127.5	126.5	Yield strength of steel tube:
		5					125			f_y =330 MPa
		7					300			Cylinder compressive strength of
		3					750			concrete:
	B	2	127.5	177.8			0	63.8	126.5	f_c' =43.3 MPa
		6					125			
		8	127.8				300			
		4					750			
	C	9	177.8	127.8			0	63.8	126.5	
		12					125			
		13					300			
		10					750			
	D	14	177.8	127.8			300	126.5	63.8	
		11					750			

4.2.2 Experimental method

4.2.2.1 Test setup and loading scheme

The transverse local compressive load tests of the CFRST were carried out on a universal testing machine, as shown in Fig. 4.2. The load values were directly read from the testing machine.

A monotonic stepwise loading scheme was adopted. Before reaching 60% of the estimated ultimate load, loading was applied in increments of 10% of the estimated ultimate load. After this stage, or once obvious nonlinear deformation appeared, the load increment was reduced to 5% of the estimated ultimate load, continuing until the load could no longer increase.

The testing procedure was as follows: (1) Install the specimen and perform geometric alignment. (2) Install the dial gauges, reset them to zero, and record the initial values from the strain gauges. (3) Apply a preload to perform physical alignment. (4) Remove the preload, reset the dial gauges to zero, and record the initial values from the strain gauges. (5) Apply stepwise loading until the load could no longer increase, that is, until the joint reached its ultimate load. (6) Unload, marking the end of the test.

4.2.2.2 Test measurements

Digital dial gauges were used to measure the vertical deformation between the upper and lower bearing plates of the specimens, as well as the horizontal deformation at the midpoint of the steel tube side plate at the specimen center under each load level (measurement point arrangement shown in Fig. 4.3). These measurements were used to obtain the load-

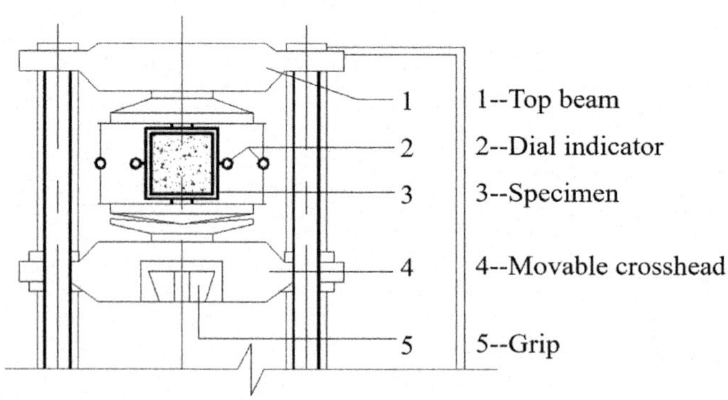

Figure 4.2 Test setup for transverse local bearing of concrete-filled rectangular steel tubular.

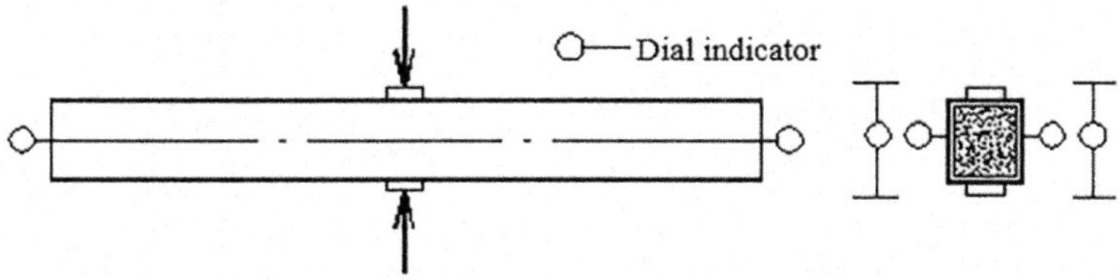

Figure 4.3 Layout of displacement measurement points.

deformation curves of the specimens, in order to analyze the deformation characteristics, load-bearing capacity, and influencing factors of the transverse local compression of CFRSTs.

For specimens Xc-40-000 and Xc-40-750, strain gauges were attached to the steel tube surface (measurement point arrangement shown in Fig. 4.4). The surface strain of the steel tube under each load level was measured using a YJ28 static resistance strain gauge, so as to understand the stress distribution characteristics of CFRSTs under transverse local compression and to analyze the effect of infilled concrete on the mechanical performance of compressed joints in CFRSTs.

4.2.3 Test results and analysis

4.2.3.1 Failure modes and failure loads

The external appearance of specimen failures is shown in Fig. 4.5. The failure mode of the specimens was closely related to the β value. When $\beta = 1$, the failure mode of the hollow steel tube specimen (Xc-80-000) was local buckling of the steel tube side plate, resulting in local instability failure. For the CFST specimen with a concrete filling length of Lc=160 mm (Xc-80-150), the side plate of the steel tube bulged outward. For specimen Xc-80-750, with a concrete filling length of 750 mm, when the ultimate load was reached, concrete at both ends was extruded 1–3 mm beyond the tube ends.

When $\beta < 1$, the failure mode of the hollow steel tube specimens (Xc-40-000 and Xc-60-000) was bending deformation failure of the steel tube flange plates and side plates. In contrast, the failure mode of the CFST specimens was compressive failure of the concrete inside the tube. For specimen Xc-40-150, an initial cracking sound occurred at about 43% (90 kN) of the ultimate load, and a louder cracking sound was heard when approaching the ultimate load (195 kN). For specimen Xc-40-750, an initial cracking sound appeared at 76% (220 kN) of the ultimate load, and a louder cracking sound occurred at the ultimate load. Similarly, the

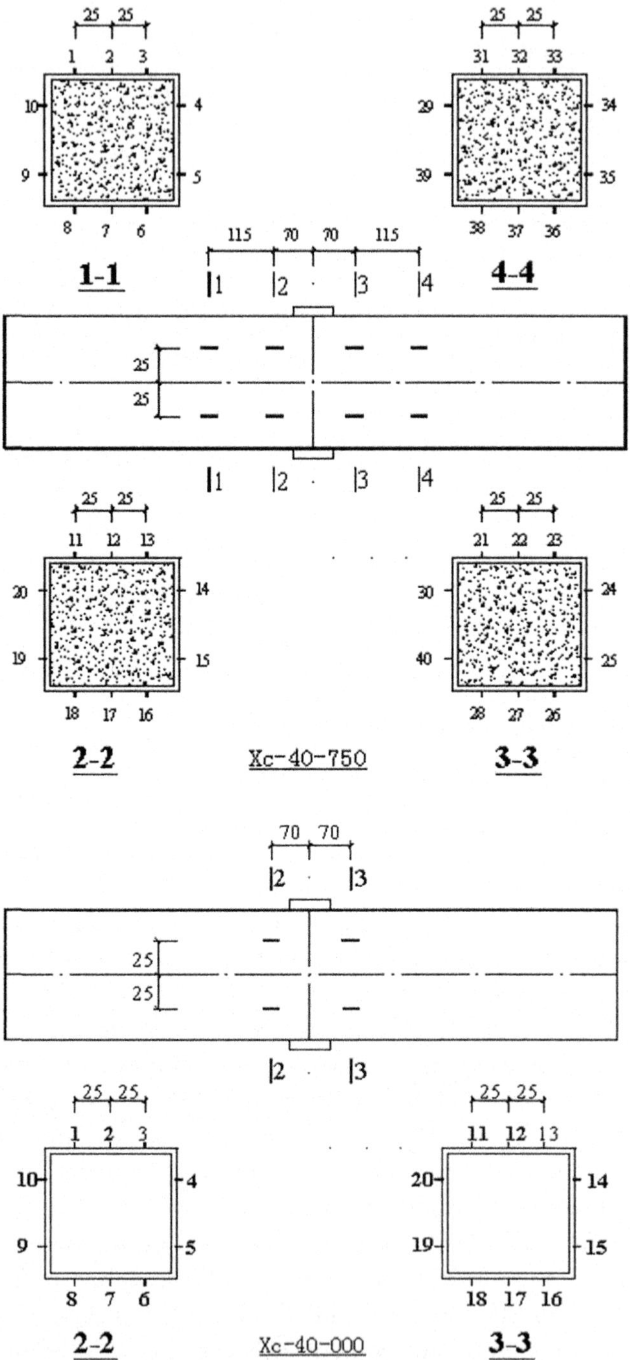

Figure 4.4 Layout of strain measurement points (mm).

Figure 4.5 External view of specimen failure condition.

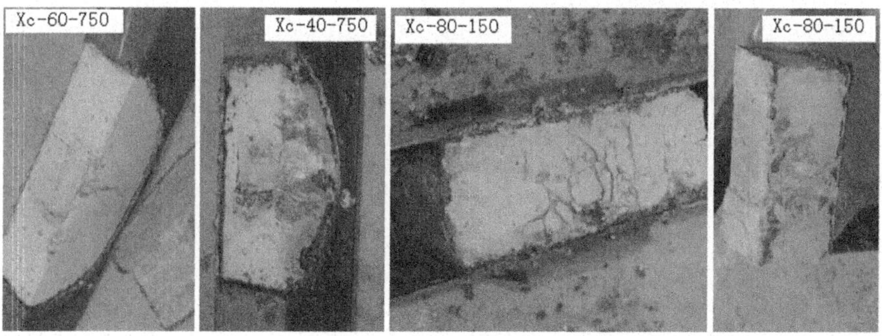

Figure 4.6 Condition of infill concrete failure in specimen.

CFST specimens in the Xc-30 and Xc-60 series also exhibited compressive failure of the infilled concrete, with the initial cracking load following a pattern similar to that of the Xc-40 series: the initial cracking load was approximately 40%–80% of the ultimate load. As the length of the infilled concrete increased, the initial cracking load approached the ultimate load. This indicates that the concrete outside the local compression zone effectively restrained cracking of the concrete within the local compression zone, delaying the onset of cracking.

Fig. 4.6 shows the concrete failure inside specimens Xc-60-750, Xc-40-750, and Xc-80-150, all of which exhibited indentation failure of the concrete beneath the steel bearing plate. The difference is that in

Xc-60-750 and Xc-40-750, due to the effective confinement provided by the concrete outside the local compression zone, only fine cracks appeared on the concrete surface. In contrast, for Xc-80-150, a larger amount of concrete fragments was observed on the side of the infilled concrete, but no cracks were visible after the spalling.

The measured failure loads of the specimens are listed in Table 4.2, which also includes the test results of Packer. Fig. 4.7 shows the relationship curve between the failure load of the specimens in this study and the length of the infilled concrete. It can be observed that the transverse local bearing strength of CFRST members increases with the filling length of the infilled concrete. However, this strength cannot increase indefinitely with concrete length; once the concrete length exceeds a certain limit, the improvement in local bearing strength becomes minimal. This observation is consistent with Packer's test results.

In the literature, Packer adopted the transverse local bearing strength enhancement factor of CFRST members as equal to the enhancement factor of plain concrete local bearing strength calculated according to the European Code for Concrete Structures (CEB-FIP). Based on this, the transverse local bearing strength of CFRST members was obtained, and the formula for calculating the bearing capacity of X-type compression joints was expressed as:

$$N_{uP} = k_c f_c' A_b \tag{4.1}$$

If the enhancement factor for the local bearing strength of plain concrete is determined according to the Chinese Code for Design of Concrete Structures (GB50010-2010), the corresponding calculation formula is obtained:

$$N_{uC} = k_c f_c A_b \tag{4–2}$$

In Eqs. (4.1) and (4.2), $k_c = \sqrt{A_d/A_b}$ denotes the enhancement factor for the local bearing strength of plain concrete, calculated according to Packer's recommended European concrete design code method and the Chinese Code for Design of Concrete Structures (GB50010-2010) method, respectively. fc' and fc represent the cylinder compressive strength and axial compressive strength of concrete, respectively. A_b=b_1h_1 is the actual bearing area of the concrete, while A_d is referred to as the dispersion bearing area or the calculated base area. The calculation methods for A_d differ significantly, as shown in Fig. 4.8.

Table 4.2 Comparison of experimental and theoretical calculation results.

Specimen number		Test results				Packer		GB50010-2002		
		N_u (kN)	N_u/A_bf_c	N_u/A_bf_c'	k_c	N_{uP} (kN)	N_u/N_{uP}	k_c	N_{uc} (kN)	N_u/N_{uC}
Xc-30-150		160	4.23	4.02	2.03	81	1.98	2.45	93	1.73
Xc-30-350		153	4.05	3.84	2.24	89	1.72	2.45	93	1.65
Xc-30-750		205	5.42	5.15	2.24	89	2.30	2.45	93	2.21
Xc-40-000		40	/	/	/	/	/	/	/	/
Xc-40-150		207	3.50	3.33	1.79	111	1.86	2.19	129	1.60
Xc-40-750		290	4.91	4.66	2.05	127	2.28	2.19	129	2.24
Xc-60-000		60	/	/	/	/	/	/	/	/
Xc-60-350		358	3.09	2.94	1.81	221	1.62	1.85	214	1.67
Xc-60-750		440	3.80	3.61	1.81	221	1.99	1.85	214	2.05
Xc-80-000		125	/	/	/	/	/	/	/	/
Xc-80-150		470	2.76	2.62	1.33	239	1.97	1.33	227	2.07
Xc-80-750		595	3.50	3.32	1.67	299	1.99	1.67	284	2.10
A	1	393	/	/	/	/	/	/	/	/
	5	902	1.41	1.29	1.00	694	1.30	1.00	639	1.41
	7	1 266	1.98	1.81	1.54	1 075	1.18	1.54	916	1.38
	3	1 421	2.22	2.03	1.95	1363	1.04	1.73	1 030	1.38
B	2	118	/	/	/	/	/	/	/	/
	6	521	1.63	1.49	1.00	347	1.50	1.41	418	1.25
	8	660	2.06	1.89	1.54	538	1.23	2.01	597	1.11
	4	721	2.25	2.06	1.95	682	1.06	2.00	596	1.21

(*Continued*)

Table 4.2 (Continued)

Specimen number		Test results			Packer			GB50010-2002		
		N_u (kN)	N_u/A_bf_c	N_u/A_bf_c'	k_c	N_{uP} (kN)	N_u/N_{uP}	k_c	N_{uc} (kN)	N_u/N_{uC}
C	9	73	/	/	/	/	/	/	/	/
	12	856	2.68	2.45	1.00	347	2.47	1.66	494	1.73
	13	1 442	4.51	4.13	1.54	538	2.68	2.37	704	2.05
	10	1 549	4.84	4.44	1.74	607	2.55	2.37	704	2.20
D	14	784	2.45	2.25	2.17	757	1.03	2.05	611	1.28
	11	991	3.10	2.84	2.24	782	1.27	2.05	611	1.62
The ratio of measured to calculated					Average value 1.75Coefficient of variation 29.8%			Average value 1.70Coefficient of variation 21.3%		

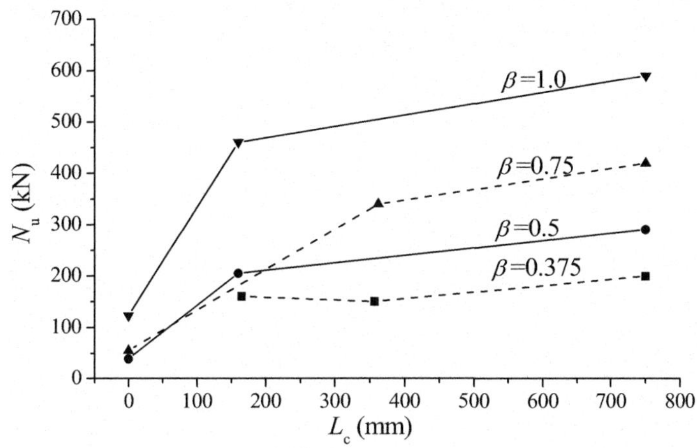

Figure 4.7 Influence of infill concrete length on transverse local compression strength.

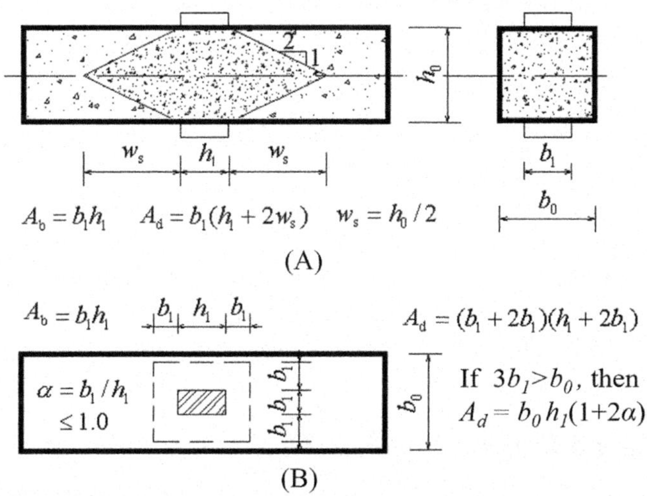

Figure 4.8 Calculation method for A_b and A_d. (A) Packer-suggested European Code for Concrete Structure diffusion bearing area. (B) Calculated base area in China's Code.

Based on the specimen parameters in Table 4.1, the bearing capacities NuP and NuC of the specimens from this test and Packer's test were calculated and are listed in Table 4.2. Table 4.2 also presents the experimentally measured enhancement factors for local bearing strength, $N_u/A_b f_c'$ and $N_u/A_b f_c$, calculated respectively using the concrete cylinder compressive strength and the axial compressive strength. It can be seen that the experimentally measured enhancement factors for local bearing strength

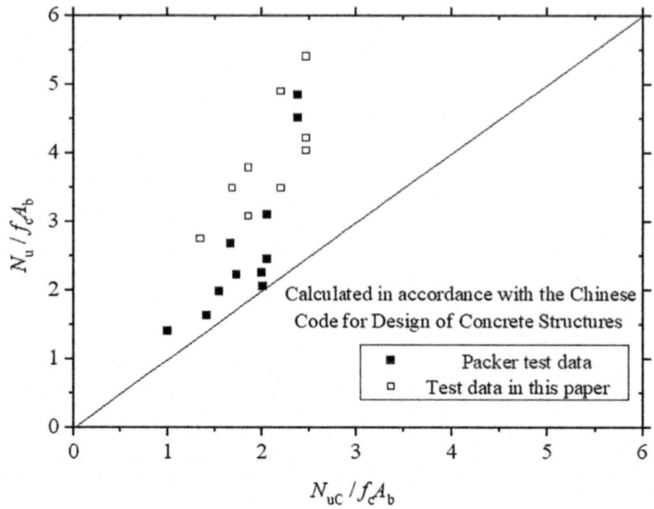

Figure 4.9 Comparison between local compression enhancement factors calculated according to Chinese code and test results.

and the bearing capacities of the specimens are both higher than the calculated values for plain concrete.

Figs. 4.9 and 4.10 compare the measured values with the theoretical values calculated by the two methods. It can be seen that, regardless of the method used, there is a significant discrepancy between the enhancement factors of transverse local bearing strength of CFRST members calculated based solely on plain concrete and the measured values. This indicates that the contribution of the steel tube cannot be neglected.

4.2.3.2 Load-deformation relationship

Fig. 4.11 shows the load-deformation curves of the tested specimens. In the figure, v represents the vertical displacement in the loading direction, which is mainly caused by the compressive deformation of the infilled concrete; u represents the lateral displacement measured at the midpoint of the steel tube side plate, which results from the bending deformation of the side plate and is related to the lateral expansion of the infilled concrete.

It can be seen that near failure the vertical displacement and lateral displacement values are nearly the same from Fig. 4.11A. However, compared with the vertical displacement, the lateral displacement is relatively small in the early stage and develops rapidly in the later stage.

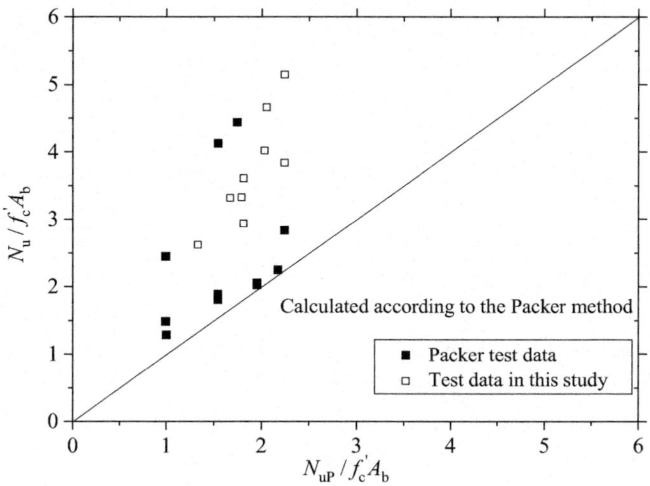

Figure 4.10 Comparison between local compression enhancement factors calculated using Packer's method and test results.

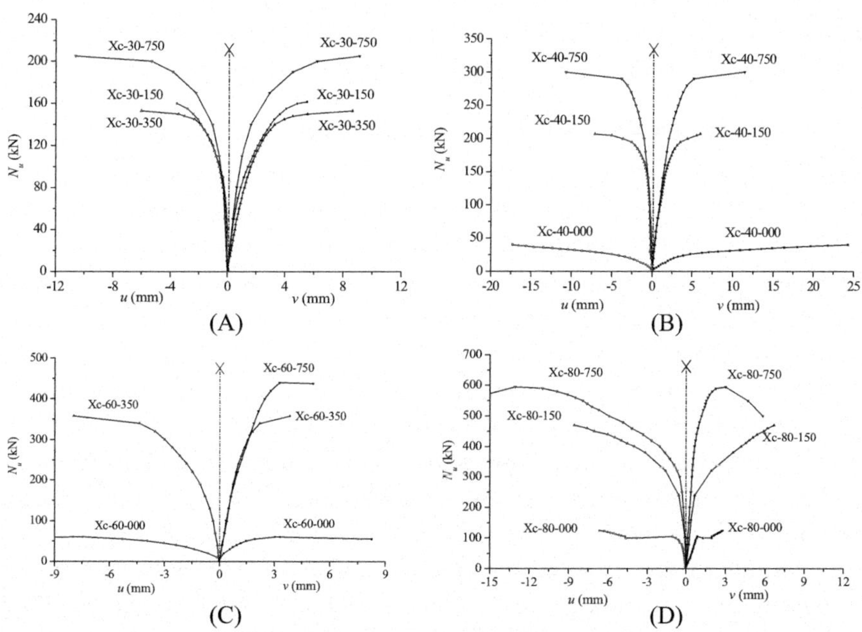

Figure 4.11 Measured load-deformation curves of specimens. (A) Measured load-deformation curve of the specimen with $\beta = 0.375$. (B) Measured load-deformation curve of the specimen with $\beta = 0.5$. (C) Measured load-deformation curve of the specimen with $\beta = 0.75$. (D) Measured load-deformation curve of the specimen with $\beta = 1.0$.

By comparing the deformation curves of hollow steel tube specimens and CFST specimens in Fig. 4.11B–D, it can be observed that the deformation at failure of the CFST specimens can reach the deformation value corresponding to the yielding of the tube wall in the hollow steel tube specimens. This indicates that under transverse local bearing loads, the steel tube wall of CFST members can also reach yielding. For specimens with $\beta = 1.0$, due to the effect of the infilled concrete, the side plates of CFST specimens avoided local instability failure, and their deformation capacity was significantly better than that of the hollow tube specimens.

Test results indicate that the failure characteristics of CFST members under transverse local bearing loads are local indentation of the infilled concrete and bending failure of the steel tube side plates. Due to the Poisson effect of the concrete, the steel tube side plates bend; in turn, the loading of the side plates provides confinement to the concrete, effectively enhancing the compressive strength of the infilled concrete. Furthermore, because of the Poisson effect, the infilled concrete tends to be extruded, but the bond force between the concrete and the inner wall of the steel tube restrains the concrete in the local compression zone, thereby further improving the compressive strength of the infilled concrete.

4.2.3.3 Load–longitudinal strain relationship

To compare and analyze the influence of the infilled concrete inside the chord on the strain distribution pattern of the chord section in the joint region and to further investigate the working mechanism of transverse local bearing in CFRST members, the longitudinal strains of the chord section were measured for the CFRST local bearing specimen Xc-40-750 and the hollow RST local bearing specimen Xc-40-000.

The measurement point arrangement of the hollow steel tube transverse local bearing specimen Xc-40-000 is shown in Fig. 4.4. Figs. 4.12 and 4.13 present the measured load–longitudinal strain curves of cross sections 2-2 and 3-3, located 70 mm away from the specimen center on both sides, for specimen Xc-40-000. In the figure, the numbering of strain gauges corresponds to the measurement point numbers in Fig. 4.4. The vertical axis represents the applied load on the specimen, and the horizontal axis shows the longitudinal strain of each measurement point along the steel tube, with tensile strain taken as positive and compressive strain as negative.

From Figs. 4.12 and 4.13, it can be seen that the strain at each measurement point on the flange plates along the steel tube longitudinal direction is essentially equal, indicating uniform tension; similarly, the

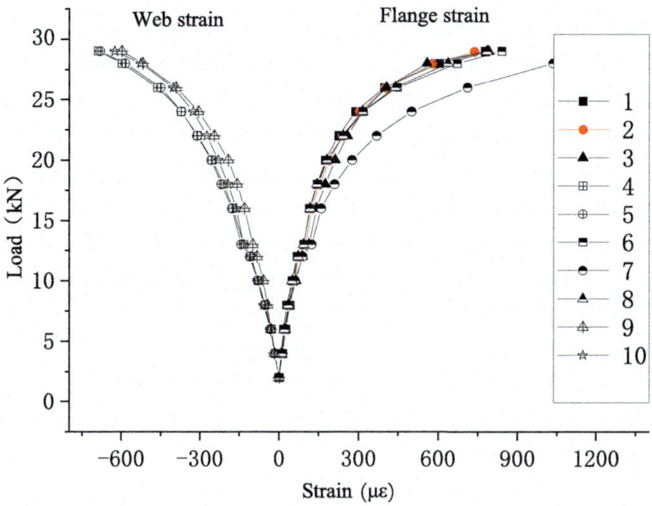

Figure 4.12 Longitudinal strain at measurement points of section 2-2 for specimen Xc-40-000.

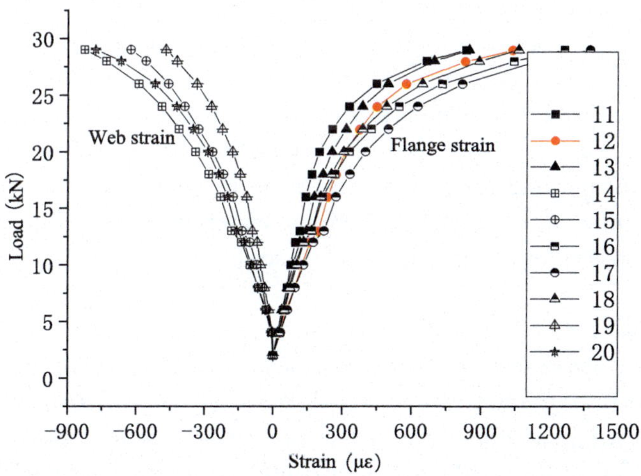

Figure 4.13 Longitudinal strain at measurement points of section 3-3 for specimen Xc-40-000.

strain at each measurement point on the brace plates is also essentially equal, indicating uniform compression. When the applied load is less than about 20 kN, the load-strain curves are nearly linear; when the load exceeds about 20 kN, the strain increases rapidly, showing a clear nonlinear relationship. The failure load of this specimen was 40 kN, but when the applied load reached 29 kN, most strain readings had already become highly unstable, and strain measurements were terminated.

Under transverse local bearing loads, local indentation occurred at the bearing area of the flange plate of the hollow steel tube. The longitudinal strains of cross sections 2-2 and 3-3 consisted of bending strains and membrane strains of the tube wall. At low loads, since cross sections 2-2 and 3-3 were outside the indentation zone, the longitudinal strain of the flange plate was mainly governed by the membrane strain. The membrane effect caused the chord surface at cross sections 2-2 and 3-3 to be subjected to longitudinal tension. As the steel tube was not subjected to longitudinal force under transverse load, according to the equilibrium condition, the brace at cross sections 2-2 and 3-3 had to bear longitudinal compressive stress of equal magnitude.

At higher loads, the indentation zone gradually expanded and approached cross sections 2-2 and 3-3. The flange plate at cross sections 2-2 and 3-3 developed significant negative bending strain, with the outer surface of the tube wall in longitudinal tension. At the same time, due to the outward bulging of the brace, significant positive bending strain occurred at cross sections 2-2 and 3-3, with the outer surface of the tube wall in longitudinal compression. Therefore, at cross sections 2-2 and 3-3, the outer surface of the flange plate was in longitudinal tension, while the outer surface of the brace was in longitudinal compression. As the load increased further and the indentation zone reached cross sections 2-2 and 3-3, the tube wall at these sections experienced significant bending deformation, leading to highly unstable strain readings at the measurement points.

For the CFRST transverse local bearing specimen Xc-40-750, in addition to arranging the same number of measurement points at the sections corresponding to sections 2-2 and 3-3 of specimen Xc-40-000, the same numbers of strain gauges were also arranged at sections 1-1 and 4-4, located 185 mm from the specimen center, as shown in Fig. 4.4.

The load-strain curves of the measurement points near the flange and brace surfaces at the corners of sections 2-2 and 3-3 of specimen Xc-40-750 are shown in Figs. 4.14–4.18. From these figures, it can be observed that when the load is below approximately 50 kN, the stresses at all measurement points are close to zero, indicating that at the initial stage of loading, the load is mainly carried by the infilled concrete. As the load increases, the infilled concrete undergoes vertical compression and, due to the Poisson effect, expands both longitudinally and laterally. The lateral

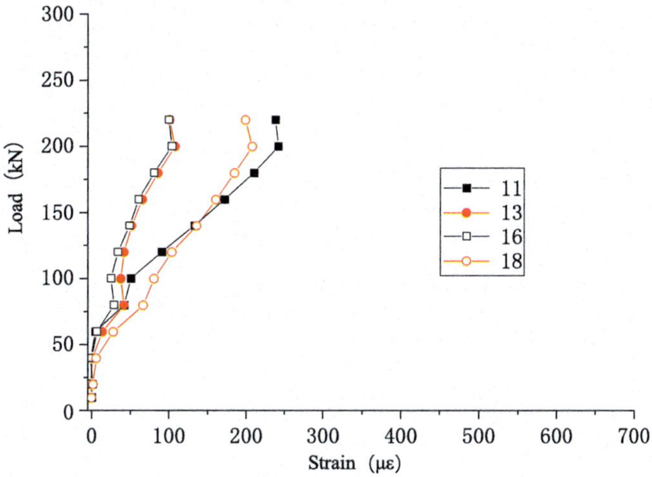

Figure 4.14 Longitudinal strain at flange edge measurement points of section 2-2 for specimen Xc-40-750.

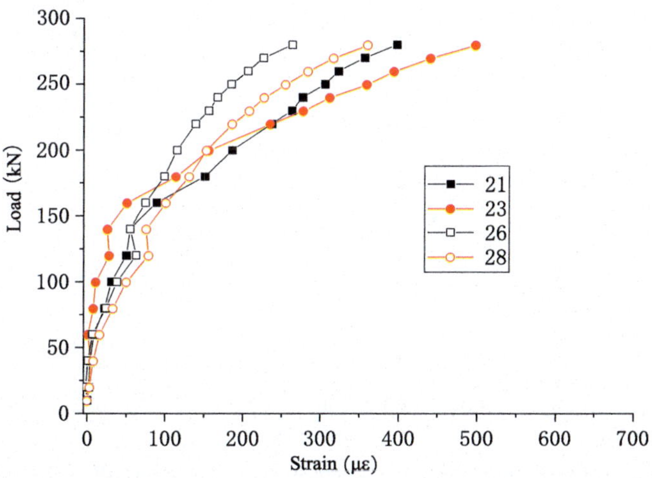

Figure 4.15 Longitudinal strain at flange edge measurement points of section 3-3 for specimen Xc-40-750.

expansion of the concrete is restrained by the braces, while its longitudinal expansion is restrained by the bond and friction between the concrete and the inner wall of the steel tube. As a result, longitudinal tensile stress develops in the steel tube wall.

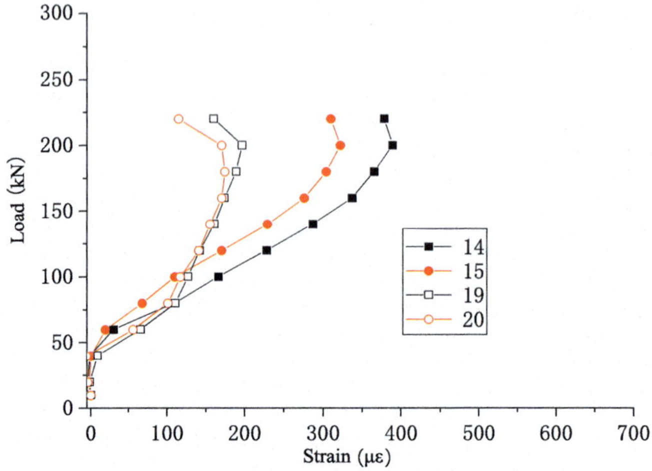

Figure 4.16 Longitudinal strain at brace measurement points of section 2-2 for specimen Xc-40-750.

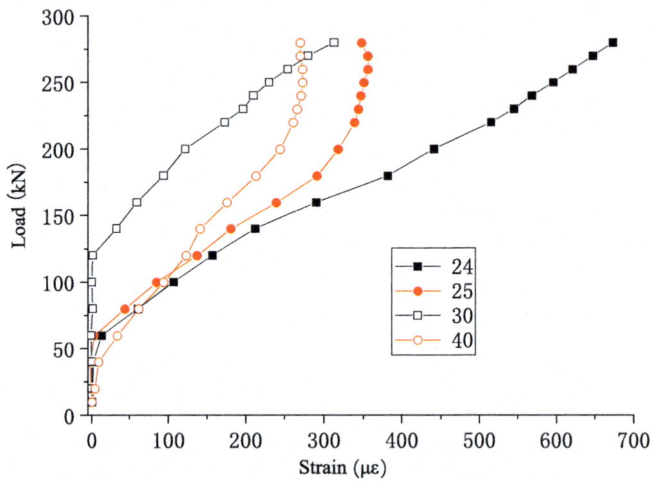

Figure 4.17 Longitudinal strain at brace measurement points of section 3-3 for specimen Xc-40-750.

A comparison of the strains of the flange and brace near the corners of sections 2-2 and 3-3 shows that the strain in the brace is significantly greater than that in the flange. This is due to the stronger frictional resistance at the brace-concrete interface caused by the brace's constraint on the lateral expansion of the infilled concrete, as well as the bending of the brace itself. When the load reaches 200 kN, the strains at each measurement point of

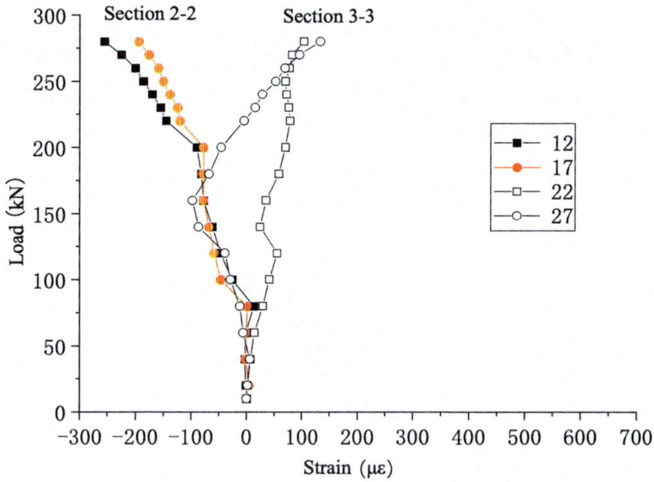

Figure 4.18 Longitudinal strain at mid-flange measurement points of section 2-2 and section 3-3 for specimen Xc-40-750.

section 2-2 decrease, while those at section 3-3 continue to increase, indicating that the bond and friction between the infilled concrete and the inner wall of the steel tube at the 2-2 end have been overcome, causing the infilled concrete to be extruded at one end. The longitudinal strain values of the measurement points at the two sections show some scatter, which can be attributed to factors such as the nonhomogeneous nature of the infilled concrete. Overall, when the load is below 200 kN, the strains at the flange corners of sections 2-2 and 3-3 are approximately equal, and the brace strains near the corners follow the same pattern.

Since the steel tube wall at the midpoints of the flanges provides relatively weak restraint on the infilled concrete, the bond friction between the inner wall of the steel tube and the concrete at these locations is significantly smaller than that at the corners. Therefore the longitudinal strain at the flange midpoints is relatively small. Fig. 4.18 shows the measured load-strain curves at the flange midpoints of sections 2-2 and 3-3, which are clearly smaller compared to those at the corner measurement points. Among them, measurement points 12 and 17 at section 2-2 show compressive strain, while measurement point 22 at section 3-3 also shows compressive strain before the load reaches approximately 200 kN, and then shifts to tensile strain, though the value remains small. Measurement point 27 remains in tension throughout, but with relatively small values. The stress distribution exhibits a certain degree of scatter.

Figs. 4.19 and 4.20 show the longitudinal load–strain curves at sections 1-1 and 4-4, located 185 mm away from the center of the bearing plate. At these sections, all measurement points are under longitudinal tension, and the strain magnitudes at the same section are essentially equal, indicating

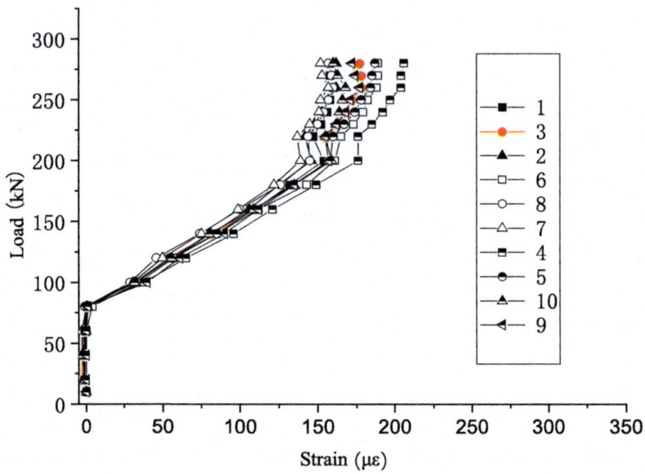

Figure 4.19 Longitudinal strain at measurement points of section 1-1 for specimen Xc-40-750.

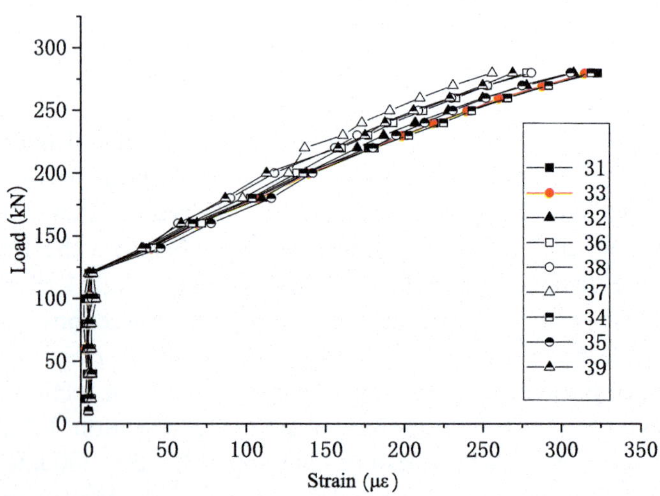

Figure 4.20 Longitudinal strain at measurement points of section 4-4 for specimen Xc-40-750.

relatively uniform strain distribution. When the load is below 75 kN, the longitudinal strain at section 1-1 is nearly zero; thereafter the strain increases approximately linearly, indicating a tendency of the infilled concrete to be extruded. When the load approaches 200 kN, the load-strain curves at the measurement points of section 1-1 exhibit a turning point, with the strain no longer increasing and stabilizing around 170 με, suggesting that the infilled concrete on that side has been extruded. At section 4-4, the longitudinal strain remains nearly zero before the load reaches 120 kN, and then increases approximately linearly until specimen failure, reaching a maximum strain of about 300 με.

Analysis of the load-strain curves of this specimen shows that the section farther from the bearing plate is subjected to tensile stress with a relatively uniform distribution, while the stress distribution near the bearing plate is more scattered. For comparison, the average strains of sections 1-1 and 4-4, as well as the average flange strain and average brace strain at the corners of sections 2-2 and 3-3, were calculated, and the corresponding load-strain curves are plotted in Figs. 4.21 and 4.22.

A comparison of the longitudinal strain between the far-from-bearing-plate section 1-1 and the near-bearing-plate section 2-2 in Fig. 4.21 shows that, after the infilled concrete tends to be extruded, the average flange strain at the corners is nearly the same as the average strain of section 1-1,

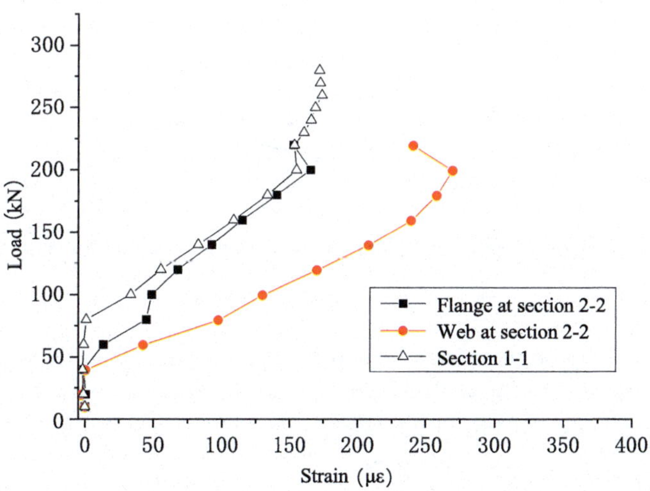

Figure 4.21 Average longitudinal strain at measurement points of section 1-1 and section 2-2 for specimen Xc-40-750.

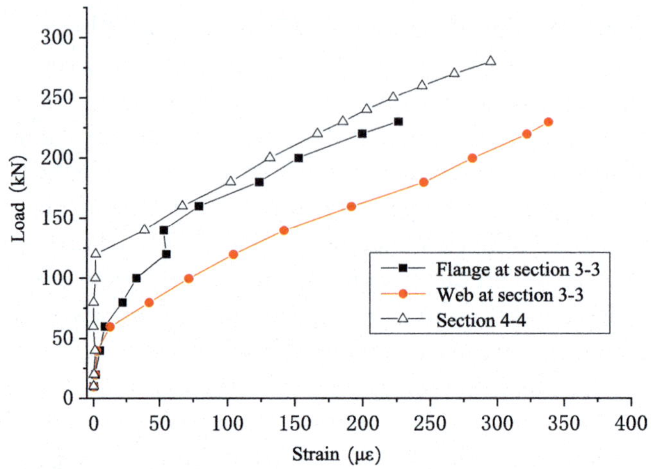

Figure 4.22 Average longitudinal strain at measurement points of section 3-3 and section 4-4 for specimen Xc-40-750.

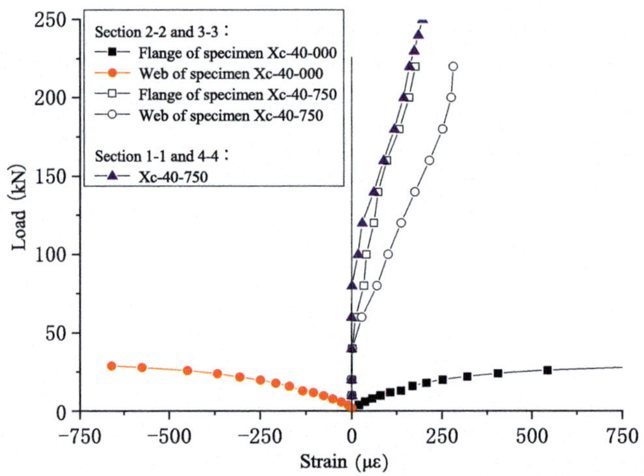

Figure 4.23 Average longitudinal strain at measurement points for specimens Xc-40-000 and Xc-40-750.

while the average brace strain near the corners is significantly larger. When the load reaches 200 kN, the extrusion of the infilled concrete causes the strain to decrease. In Fig. 4.22, the average strains of sections 3-3 and 4-4 exhibit a similar variation pattern.

Fig. 4.23 compares the longitudinal average strains on the steel tube surface between the hollow tube specimen and the CFST specimen. Near the bearing plate, the hollow tube specimen shows tensile strains along the flange plate and compressive strains along the brace plate, with longitudinal stresses being self-balanced, and the strain values are relatively large.

For the CFST specimen, the load is mainly carried by the infilled concrete, and the longitudinal strain on the steel tube surface remains nearly zero when the applied load is below a certain value. Under transverse loading, the infilled concrete undergoes longitudinal expansion; the concrete outside the local bearing zone is subjected to longitudinal compression, which is balanced by longitudinal tensile stresses in the steel tube flanges and braces through the bond friction between the inner steel surface and the concrete. As the longitudinal expansion of the concrete in the local bearing zone increases, the bond friction gradually extends toward both ends of the specimen, causing the longitudinal tensile stresses in the flange and brace plates to increase until the bond friction is completely overcome and the infilled concrete is extruded from the specimen ends.

Due to this difference in the load-transfer mechanism, the longitudinal strain on the steel tube surface of the CFST specimen is significantly smaller than that of the corresponding hollow tube specimen. Among them, the brace stress near the local bearing zone is larger than the flange stress, while at positions farther away from the local bearing zone, the stress distribution becomes more uniform and smaller than that near the local bearing zone.

The phenomenon that the brace stress of the steel tube near the local bearing zone is greater than the flange stress is related to the wedge effect. Under transverse local bearing load, the flange near the local bearing zone of the CFST specimen deflects inward, forming an inverted wedge shape as shown in Fig. 4.24, which reduces the bond friction between the concrete and the inner wall of the steel tube. In contrast, the brace near the local bearing zone bulges outward due to the lateral expansion of the infilled concrete, creating the positive wedge effect illustrated in Fig. 4.24. This results in larger bond friction. Therefore, from the onset of bond friction, the longitudinal stress in the steel tube brace near the local bearing zone is consistently greater than that in the flange, as shown in Fig. 4.23.

4.2.4 Calculation formula for the transverse local bearing strength of concrete-filled rectangular steel tubular

The above experimental results indicate that, due to the Poisson effect of concrete, when a CFST member is subjected to transverse local compressive loading, the concrete inside the tube not only indents but also expands laterally and longitudinally. The lateral expansion causes the steel tube

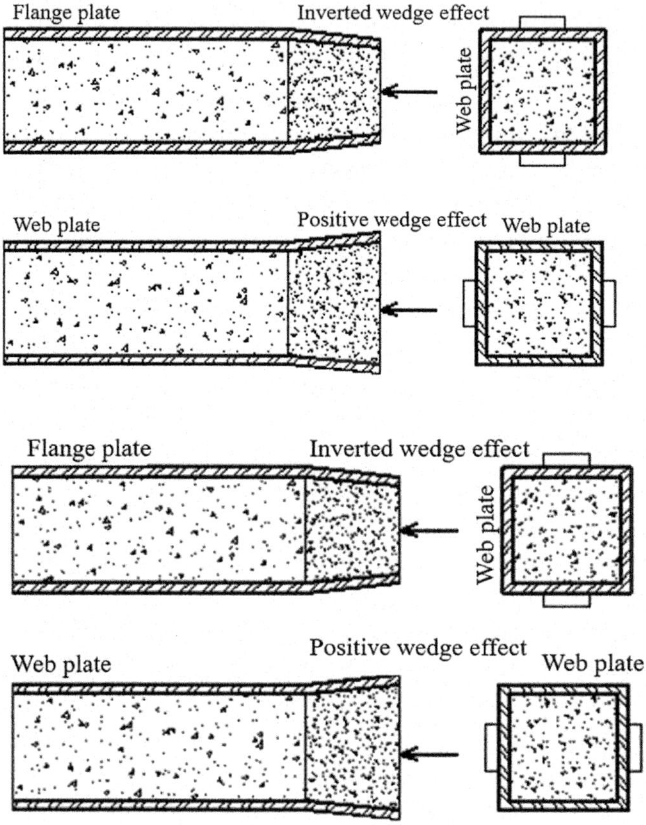

Figure 4.24 Wedging effect in local compression zone.

brace to bend, while the longitudinal expansion is restrained by the bond and friction between the concrete and the inner wall of the steel tube, thereby inducing longitudinal tension in the steel tube. Through the bond and friction, the lateral loading of the steel tube brace and the longitudinal tension of the steel tube in turn impose a confining effect on the concrete in the local bearing zone. As a result, the concrete in this region is subjected to a triaxial compressive stress state, which effectively enhances its compressive strength. Therefore the transverse local bearing strength of CFRSTs is higher than that of plain concrete.

4.2.4.1 Participation coefficient of steel tube in transverse local bearing

To investigate the steel tube's contribution to transverse local bearing performance and to reflect the restraining action of the steel tube brace on

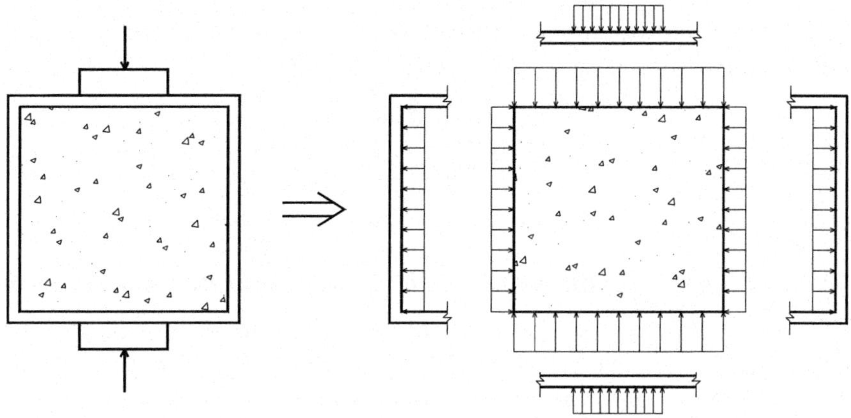

Figure 4.25 Simplified model of the working mechanism of the steel tube's participation.

the concrete inside the tube, a simplified model shown in Fig. 4.25 was established based on the above experimental failure modes and deformation characteristics of CFRSTs under transverse local compression.

In Fig. 4.25, the plastic yield moment of the steel tube brace under horizontal load is given by $m_s = f_y t_0^2/4$. When the concrete inside the tube is subjected to local bearing load, due to the Poisson effect, the concrete applies a load ppp on the steel tube brace. Considering the brace as a fixed-ended beam, the maximum bending moment produced by ppp is $m_c = p h_0^2/12$. Similar to the confinement coefficient in axially compressed circular CFSTs, the participation coefficient of the steel tube in transverse local bearing is defined as $k_s = m_s/m_c$. Assuming the concrete inside the tube is an ideal isotropic material, at failure, the maximum value of the load ppp applied on the steel tube brace due to the Poisson effect of the concrete should be $v_c f_c$, where v_c is the Poisson's ratio of the concrete. Considering that p is nonuniformly distributed along the inner surface of the steel tube brace, the following expression is obtained:

$$p = \phi v_c f_c \tag{4.3}$$

where ϕ—coefficient of nonuniform load distribution; based on trial calculations and comparative analysis, it is recommended to adopt:

$$\phi = \frac{A_b/A_d}{(h_1/h_0)^2} \tag{4.4}$$

Thus the participation coefficient of the steel tube in transverse local bearing under transverse local compressive loading of CFRSTs is:

$$k_s = \frac{3f_y t_0^2 (h_1/h_0)^2}{\nu_c f_c h_0^2 (A_b/A_d)} \tag{4.5}$$

where $\nu_c = 0.16 \sim 0.23$; in this book, $\nu_c = 0.23$ is adopted.

4.2.4.2 Influence coefficient of the length of concrete inside the tube

The restraining effect of the steel tube on the concrete inside the tube through bond and friction depends on the bond-friction resistance along the inner wall of the steel tube, which is mainly determined by the filling length of the concrete. Packer suggested that for X-type compression joints, the concrete length inside the tube should satisfy $L_c \geq h_1 + 2h_0$. Therefore, in this book, the local bearing region of concrete is taken as $h_1 + 2h_0$. Based on statistical analysis of the experimental data from this book and from Packer, the influence coefficient of the length of concrete inside the tube is obtained as:

$$k_L = 0.12(L_c - h_1 - 2h_0)/L_c \tag{4.6}$$

In practice, the local bearing strength cannot increase indefinitely with the length of the concrete. Therefore it is recommended that $k_L \leq 0.1$.

4.2.4.3 Calculation formula for the transverse local bearing strength of concrete-filled rectangular steel tubulars

By applying the concept of the confinement effect of circular CFSTs in the Chinese Code CECS 28:2012 Design and Construction Specification for Concrete-Filled Steel Tubular Structures, the calculation formula for the transverse local bearing strength of CFRSTs is established as follows:

$$N_{ucs} = k_{cs} f_c A_b \tag{4.7}$$

where k_{cs} is the enhancement coefficient of the transverse local bearing strength of CFRSTs, given by $k_{cs} = (1+k_s)(1+k_L)k_c$. k_c is the enhancement coefficient of the local bearing strength of plain concrete, calculated as $k_c = \sqrt{A_d/A_b}$ in accordance with the calculation method of the *Code for Design of Concrete Structures* of China, while neglecting the provision that the local bearing height should be greater than the width of the local bearing area. When the concrete inside the tube is high-strength concrete, based on the series of experimental results conducted by the China Academy of Building Research on concretes of different strengths, the enhancement coefficient of the local bearing strength of plain concrete is calculated as follows:

$$k_c = \frac{190 - f_{cu}}{150}\sqrt{A_d} \text{ and } k_c \geq 1 \tag{4.8}$$

The enhancement coefficients of transverse local bearing strength and the corresponding bearing capacities of the test specimens, calculated using Eq. (4.7), are listed in Table 4.3. In the table, the cylindrical compressive strength of the concrete inside the steel tube in Packer's test specimens is $f_c' = 43.3$ MPa, which is converted to a 150 mm cube compressive strength of 52.17 MPa. The value of k_c is calculated according to Eq. (4.8). The ratios of the measured to the calculated local bearing strengths in the table are all greater than 1, with an average of 1.23 and a coefficient of variation of 12.0%. This indicates that Eq. (4.7) agrees well with both the experimental results in this book and those of Packer.

Using the specimen parameters, the calculated bearing capacities were taken as the horizontal axis and the measured bearing capacities as the vertical axis, and the experimental data were plotted in Fig. 4.26. The test points in the Figure are all located above the isopleth and generally close to it, indicating that the calculated values are in good agreement with the experimental results. Therefore considering the participation of the steel tube and the influence of the concrete length inside the tube can effectively reflect the load-bearing mechanism of the transverse local compression of CFRST. The enhancement coefficient of transverse local bearing strength calculated by the method presented in this book is thus both safe and appropriate.

4.3 Experimental study on the tensile and bending performance of concrete-filled rectangular steel tubular truss joints

In this section, X-type CFRST tension joints are taken as test specimens to study their mechanical performance under axial tensile loading of the brace members, with comparison to Packer's test results on T-type tension joints. X-type CFRST bending joints are also tested to investigate the mechanical performance of CFRST truss joints under brace member bending loads. The experiments focus on the influence of factors such as the width ratio between brace and chord members, the width-to-thickness ratio of chord members, and the presence of infilled concrete on joint performance. Comparative tests were conducted on four CFRST truss tension joints and one hollow tube tension joint, as well as four CFRST truss bending joints and one hollow tube bending joint.

Table 4.3 Comparison between experimental results and the calculated results of Eq. (4.18).

Tests described in this book		Calculation using Eq. (4.18)			Packer test			Calculation using Eq. (4.18)		
Specimen number	N_u (kN)	$k_{c\ s}$	N_{ucs} (kN)	Specimen number		N_u (kN)	$k_{c\ s}$	N_{ucs} (kN)	Specimen number	N_u (kN)
Xc-30-150	160	3.19	121	1.33	A	5	902	1.04	664	1.36
Xc-30-350	153	3.45	130	1.17		7	1 266	1.49	951	1.33
Xc-30-750	205	3.56	135	1.52		3	1 421	1.88	1 201	1.18
Xc-40-150	207	3.00	177	1.17	B	6	521	1.34	428	1.22
Xc-40-750	290	3.38	200	1.45		8	660	2.05	656	1.01
Xc-60-350	358	3.07	356	1.01		4	721	2.25	719	1.00
Xc-60-750	440	3.19	369	1.19	C	12	856	2.06	659	1.30
Xc-80-150	470	1.91	324	1.45		13	1 442	3.88	1 241	1.16
Xc-80-750	595	3.22	548	1.09		10	1 549	4.25	1 358	1.14
Ratio of measured-to-calculated values:					D	14	784	2.19	701	1.12
Average value: 1.23						11	991	2.36	755	1.31
Coefficient of variation: 12.0%										

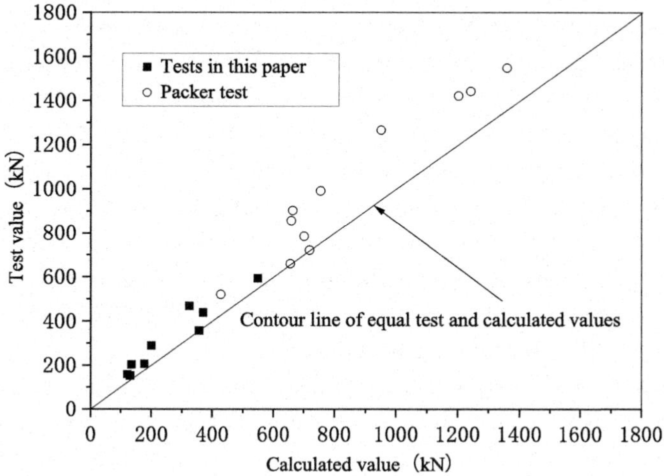

Figure 4.26 Comparison between the experimental and calculated values of specimen bearing capacity.

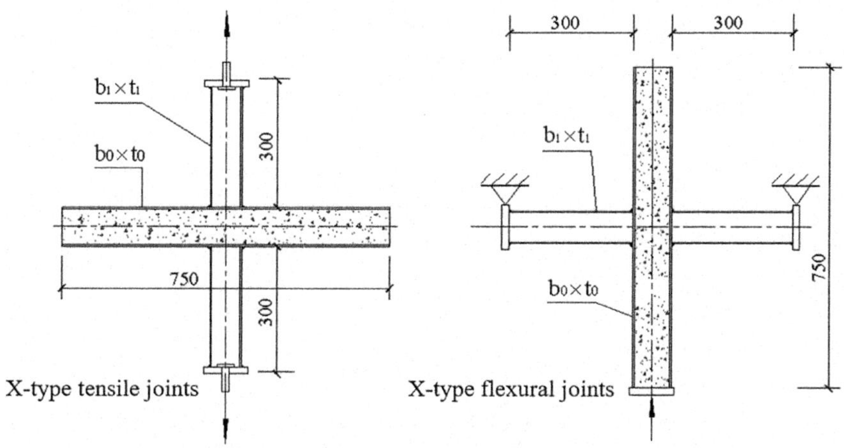

Figure 4.27 Joint configuration in the test.

4.3.1 Test design and fabrication

In this experimental design, the angle between the brace and chord members of the X-type joints is 90 degrees, and the specimen configuration and loading conditions are shown in Fig. 4.27. The brace and chord members are all made of high-frequency welded tubes, with only the chord members filled with concrete, which was manually placed and compacted.

The chord members are 750 mm long, and the brace members are 300 mm long. For ease of loading, the ends of the brace members were sealed with 12 mm thick steel plates, and one end of the chord member in the bending joint was also sealed with a 12 mm thick steel plate. Both tensile and bending joint groups included one hollow tube joint. The welding electrodes used for the test joints were J422 type produced by Zhuzhou Electrode Factory, with weld thickness equal to twice the tube wall thickness, and manual DC welding was adopted. After welding, the concrete infilling of the chord members of the CFRST truss joints was completed, with a filling length of 750 mm.

In Packer's tests, six T-type tensile joints were fabricated, including three T-type RST tensile joints and three T-type CFRST tensile joints. The width ratio between brace and chord members (β) was taken as 0.4 and 0.7, and the chord member width-to-thickness ratio (b_0/t_0) was taken as 17 and 27. In this study, a total of five X-type tensile joint specimens were fabricated, including four CFST joints (β = 0.4, 0.6, and 1.0; b_0/t_0=25.0 and 36.36) and one hollow steel tube joint (β = 0.4, 0.6, and 1.0; b_0/t_0 = 25.0 and 36.36). For the X-type bending joints, a total of five specimens were fabricated, including four CFST joints (β = 0.4, 0.6, and 1.0; b_0/t_0 = 25.0 and 36.36) and one hollow steel tube joint (β = 0.8; b_0/t_0 = 25.0).

The measured geometric parameters of the joints are shown in Table 4.4, which also provides the geometric parameters of Packer's T-type tensile joints. In Table 4.1, the specimen numbering of the joints tested in this study represents:

X—joint type; t—tension, b—bending; 0—hollow steel tube joint; c—CFST joint;

1—b_0/t_0=36.36; 2—b_0/t_0=25.0;

4—β=0.4; 6—β=0.6; 8—β=0.8; 1—β=1.0

The mechanical properties of the steel tubes were obtained through tensile tests conducted in accordance with the Method of Tensile Test for Metals (GB228-87). The cube compressive strength of concrete f_{cu} was determined according to the Standard Test Method for Mechanical Properties of Ordinary Concrete (GBJ81-85). The measured mechanical properties of the materials are shown in Table 4.5, while Table 4.6 presents the mechanical properties of the materials used in Packer's joint tests.

Table 4.4 Geometric parameters of the test joints.

Joint type	Joint number	b_0	h_0	t_0	L_c	b_1	h_1	t_1	β	γ
		/	/	/	(mm)	/	/	/	(b_1/b_0)	$(b_0/2t_0)$
X-type tensile joints in this section	Xt-018	100	100	2.75	0	80	80	2.83	0.8	18.18
	Xt-c16	100	100	2.75	750	60	60	2.83	0.6	18.18
	Xt-c14	100	100	2.75	750	40	40	2.83	0.4	18.18
	Xt-c21	100	100	4.0	750	100	100	2.75	1.0	12.5
	Xt-c26	100	100	4.0	750	60	60	2.83	0.6	12.5
X-type flexural joints in this section	Xb-018	100	100	2.75	0	80	80	2.83	0.8	18.18
	Xb-c16	100	100	2.75	750	60	60	2.83	0.6	18.18
	Xb-c14	100	100	2.75	750	40	40	2.83	0.4	18.18
	Xb-c21	100	100	4.0	750	100	100	2.75	1.0	12.5
	Xb-c26	100	100	4.0	750	60	60	2.83	0.6	12.5
Packer T-type tensile joints	T1	127.2	178.1	4.74	0	50.9	51.1	6.32	0.4	13.42
	T1C	127.2	178.1	4.74	/	50.9	51.1	6.32	0.4	13.42
	T2	127.2	178.1	4.74	0	88.9	89.3	7.6	0.7	13.42
	T2C	127.2	178.1	4.74	/	88.9	89.3	7.6	0.7	13.42
	T3	127.1	177.9	7.54	0	50.9	51.1	6.32	0.4	8.43
	T3C	127.1	177.9	7.54	/	50.9	51.1	6.32	0.4	8.43

Table 4.5 Mechanical properties of materials used in this section's tests.

Material	Steel tube		Concrete
Wall thickness of steel tube (mm)	3	4	Wall thickness of steel tube (mm)
Tensile strength f_u (MPa)	580	580	Tensile strength f_u (MPa)
Yield strength f_y (MPa)	325	320	Yield strength f_y (MPa)
Elongation δ_5 (%)	13	20	Elongation δ_5 (%)
Cube strength f_{cu} (MPa)			31.1

Table 4.6 Packer mechanical properties of the test materials.

Material	Steel tube		Concrete
Wall thickness of steel tube (mm)	4.74	7.54	Wall thickness of steel tube (mm)
Yield strength f_y (MPa)	400	391	Yield strength f_y (MPa)
Wall thickness of steel tube (mm)	6.32	7.6	Wall thickness of steel tube (mm)
Yield strength f_y (MPa)	443	426	Yield strength f_y (MPa)
Concrete cylinder strength f_c' (MPa)			30.0

4.3.2 Test method

4.3.2.1 Test setup and loading scheme

The load tests of the X-type tensile joints were carried out on a universal testing machine, with the test setup shown in Fig. 4.28A. The anchor bolts welded to the end plates of the brace members in the test joints were connected to the self-anchoring grips of the testing machine. The tensile force generated by the machine was transferred through the self-anchoring grips to the tensile brace members of the joints, and loading was applied until joint failure occurred.

The X-type bending joint tests were conducted using a self-fabricated loading device, as shown in Fig. 4.28B. The steel plates welded to the ends of the brace members were supported on bearings, and a jack was used to apply load to the end of the chord member, thereby inducing bending in the brace members. The load magnitude was determined by a precalibrated load sensor, and loading was applied until joint failure occurred.

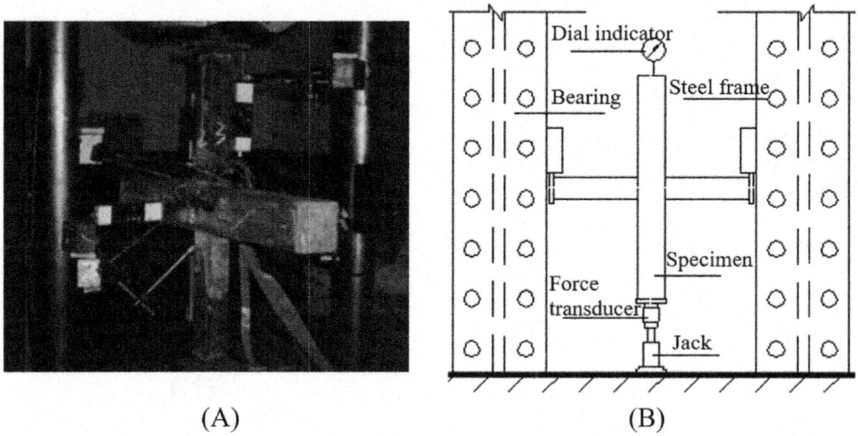

Figure 4.28 Test setup. (A) Test setup for X-type tensile joints. (B) Test setup for X-type flexural joints.

A monotonic stepwise loading scheme was adopted, with load levels and test procedures identical to those described in Section 4.1.

4.3.2.2 Test measurements

A digital dial indicator was used to measure the displacement at the brace ends of the joints and the deformation values in the joint region under each load level, in order to analyze the deformation characteristics, bearing capacity, and influencing factors of CFRST truss joints, as well as to compare and evaluate the effect of infilled concrete on joint bearing capacity and stiffness.

For the rectangular steel tubular (RST) tensile joint Xt-018 and the CFRST tensile joint Xt-c26, as well as the RST bending joint Xb-018 and the CFRST bending joint Xb-c16, strain gauges were affixed to the surfaces of the steel tubes. A YJ28 static resistance strain gauge was used to measure the strain values on the steel tube surface in the joint region under each load level, in order to understand the stress distribution characteristics and load transfer mechanism of CFRST truss joints, and to analyze the influence of infilled concrete on their mechanical performance.

The arrangement of displacement measuring points is shown in Fig. 4.29, while the arrangement of strain measuring points is shown in Figs. 4.30 and 4.31.

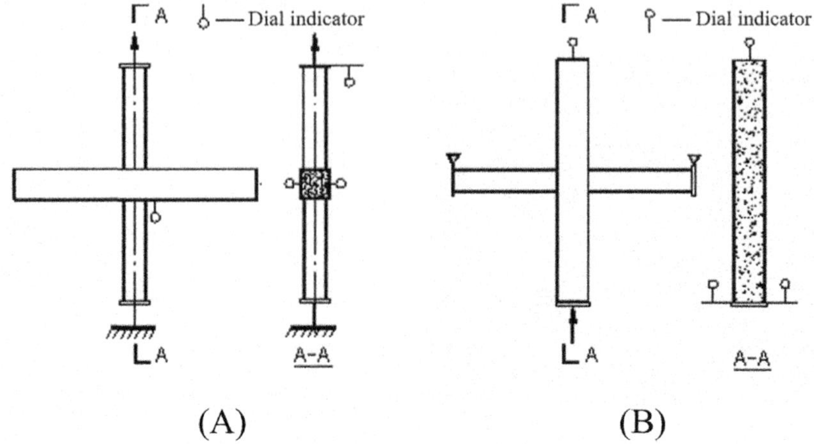

Figure 4.29 Arrangement of displacement measuring points. (A) Layout of displacement measurement points for X-type tensile joints. (B) Layout of displacement measurement points for X-type flexural joints.

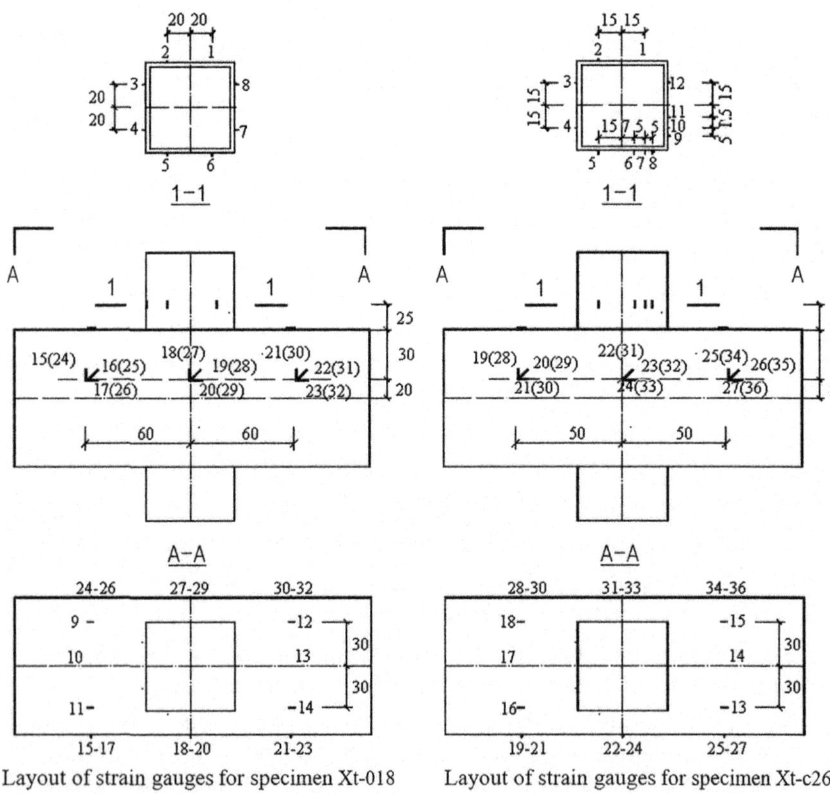

Figure 4.30 Arrangement of strain measuring points for X-type tensile joints.

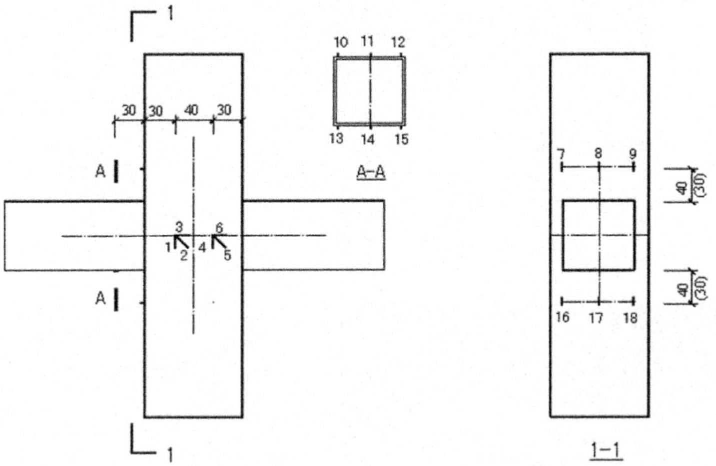

Layout of strain gauges for specimen Xb-c16(Xb-018)

Figure 4.31 Arrangement of strain measuring points for X-type bending joints.

4.3.3 X-type tensile joint test results and analysis

4.3.3.1 Failure modes and failure loads

The deformation characteristics of the tested steel tube tensile joint (Xt-018) were outward bulging of the chord surface and inward concavity of the sidewalls. At failure, the outward bulging deformation of the chord became severe, and failure initiated locally at the chord surface near the brace corner, then rapidly propagated along the sidewall of the brace, leading to shear failure of the chord surface near the brace sidewall.

The deformation characteristics of the tested X-type CFST tensile joints were also outward bulging of the chord surface; however, the inward concave deformation of the sidewalls did not occur due to the confinement of the infilled concrete. The failure modes were excessive outward bulging of the chord surface, eventually leading to shear rupture of the chord surface near the brace sidewall (Xt-c14, Xt-c26) or weld failure at the brace-chord connection (Xt-c16). For joint Xt-c21, when the load approached 200 kN, the weld of the screw connection in the loading device suddenly fractured, so the ultimate load and deformation of the joint were not measured, and joint failure was not reached. Fig. 4.32 shows the failure modes of the tested joints. After flame cutting the chord of joint Xt-c14, the concrete inside the tube remained basically intact.

Figure 4.32 Failure modes of X-type tensile joints.

Before failure, the joints exhibited large deformations. Failure occurred suddenly and was accompanied by loud fracturing sounds. Table 4.7 presents the test results of the five joints. In the table, the ultimate load NuT refers to the maximum load attained during the test, while N3 is the load corresponding to a joint deformation equal to 3% of the chord width. Table 4.7 also provides the test results of Packer's T-type tensile joints, as well as the calculated bearing capacities according to the formulas for RST given in Chapter 2.

In Table 4.7, the ratio of the experimental to calculated bearing capacity of CFRST truss joints based on the punching-shear failure mode is 1.57, with a coefficient of variation of 14.3%. When calculated according to the yield-line failure mode, the ratio is 4.50, with a coefficient of variation of 27.7%. This indicates that the punching-shear failure mode is closer to the actual failure mode of CFRST tensile joints, which is consistent with the observed failure modes in the tests.

In Table 4.7, for both punching-shear and yield-line failure modes, the ratios of experimental to calculated bearing capacity are higher for CFST joints than for hollow steel tube joints, showing that the ultimate bearing capacity of CFRST tensile joints is greater than that of their corresponding hollow steel tube joints.

Table 4.7 Comparison of failure loads of test joints with theoretical bearing capacity calculations for rectangular steel tubular.

Node type	Joint number	Ultimate bearing capacity (kN)						Joint geometric parameters	
		Test value		Punching-shear failure mode		Yield-line failure mode			
		N_u^T	N^3	N_u	N_u^T/N_u	N_u^T	N^3	β	γ
X-type tensile	Xt-018	160	91	X-type tensile	Xt-018	160	91	X-type tensile	Xt-018
	Xt-c14	90	23	54.7	Xt-c14	90	23	0.4	Xt-c14
	Xt-c16	140	57	82.0	Xt-c16	140	57	0.6	Xt-c16
	Xt-c21	>200	/	/	Xt-c21	>200	/	1.0	Xt-c21
	Xt-c26	149.5	140	86.0	Xt-c26	149.5	140	0.6	Xt-c26
T-type tensile	T1	216	/	T-type tensile	T1	216	/	T-type tensile	T1
	T2	395	/	278.5	T2	395	/	0.7	T2
	T3	323	/	287.6	T3	323	/	0.4	T3
	T1C	240	/	159.4	T1C	240	/	0.4	T1C
	T2C	484	/	278.5	T2C	484	/	0.7	T2C
	T3C	317	/	287.6	T3C	317	/	0.4	T3C

(*Continued*)

Table 4.7 (Continued)

Node type	Joint number	Ultimate bearing capacity (kN)						Joint geometric parameters	
		Test value		Punching-shear failure mode		Yield-line failure mode			
		N_u^T	N^3	N_u	N_u^T/N_u	N_u^T	N^3	β	γ
Steel tubular joints		Average value		/	1.34	Steel tubular joints	Average value	/	
		Coefficient of variation		/	9.8%	/	Coefficient of variation	/	
Concrete-filled steel tubular (CFST) joints		Average value		/	1.57	Concrete-filled steel tubular (CFST) joints	Average value	/	
		Coefficient of variation		/	14.3%	/	Coefficient of variation	/	

For the tested CFST joints Xt-c14 and Xt-c16, the load corresponding to 3% of the chord width deformation was lower than the bearing capacity calculated using the punching-shear failure mode, but higher than that calculated using the yield-line failure mode. Therefore, if the bearing capacity of a joint is defined as the load N^3 corresponding to a deformation equal to 3% of the chord width, it is inappropriate to adopt the punching-shear failure mode for calculating the bearing capacity of CFRST X-type tensile joints when β is small and γ is large. In such cases, it is more appropriate, for safety reasons, to calculate the joint bearing capacity according to the yield-line failure mode.

4.3.3.2 Load-deformation relationship

Fig. 4.33 shows the load-deformation curves of all specimens. Since the branch of the test node was subjected to tension at one end, the outward bulging deformation of the main chord flange in the figure was taken as half of the dial gauge readings measured at the branch end. For the hollow section node Xt-018 under vertical tension of the branch, the outward bulging of the main chord flange was accompanied by inward buckling of the brace. In Fig. 4.33A, the inward buckling deformation of the brace was taken as the average value of the dial gauge readings on both sides of the brace. During the test, the inward displacement of the main chord brace in CFST nodes was essentially zero. As shown in Fig. 4.33A, the inward buckling deformation of the brace in RST tension nodes mainly occurred before the bending yield load of the main chord flange, that is, before the load corresponding to the end of the linear segment of the flange outward bulging load-deformation curve. With further increase of the load, membrane tension developed in the main chord flange, and simultaneously, membrane tension also arose in the brace, which progressively restored the inward deformation. After the load reached 80–90 kN, the inward deformation basically ceased to increase.

Fig. 4.33B shows an enlarged view of the load-deformation curves of the test nodes where the outward bulging deformation of the main chord flange was less than 3% of the chord width. It can be seen that the stiffness of the nodes increases with increasing branch-to-chord width ratio (β) and chord wall stiffness (i.e., decreasing γ). In Fig. 4.33B, the stiffness of Xt-c16 before reaching the yield-line load (22.9 kN) was basically the same as that of Xt-018, indirectly indicating that the stiffness of CFRST tension nodes is higher than that of RST tension nodes. This result is consistent with Packer's T-type node tests.

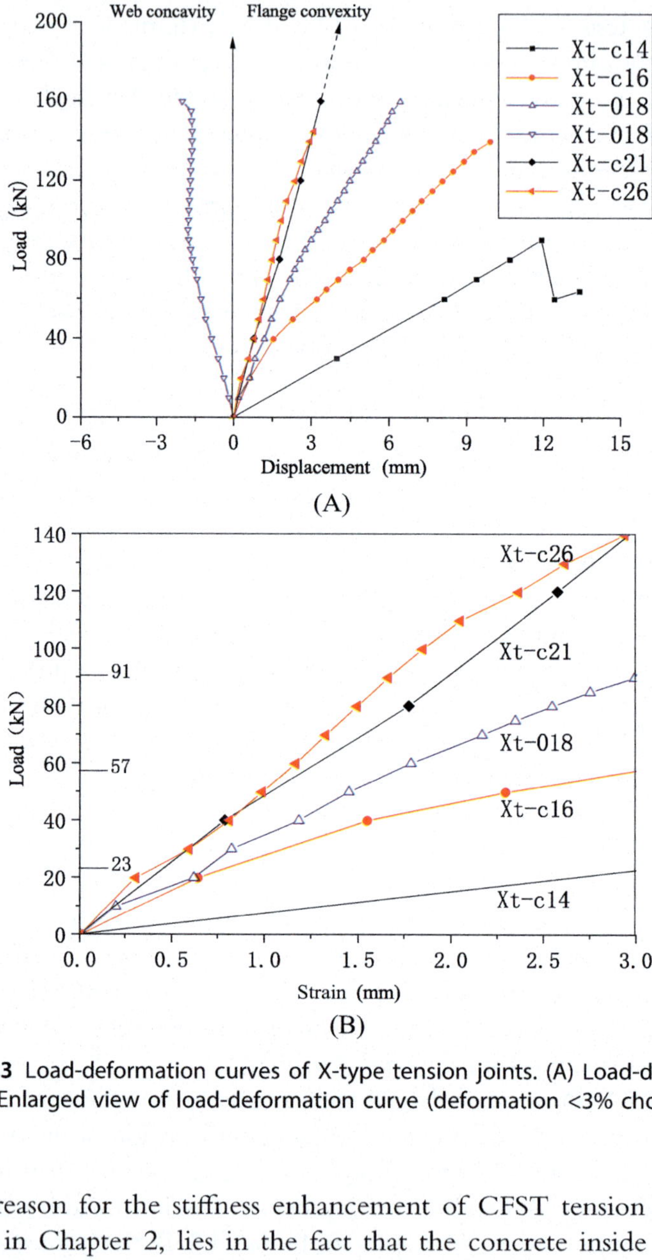

Figure 4.33 Load-deformation curves of X-type tension joints. (A) Load-deformation curve. (B) Enlarged view of load-deformation curve (deformation <3% chord width).

The reason for the stiffness enhancement of CFST tension nodes, as analyzed in Chapter 2, lies in the fact that the concrete inside the main chord prevents inward buckling of the chord braces and rotation at the chord corners, thereby allowing the chord flange plates to develop greater membrane tension. However, when the main chord wall of the CFST tension node is relatively flexible and the branch-to-chord width ratio is small, relatively large deformations can still occur under loading. Therefore Packer suggested using the load-bearing capacity formula of RST tension nodes to design CFST tension nodes. Nevertheless, applying this method to calculate the bearing capacity of CFST nodes with larger β values and smaller γ values results in overly conservative estimates.

In this study, as well as in Packer's three T-type CFST nodes, the failure mode was punching shear. As shown in Table 4.7, the ratios of the experimental bearing capacity to the calculated value based on the punching-shear failure mode for Xt-c16, Xt-c26, and T2c RST tension nodes were in the range of 1.71–1.75. Therefore, for CFRST tension nodes with larger β values and smaller γ values, it is more appropriate to calculate the node bearing capacity using the punching-shear failure mode.

4.3.3.3 Load-strain relationship

During the test, the strains of the brace and chord sections of the CFRST X-type tension joint (Xt-c26) and the RST tension joint (Xt-018) were measured.

Fig. 4.34 shows the relationship curves between strain at the brace root and load, where the strain values are taken as the averages of the corresponding measuring points. For the RST Xt-018, the strain at the brace flange edge measuring point I is smaller than that at the brace edge measuring point H. For the CFRST truss joint Xt-c26, the strain values at the brace flange measuring

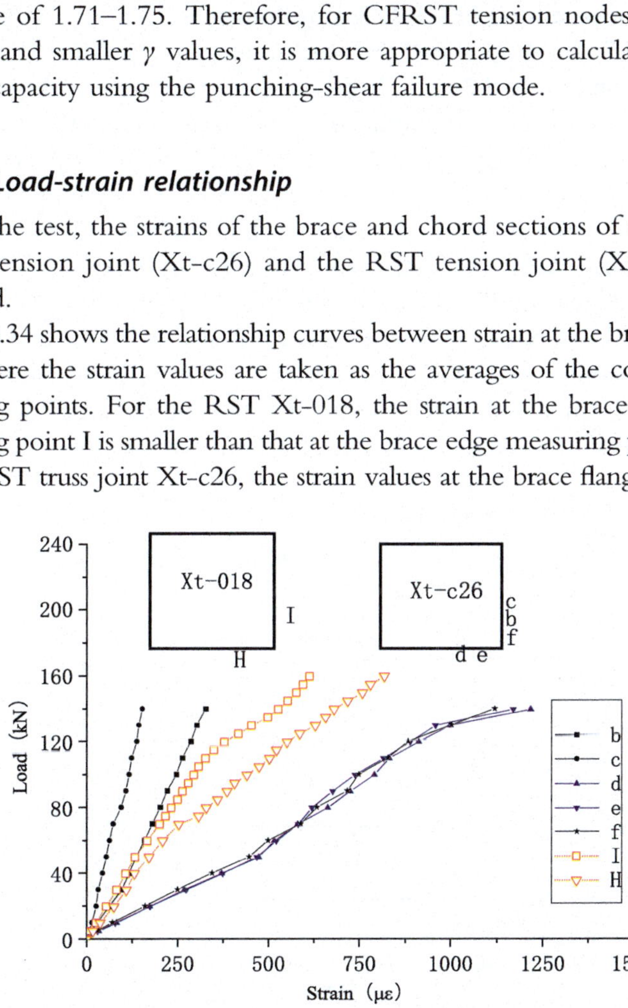

Figure 4.34 Load-strain curves at the brace measuring points for Xt-c26 and Xt-018.

points decrease as the location moves closer to the center, that is, the strain values at points *f*, *b*, and *c* decrease successively. The strain values at brace measuring points d and e and at flange edge measuring point f are approximately equal.

It can be seen that, whether for the RST or the CFRST truss joint, the strain distribution across the brace root section under tension is nonuniform. Among them, the brace parallel to the chord side plate bears greater stress than the flange plate. Therefore the concept of the effective width of the brace under tension in RST, as discussed in Chapter 2, can also be applied to the brace under tension in CFRST truss joints.

Figs. 4.35 and 4.36 show the load-strain curves at the chord flange measuring point located 20 mm from the brace flange. From Figs. 4.35 and 4.36 the chord flange midpoint (C) and edge-point (B) strains were obtained as indicated. In Figs. 4.36 and 4.37, the strain at the chord flange midpoint C of the RST Xt-018 remained tensile throughout, mainly due to membrane tension; when the load reached the theoretically calculated punching-shear load, the load-strain curve began to decline. The edge point B exhibited tensile strain when the load was below roughly 60 kN, but the strain magnitude was small (on the order of 100 $\mu\varepsilon$); thereafter it turned into compressive strain and remained so until joint failure.

The appearance of compressive strain at the chord flange edge is caused by restraint of the flange's outward bulging deformation: the flange is

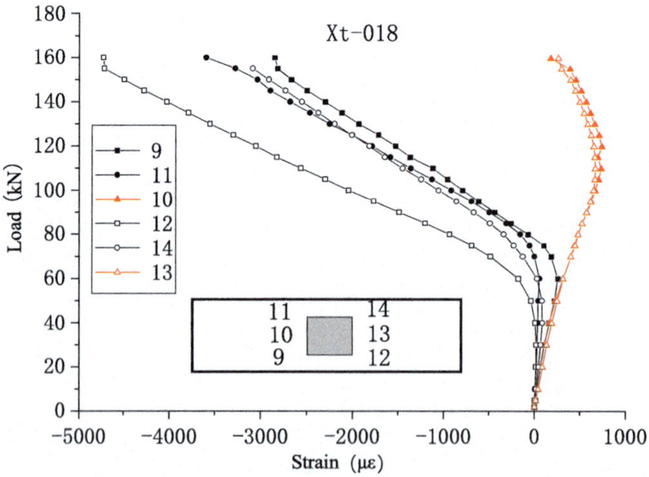

Figure 4.35 Load-strain curves at the chord flange measuring points for Xt-018.

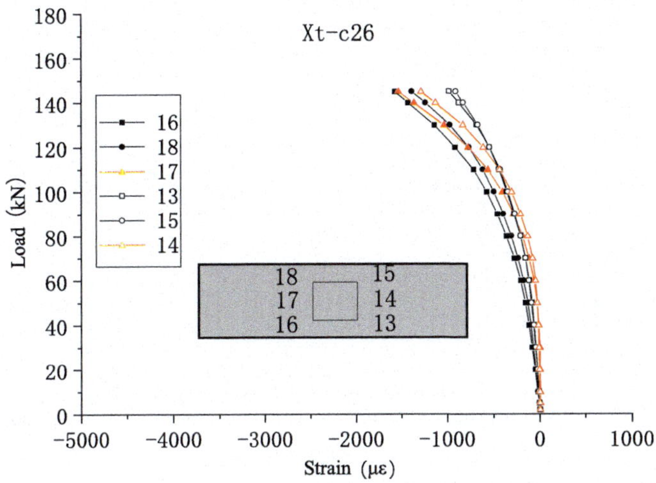

Figure 4.36 Load-strain curves at the chord flange measuring points for Xt-c26.

subjected to bending, and the closer the flange is to the brace, the greater the restraining effect. Consequently, once the outward bulging of the chord flange extends to the edge point, that point develops bending-induced compressive strain. Because the outward bulging cannot continue further in that direction, the compressive strain at the edge point increases with increasing load. By contrast, the chord flange midpoint is less restrained, so its strain is dominated by membrane stress.

In Figs. 4.36 and 4.37, all measurement points of the CFRST truss joint Xt-c26 show compressive strain. The strain at the midpoint C remains nearly zero when the load is below 40–60 kN, and then develops more rapidly than the strain at the edge point B, indicating that the infilled concrete in the chord effectively restrains the outward bulging deformation of the chord flange.

Fig. 4.38 presents the principal tensile strain curves of the chord brace under load, calculated from the measured strains. For the CFST truss joint Xt-c26, the strains are tensile throughout; at failure load, the strain at point a reaches about 1250 με, corresponding to a stress of approximately 250 MPa. In contrast, for the steel tube joint Xt-018, the brace surface shows compressive strain due to the simultaneous outward bulging of the flange and inward buckling of the brace. When the load reaches 90–100 kN, the membrane stress in the brace exceeds the bending

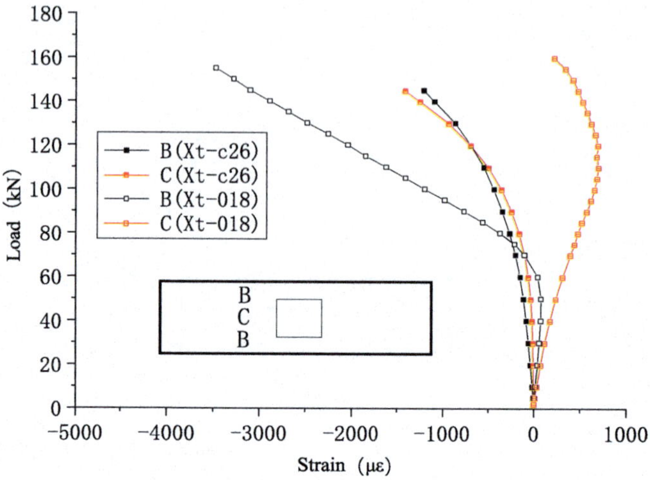

Figure 4.37 Average load-strain curves at the chord flange measuring points for Xt-c26 and Xt-018.

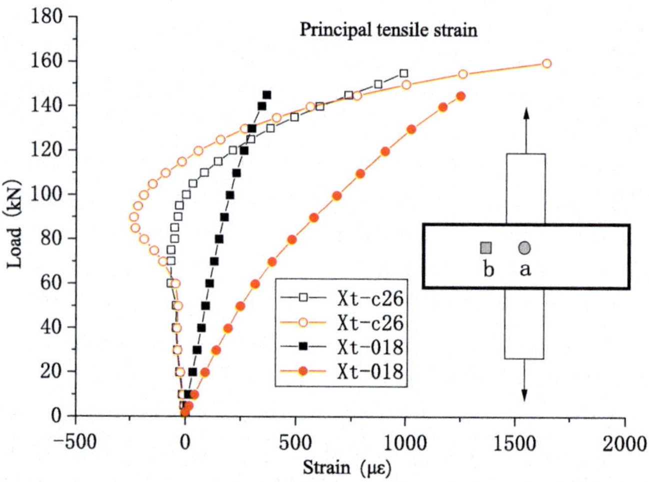

Figure 4.38 Principal tensile strain curves of the chord brace under load for Xt-c26 and Xt-018.

compressive stress, after which the strain gradually shifts to tension and increases rapidly, leading to quick joint failure.

A comparison of the strain test results of Xt-c26 and Xt-018 reveals that the stress distribution on the surface of the CFST truss joint is more uniform and develops more smoothly, whereas the steel tube joint behaves oppositely. This indicates that the infilled concrete in the chord improves the joint's load-bearing performance and enhances the load transfer efficiency of the branch tube.

4.3.4 Test results and analysis of X-type bending joints

4.3.4.1 Failure modes and failure loads

For the tested RST bending joint (Xb-018), the failure mode was characterized by bulging and inward denting of the chord sidewall at the compressed flange of the branch, while the chord surface at the tension flange of the branch exhibited slight outward bulging. For the CFST bending joints, due to the infilled concrete in the chord, the chord surface at the tension flange of the branch showed more significant outward bulging. When the β value was relatively small, this led to excessive rotational or bending deformation of the branch as a whole, or tensile failure of the weld at the branch root. In specimens Xb-c26, Xb-c16, and Xb-c14, cracks first appeared at the two corner points on the tension edge of the branch root, followed by complete fracture of the entire tension edge. When the β value was relatively large, the specimen Xb-c21 experienced local buckling failure at the compressed edge of the branch root. Fig. 4.39 illustrates typical failure modes of the tested joints, where for Xb-c16, the chord concrete at the compressed edge of the branch showed only slight indentation.

The failure loads and failure modes of the tested joints are presented in Table 4.8. In the table, N_u^T denotes the failure load applied at the chord end. Given that the branch length is 0.3 m, the corresponding joint failure bending moment applied to the branch is $M_u^T = 0.3\ N_u^T/2$.

Figure 4.39 Failure modes of X-type bending joints.

Table 4.8 Failure loads and failure modes of X-type bending joint specimens.

Joint number	Test value		Yield-line failure		Effective width failure		Failure pattern
	N_u^T (kN)	M_u^T (kN·m)	M_u (kN·m)	M_u/M_u^T	M_u (kN·m)	M_u/M_u^T	
Xb-018	19.93	2.99	1.79	1.67	4.06	0.74	CY
Xb-c14	13.41	2.01	0.44	4.55	0.91	2.22	BS
Xb-c16	24.64	3.70	0.81	4.56	2.20	1.68	BS
Xb-c21	81.88	12.28	/		9.25	1.33	BB
Xb-c26	34.06	5.11	1.71	2.98	3.18	1.61	BS

Notes: *BB*, Brace compression-side buckling; *BS*, brace tension-side failure; *CY*, chord flange yielding and brace buckling.

As shown in Table 4.8, the failure load of CFRST bending joints is closely related to the chord width-to-thickness ratio and the branch-to-chord width ratio. The larger the β value and the smaller the γ value, the higher the failure load of the joint. A comparison between specimens Xb-018 and Xb-c16 indicates that infilled concrete enhances the flexural capacity of the joints. In Table 4.8, the theoretical flexural capacities of joints based on yield-line failure and effective width failure modes are calculated using the corresponding flexural capacity formulas for RST. By comparing the theoretical values with the experimental results, it can be observed that calculating the flexural capacity of CFRST truss joints using the effective width failure mode is more reasonable, as it better reflects the actual failure modes of the joints.

4.3.4.2 Load-deformation relationship

Fig. 4.40 shows the load-displacement curves at the chord end of the tested joints. Comparing the load-displacement curves of specimens Xb-c16 and Xb-018 in the figure, it can be seen that the flexural capacity of CFRST bending joints is improved, while their deformation capacity is also enhanced. A comparison of the load-displacement curves of Xb-c16 and Xb-c26, as well as Xb-c14 and Xb-c16, shows that as the ratio of chord wall thickness to branch-to-chord width increases, the joint flexural capacity improves, but the deformation capacity decreases.

For CFRST bending joints, the indentation deformation on the compressed side of the branch is restrained by the infilled concrete in the chord, and the deformation capacity mainly comes from the bulging deformation on the tension side of the branch. Therefore the flexural

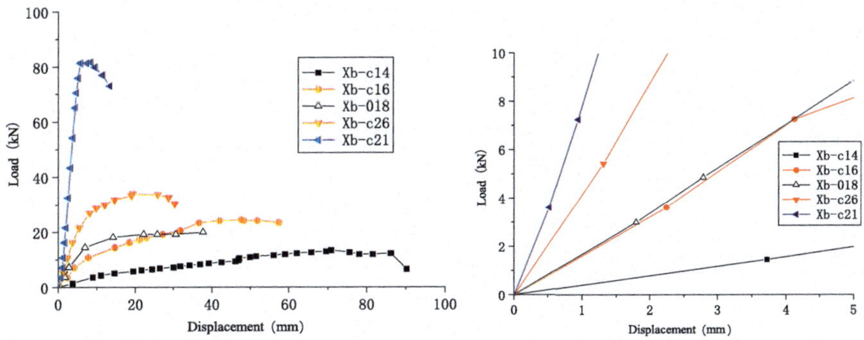

Figure 4.40 Load-displacement curve of an X-type bending node.

stiffness of CFRST truss joints in the elastic stage is higher than that of the corresponding hollow steel tube joints. This analysis is confirmed by the fact that in Fig. 4.40, the elastic portions of the load-displacement curves for Xb-c16 and Xb-018 essentially coincide.

4.3.4.3 Load-strain relationship

The strain test results of the RST bending node Xb-018 and the CFRST X-type bending node Xb-c16 at the branch section 30 mm away from the surface of the main tube are shown in Figs. 4.41 and 4.42. In the figure, D and E are measurement points on the tension flange of the branch, while B and C are measurement points on the compression flange. B and D are located at the midpoints of the flanges, and C and E are located at the flange edges near the brace.

As analyzed in Chapter 2, due to the flexibility of the main tube wall in RST, the stress distribution on the flange plate at the root of the branch is nonuniform, with higher stresses at the edges and lower stresses at the midpoint. In Fig. 4.42, the strains at the midpoint B and D of the branch's tension and compression flanges remain much smaller than those at the edge points C and E until the joint failure, indicating that the stress carried by the middle part of the branch flange is minimal.

For CFST bending joints, the stress distribution of the branch tension flange follows a similar pattern to that of the RST bending joints. As shown

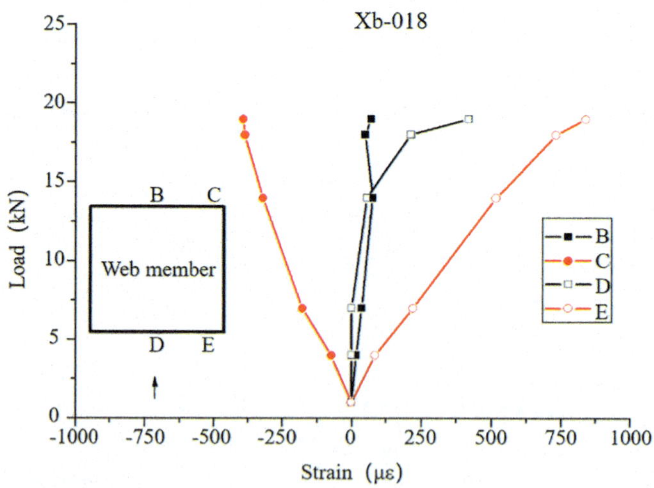

Figure 4.41 Load-strain curve of branch Xb-018.

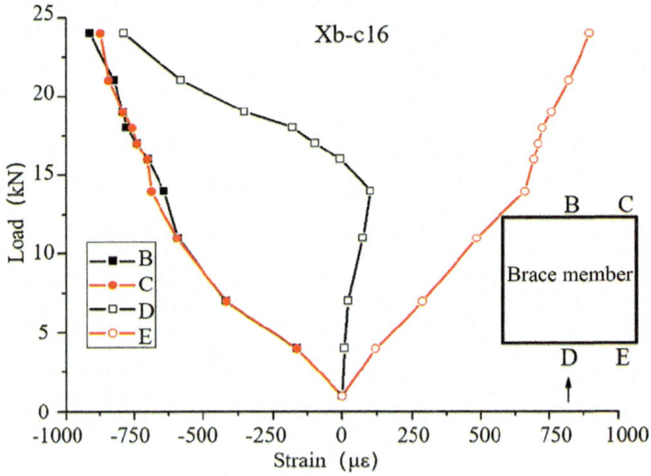

Figure 4.42 Load-strain curve of branch Xb-c16.

in Fig. 4.42, the tensile strain at point D increases slowly and is much smaller than that at point E, which is consistent with the stress distribution pattern of the branch tension flange under tension. Therefore the concept of effective width can also be applied to reflect the nonuniform stress distribution of the branch tension flange in CFST bending joints.

In Fig. 4.42, the strain at point D changes from tensile to compressive once the load exceeds 15 kN, which is due to excessive rotational deformation of the branch, causing bending in the tension flange. Unlike RST bending joints, however, the presence of infilled concrete in the main tube prevents the inward buckling of the main tube flange plate. This results in a more uniform stress distribution in the branch compression flange of the CFST bending joints. The nearly coincident load-strain curves of points B and C in Fig. 4.42 confirm this analysis.

Thus it can be concluded that the infilled concrete in the main tube improves the stress distribution of the branch compression flange in bending joints and enhances the flexural efficiency of the branch (Figs. 4.43 and 4.44).

Fig. 4.45 shows the variation curves of the principal tensile strain in the main tube brace at the 1/4 height point along the branch axis of joints Xb-018 and Xb-c16 with respect to the end load of the main tube. All strains are tensile and increase nonlinearly with increasing load. The

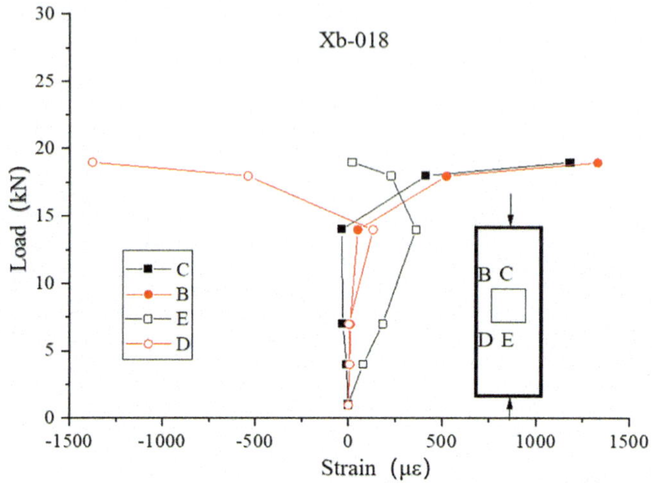

Figure 4.43 Load-strain curve of the chord flange for Xb-018.

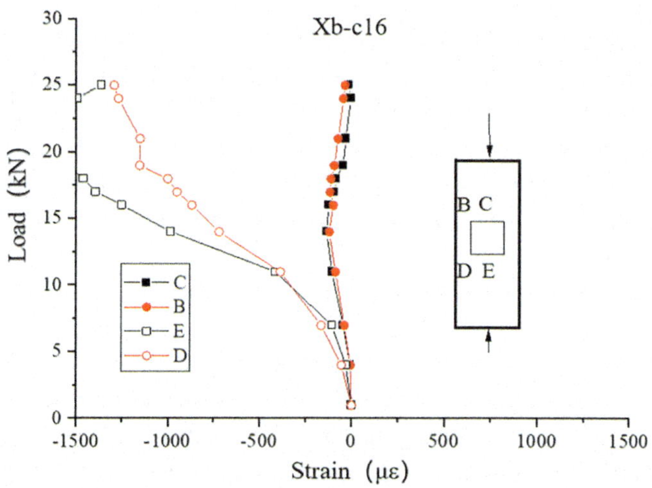

Figure 4.44 Load-strain curve of the chord flange for Xb-c16.

formation mechanisms of the principal tensile strain in the main tube brace differ between the RST bending joint and the CFST bending joint.

For the RST bending joint Xb-018, the principal tensile strain in the main tube brace is mainly generated by the bending deformation of the brace—that is, the convex deformation on the tension side and the concave deformation on the compression side of the main tube flange caused by the

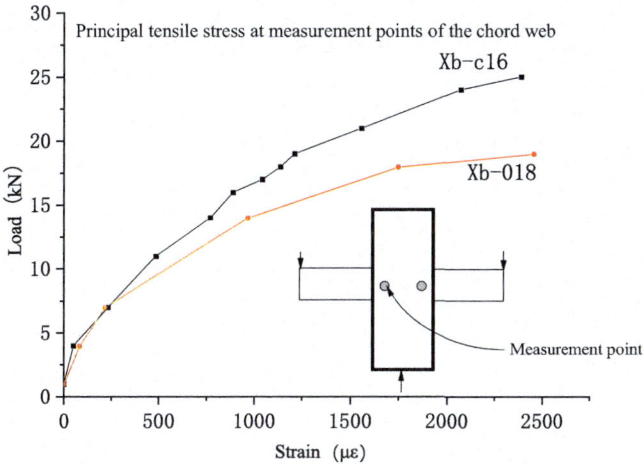

Figure 4.45 Principal tensile strain curves of the chord brace under load for Xb-018 and Xb-c16.

branch, which in turn leads to corresponding concave and convex deformations of the brace, especially near joint failure. For the CFST bending joint, the principal tensile strain in the main tube brace is mainly membrane strain.

This is further illustrated in Fig. 4.46, which shows the variation curves of the angle of the principal tensile strain in the main tube brace with respect to the load, where ω is the angle between the principal tensile strain direction and the branch axis. In joint Xb-c16, the change of ω with increasing load is small, mainly caused by the shift of the neutral axis in the branch section. In contrast, in joint Xb-018, ω changes significantly with increasing load, turning from positive to negative near joint failure.

4.3.4.4 Capacity calculation for concrete-filled steel tubular truss tensile and bending nodes

The above tensile and bending node test results both indicate that filling the main tube with concrete can improve the mechanical performance of the joint, increasing the joint stiffness and reserve bearing capacity.

The test results show that the branch-to-main width ratio and the main-tube width-to-thickness ratio are the primary factors affecting the tensile and bending capacity of CFRST. As with plain RST, the larger the branch-to-main width ratio and the smaller the main-tube width-to-thickness ratio, the higher the tensile and bending capacity and the greater

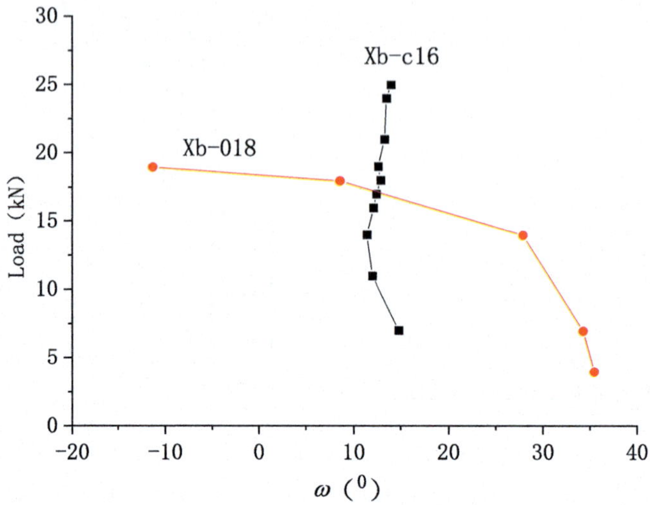

Figure 4.46 Variation of principal tensile strain angle in chord brace for Xb-018 and Xb-c16.

the stiffness of the CFST truss node. Unlike RST, however, the presence of concrete inside the main tube alters the failure mode of the joint. Based on the test results presented here, it is recommended to calculate the tensile and bending capacity of CFST nodes by applying the bearing capacity formulas corresponding to the failure modes of the (plain) RST nodes.

The tensile capacity of CFST T-type, Y-type, and X-type nodes is:

When $\beta < 0.6$, calculate using the yield-line failure mode:

$$N_{2u} = \frac{f_{y0}t_0^2}{(1-\beta)\sin\theta_2}\left[\frac{2h_2}{\sin\theta_2} + 4(1-\beta)^{0.5}\right] \tag{4.9}$$

When $0.6 \le \beta \le (1-\gamma)$ and $\gamma<13$, calculate according to the punching-shear failure mode:

$$N_{2u} = \frac{f_{y0}t_0}{\sqrt{3}\sin\theta_2}\left(\frac{2h_2}{\sin\theta_2} + 2b_{ep}\right) \tag{4.10}$$

When $\beta \ge 0.6$, it is still necessary to verify the bearing capacity using the branch effective-width failure mode. The corresponding bearing capacity calculation formula is:

$$N_{2u} = f_{y2}t_2(2h_2 - 4t_2 + 2b_e) \tag{4.11}$$

For CFST bending joints, the flexural capacity of the branch should be calculated based on the effective-width failure mode of the branch tension flange. The calculation formula is:

$$M_u = f_{y1}\left[Z_1 - (1 - b_e/b_1)b_1 t_1 (h_1 - t_1)\right] \tag{4.12}$$

4.4 Experimental study on T-type, Y-type, and K-type joints of concrete-filled steel tubular trusses

On the basis of transverse local bearing tests on the compressive performance of CFST truss joints, this section conducts experimental studies on T-type and Y-type compression branch joints of CFSTs, in order to further explore the actual failure modes and mechanical mechanisms of rectangular concrete compression joints. Through tensile joint tests on X-type nodes, the tensile performance of CFST truss joints has been investigated. The K-type joint, in which one branch is in tension and the other in compression, is the most common joint form in trusses. Therefore experimental studies on CFST K-type joints will be carried out to obtain a more systematic understanding of the mechanical behavior of such truss joints.

4.4.1 Specimen design and fabrication

The configuration of the T-type test joint is shown in Fig. 4.47, and that of the K-type test joint is shown in Fig. 4.48. The angle between the main tube and the branch in the T-type joint is 90 degrees, while in the Y-type joint it is 45 degrees, with both branches subjected to compression. In the

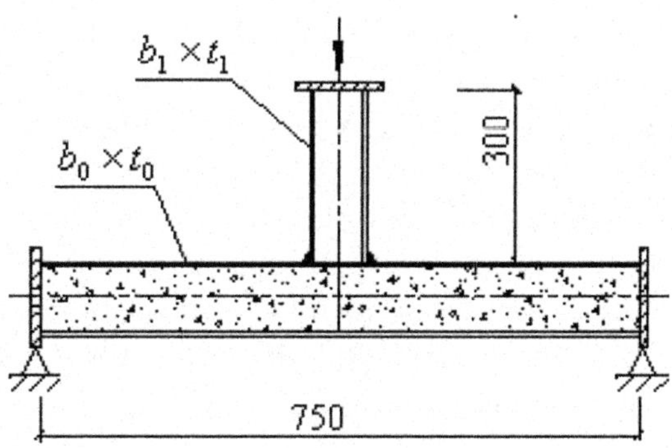

Figure 4.47 Configuration of a T-type test joint.

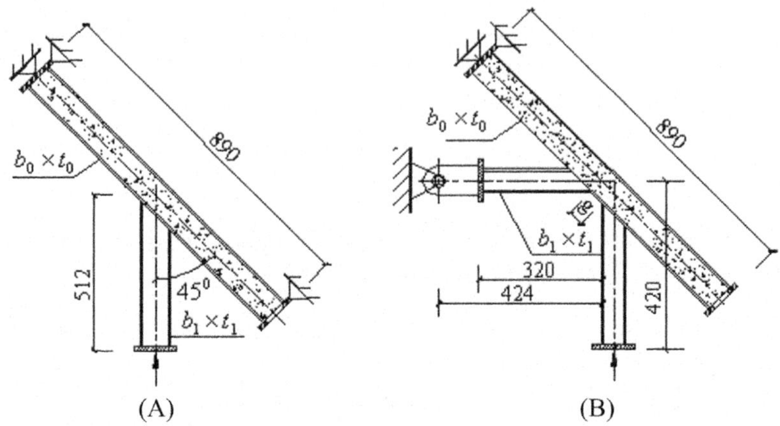

Figure 4.48 Joint configuration. (A) Structure of a Y-type compression test joint. (B) Structure of a K-type compression test joint.

K-type joint, the angle between the main tube and the branches is 45 degrees, with one branch under tension and the other under compression. For convenient loading, the ends of the steel tubes at the loading points and at the ends of the main tube were sealed with steel plates of 12 mm thickness. Additionally, for the tensile branch of the K-type joint, a steel plate with a circular hole was welded onto the end plate to facilitate connection with the tensile support.

The length of the main tube in the T-type joint is 750 mm, which meets the requirements of the loading device and allows loading under the same loading frame. The main tube length of the Y-type and K-type test joints is 890 mm. The branch and main tubes of the test joints are all square-section high-frequency welded tubes, with only the main tube filled with concrete, which was placed manually. The welds connecting the branch and main tubes were made by manual direct-current welding. After the welding of the joints was completed, concrete was cast into the main tube. The weld thickness was set to be twice the tube wall thickness.

A total of four T-type compression joint specimens were fabricated, all of which were CFST joints. The focus was on investigating the effects of the branch-to-main width ratio and the main-tube width-to-thickness ratio on the joint bearing capacity, with β = 0.4, 0.6, 0.8, 1.0 and γ = 12.5, 18.18.

A total of four Y-type compression joint specimens were fabricated, including three CFST joints (β = 0.4, 0.6, 0.8; γ = 12.5, 18.18) and one

Table 4.9 Geometric parameters of test joint.

Joint type	Joint number	b_0	h_0	t_0	L_c	b_1	h_1	t_1	g	β	γ
		mm	mm	mm	mm	mm	mm	mm		b_1/b_0	$b_0/2t_0$
The T-type compression joints in this paper	Tc-c11	100	100	2.75	750	100	100	4.0	/	1.0	18.18
	Tc-c16					60	60	2.83	/	0.6	
	Tc-c24			4.0		40	40		/	0.4	12.5
	Tc-c28					80	80		/	0.8	
The Y-type compression joints in this paper	Yc-018	100	100	2.75	0	80	80	2.83		0.8	18.18
	Yc-c16			2.75	750	60	60			0.6	
	Yc-c14					40	40			0.4	
	Yc-c28			4.0		80	80			0.8	12.5
The K-type joints in this paper	K0-c18	100	100	2.75	750	80	80	2.83	0	0.8	18.18
	K0-c16					60	60		0	0.6	
	K0-c14					40	40		0	0.4	
	K2-c18					80	80		20	0.8	
	K2-028			4.0	0				20		12.5
	K4-c18			2.75	750				40		18.18
	K4-c28			4.0					40		12.5
Packer K-type joints	K1	127.2	178.1	4.74	0	50.9	51.1	6.32	114	0.4	13.42
	K1C				1 000						
	K1X										
	K2			4.86	0				38.2		13.09
	K2C				1 000						
	K3			4.74	0	88.9	89.3	7.6	57.2	0.7	13.42
	K3C				1 000						
	K4				0				19.1		
	K4C				1 000						
	K5	127.1	177.9	7.54	0				57.2		8.43
	K5C				1 000						

plain steel tube joint ($\beta = 0.8$, $\gamma = 12.5$), in order to examine the influence of infilled concrete and the branch-to-main angle on the compressive performance of the joints.

A total of seven K-type joint specimens were fabricated to study the effects of branch spacing ggg on the joint bearing capacity. Among them, six were CFST joints ($g = 0$, 20, and 40 mm; $\beta = 0.4$, 0.6, and 0.8; $\gamma = 12.5$ and 18.18), and one was a plain steel tube joint ($g = 20$ mm; $\beta = 0.8$; $\gamma = 18.18$).

The geometric parameters of the test joints are shown in Table 4.9, where the test joint labels A, B, C, D, and E represent, respectively:
A—Joint type (Y, T, K)
B—Loading method (compression—c) or branch spacing (gap—0, 2, 4 cm)
C—Concrete infill inside the tube (0—hollow steel tube joint; c—CFST joint)
D—Main tube width-to-thickness ratio (1—$b_0/t_0=36.36$; 2—$b_0/t_0=25.0$)
E—Branch-to-main width ratio (4—$\beta=0.4$; 6—$\beta=0.6$; 8—$\beta=0.8$; 1—$\beta=1.0$)

Table 4.9 also lists the geometric parameters of the Packer test joints to compensate for the insufficient number of specimens in this study. The steel tubes used for the test joints were from the same batch of high-frequency welded tubes as those used for the X-type tensile test joints, and the concrete inside the tubes was poured from the same batch. The mechanical properties of the steel tubes and concrete are consistent with those of the X-type test joints in Section 3.8. The yield strength of the steel tubes for the Packer test joints is provided in Chapter 4, and the compressive strength of the concrete cylinders is 42.7 MPa.

4.4.2 Test method

4.4.2.1 Test setup

For the T-, Y-, and K-type joints, a self-designed loading device was used to apply load to the compressed branch through a jack. A monotonic stepwise loading scheme was adopted, with load levels and testing procedures identical to those in Chapter 3. The applied load was measured using precalibrated load sensors. The loading device is shown in Fig. 4.49.

4.4.2.2 Test measurements

A digital dial indicator was used to measure the displacement of the compressed branch and the deformation of the joint zone under each load

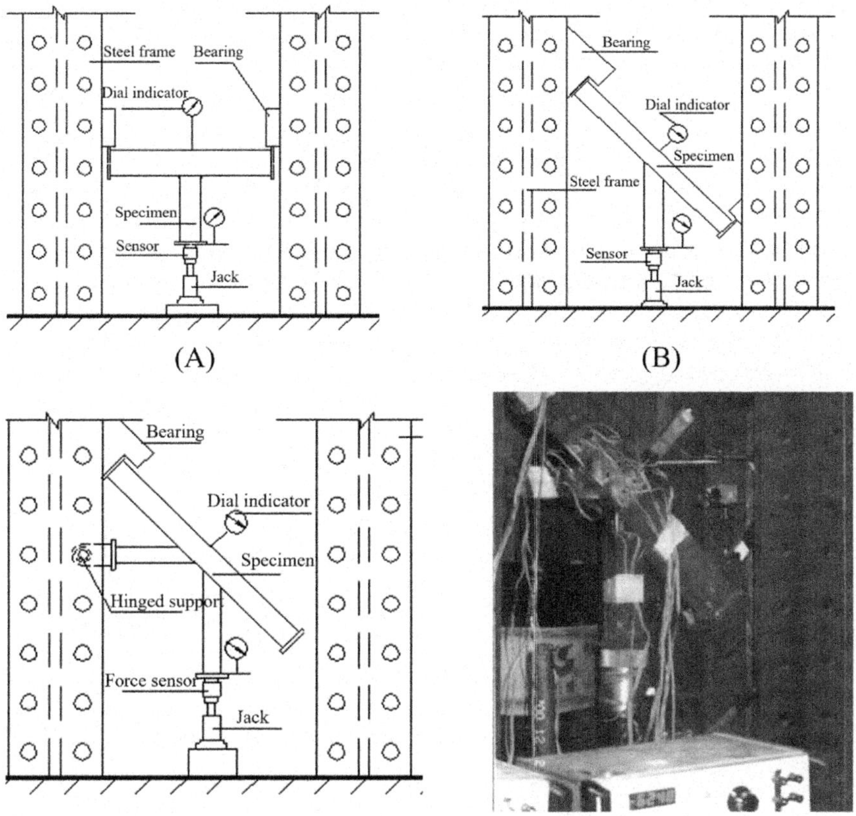

(C)

Figure 4.49 Test loading setup. (A) Test setup for T-type joints. (B) Test setup for Y-type joints. (C) Test setup for K-type joints.

level. The arrangement of displacement measurement points is shown in Figs. 4.50 and 4.51. Strain gauges were affixed to the steel tube surfaces of specimens Yc-018, Yc-c16, Tc-c16, K2-028, and K2-c18. A YJ28 static resistance strain gauge was used to measure the strain values on the steel tube surface in the joint region under each load level. The layout of strain measurement points is shown in Figs. 4.52–4.55.

4.4.3 Failure modes of T-type and Y-type compression joints

In the T-type compression joint tests, the chord members all exhibited significant bending deformation, indicating chord bending failure. Fig. 4.56

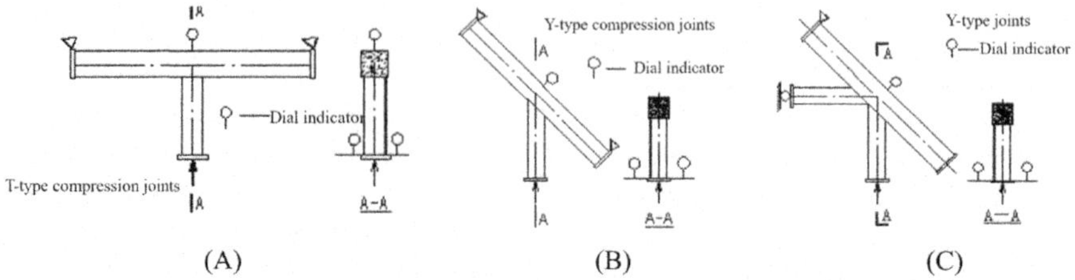

Figure 4.50 Layout of joint displacement measurement points. (A) T-type joints. (B) Y-type joints. (C) K-type joints.

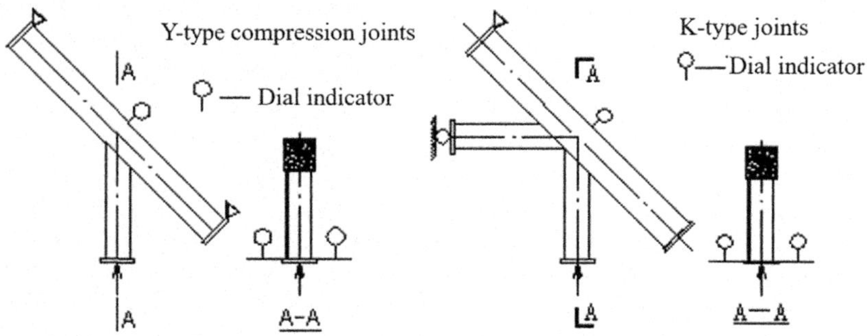

Figure 4.51 Layout of displacement measurement points for Y-type and K-type joints.

shows the failure modes of the test joints: specimen Tc-c28 failed by fracture of the tension flange of the main tube; specimens Tc-c16 and Tc-c11 showed local buckling of the compression flange on both sides of the branch, with Tc-c11 exhibiting particularly pronounced local buckling. The failure load was controlled by the wall thickness of the main tube steel plate, that is, the cross-sectional size, and was not strongly related to the branch-to-main width ratio β. The failure loads of Tc-c16 and Tc-c11 were relatively close, while those of Tc-c24 and Tc-c28 were also similar. At the branch location, the sidewalls of the main tube in the test joints all showed slight outward bulging. Fig. 4.56 also illustrates the condition of the concrete inside specimen Tc-c11: in the branch action area, the concrete surface contained a considerable amount of debris and clear imprints of the compressed branch, while outside the branch action zone the concrete surface remained intact.

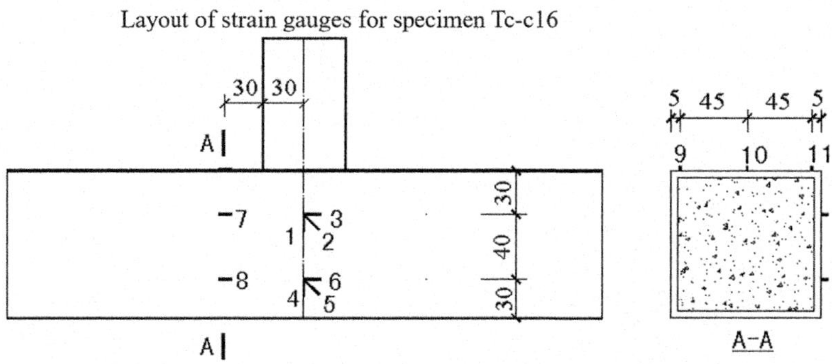

Figure 4.52 Strain measurement point layout for joint Tc-c16.

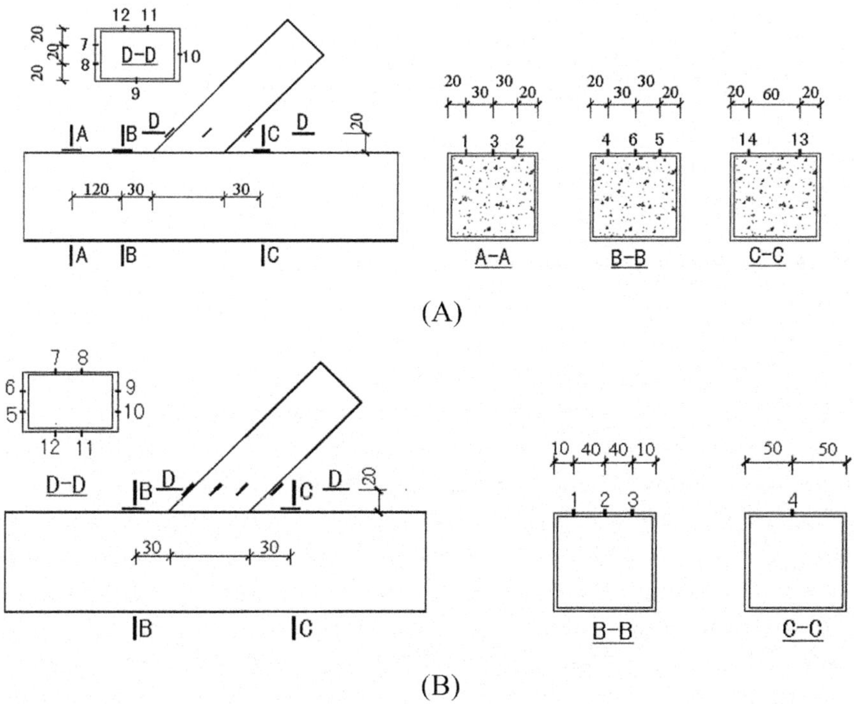

Figure 4.53 Strain measurement point layout for joints Yc-c16 and Yc-018. (A) Yc-c16. (B) Yc-018.

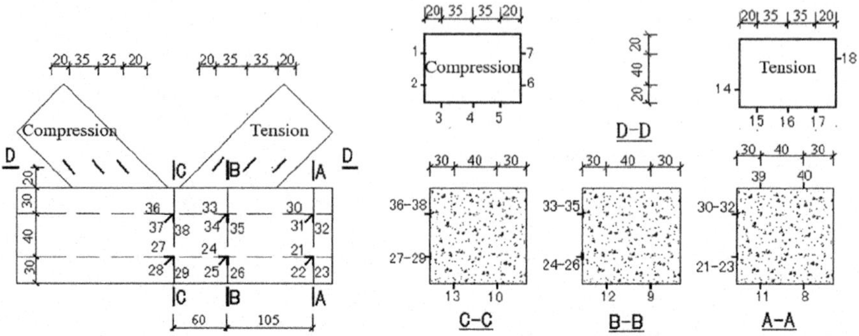

Figure 4.54 Strain measurement point layout for joint K2-c18.

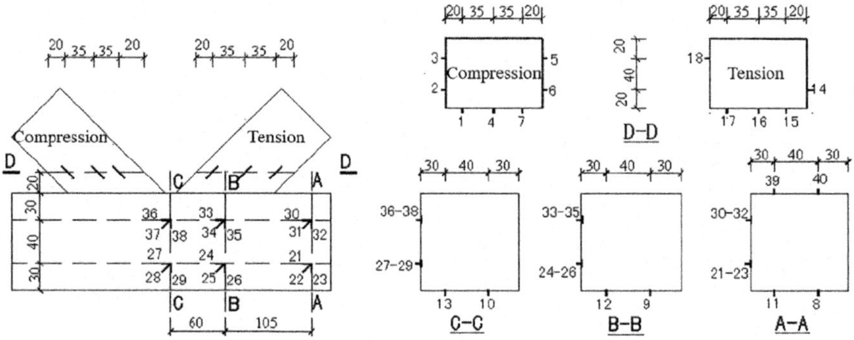

Figure 4.55 Strain measurement point layout for joint K2-028.

Fig. 4.57 shows the failure modes of the Y-type test joints. For the RST compression joint Yc-018, the failure mode was local buckling of the main tube sidewall. After the main tube was filled with concrete, this failure mode was avoided. The failure mode of Yc-c28 was local buckling of the compressed branch, accompanied by excessive bending deformation of the chord. For Yc-c16 and Yc-c14, the failure mode was local buckling on the compression side of the main tube surface at the front end of the branch, with the chord also exhibiting excessive bending deformation. The failure loads of these two specimens were relatively close. Fig. 4.57 also shows the condition of the concrete inside specimen Yc-c16, where concrete debris can be observed on the concrete surface at the branch action area.

Table 4.10 lists the failure loads and failure modes of the test joints. Since all the CFST truss joints in the table exhibited chord bending failure,

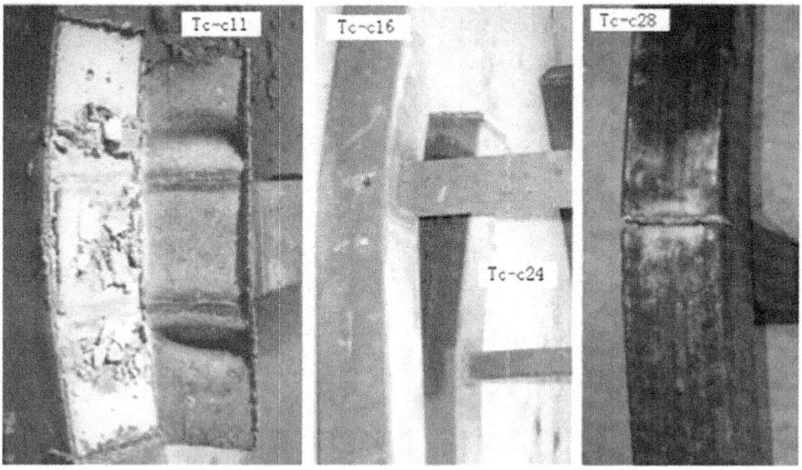

Figure 4.56 Failure condition of a T-type joint.

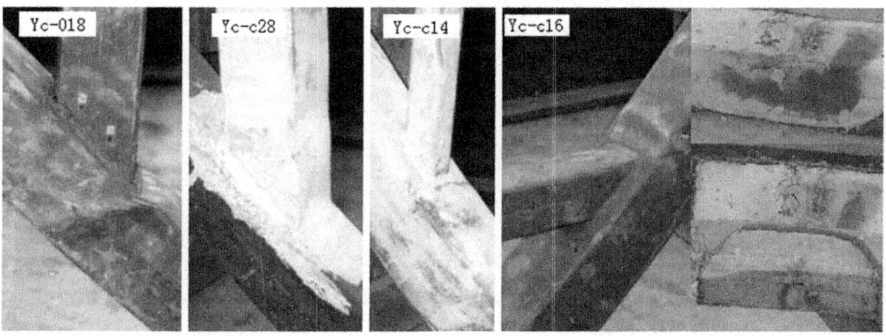

Figure 4.57 Failure condition of a Y-type joint.

which belongs to member failure, the table also includes the flexural capacity of the chord calculated using the ultimate equilibrium theory.

The calculation model for the flexural capacity of the chord in CFST truss joints based on the ultimate equilibrium theory is shown in Fig. 4.58, with the following assumptions:

1. The tensile strength of the concrete in the tension zone inside the tube is neglected.
2. The steel tube is assumed to yield across the full section, that is, the sectional stress reaches the yield strength f_y.

Table 4.10 Failure load and failure modes of T-type and Y-type joints under compression.

Joint number	Test value (kN) (kNm)		Flexural capacity of chord (kNm)		Transverse local bearing capacity (kN)		Yield-line failure bearing capacity (kN)		Failure mode
	N_u^T	M_u^T	M_u	M_u^T/M_u	N_u	N_u^T/N_u	N_u	N_u^T/N_u	
Tc-c16	103.6	19.43	14.31	1.36	237	0.44	22.9	4.52	CF
Tc-c11	112.3	21.06	14.31	1.47	576	0.19	/	/	CF
Tc-c24	152.9	28.67	19.46	1.47	139	1.1	33.3	4.60	CF+Bb
Tc-c28	163.0	30.56	19.46	1.57	492	0.33	86.8	1.88	CF
Yc-018	67.0	/	/	/	/	/	70.4	0.95	Cb
Yc-c14	123.2	20.77	14.31	1.45	149	0.83	24.5	5.03	CF+Cb
Yc-c16	123.2	20.77	14.31	1.45	289	0.43	36.7	3.35	CF+Cb
Yc-c28	186.2	31.66	19.46	1.63	574	0.33	146.7	1.27	CF+Bl

Notes: *Bb*, Global buckling of the brace in compression; *Bl*, local buckling at brace end; *Cb*, local buckling of the chord flange and sidewalls; *CF*, overall bending of the chord.

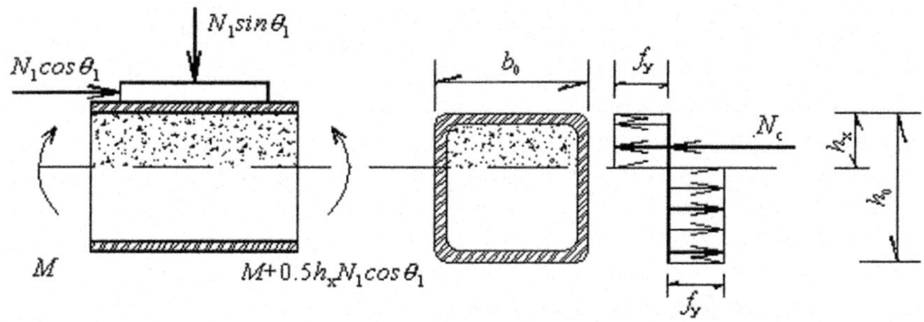

Figure 4.58 Calculation diagram for the flexural capacity of the chord member in concrete-filled rectangular hollow section truss joint.

3. The concrete in the compression zone is assumed to reach its axial compressive strength f_c; thus in Fig. 4.43, $N_c = A_c f_c$, where A_c is the cross-sectional area of the concrete in the compression zone.
4. At the front end of the branch (the right end in Fig. 4.58), the horizontal component of the branch force is assumed to act at half the height h_x of the compression zone. Therefore the maximum bending moment of the chord at the front end of the branch is $M+0.5h_xN_1\cos\theta_1$.

From static equilibrium, the height of the compression zone h_x and the flexural capacity M_u of the chord at the joint are obtained as follows:

$$h_x = \frac{(b_c + 2h_0)t_0}{b_c + 4t_0} \tag{4.13}$$

$$M_u = [(h_0 - h_x)h_0 + (b_0 - 2t_0)(h_0 - t_0)]t_0 f_y \tag{4.14}$$

where $b_c = (b_0 - 2t_0) f_c / f_y$; t_0 is the wall thickness of the main tube.

In Table 4.10, the bending capacity of the chord is given as:

$M_u^T = (N_u^T \sin\theta_1)L/4 + 0.5h_xN_u^T\cos\theta_1$, where L is the chord length. Table 4.10 also lists the calculated joint capacities based on (1) the design formula for the bearing capacity of CFRST compression joints and (2) the yield-line failure mode of the RST.

The experimental results show that for CFRST T-type and Y-type compression joints, the actual failure mode was chord bending failure rather than joint failure. The measured failure loads were mostly lower than the values calculated using the CFRST joint capacity formula. Therefore using Eq. (4.14) to evaluate the chord flexural capacity better reflects the actual failure mechanism of the joints. A comparison of the results in Table 4.10 shows that the bearing capacity of CFRST compression joints is significantly higher than that of corresponding RST. Moreover, the smaller the branch-to-chord width ratio β, the greater the increase in capacity. This indicates that for CFRST T-type and Y-type compression joints with longer chords, large bending moments develop in the chord, and the failure is usually governed by the chord flexural capacity rather than by joint failure.

4.4.3.1 Load-displacement curves

The displacement of the branch member is composed of the chord deflection, branch compression, local buckling deformation, and local deformation in the joint region, as illustrated in Fig. 4.59.

Fig. 4.60 shows the measured load-displacement curves of T-type joint branch members, which exhibit distinct elastic, elastic-plastic, and plastic stages. These stages correspond to the structural behaviors of elastic bending of the chord section, initial yielding at the section edges, and the formation and development of plastic hinges. The load-displacement curves of specimens Tc-c16 and Tc-c11 in Fig. 4.60 are quite similar, and those of Tc-c24 and Tc-c28 are also similar, indicating that the displacement of the joint branch members is mainly composed of chord deflection. However, a comparison between Tc-c24 and Tc-c28 shows that the displacement of Tc-c24 is larger than that of Tc-c28, suggesting that as the branch-to-chord width ratio decreases, local deformation in the joint region becomes more pronounced and the joint stiffness decreases. A similar trend is also observed when comparing the load-displacement curves of Tc-c16 and Tc-c11.

Fig. 4.61 shows the measured load-displacement curves of Y-type joint branch members. The curves of specimens Yc-c14 and Yc-c16 are quite similar, but before failure, the displacement of Yc-c14 is larger than that of

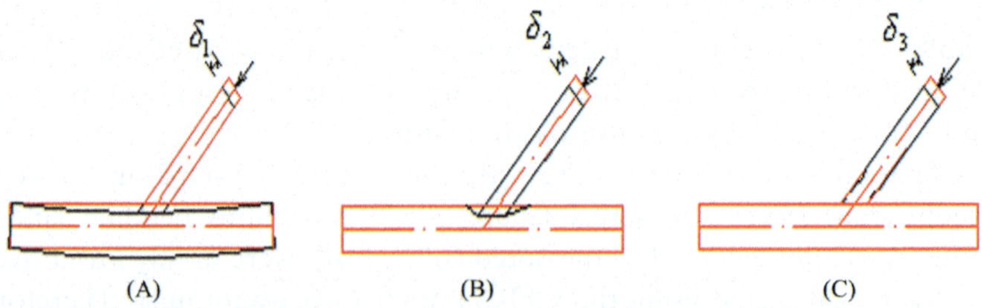

Figure 4.59 Joint deformation. (A) Flexural deformation of the chord. (B) Local deformation of the joint. (C) Compression and local buckling deformation of the brace.

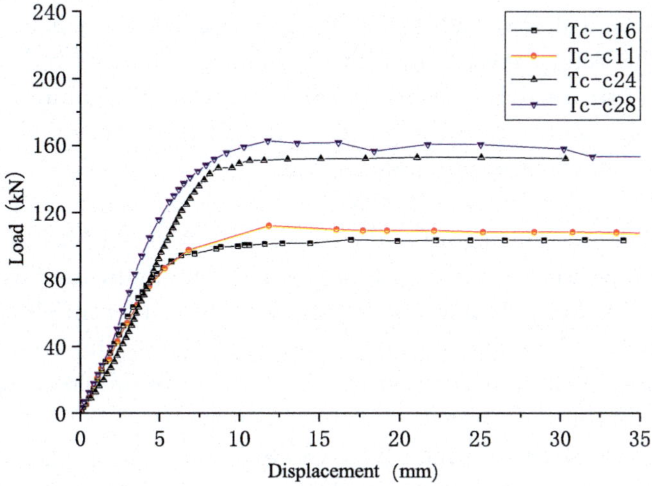

Figure 4.60 Load-displacement curve of the brace member for a T-type joint.

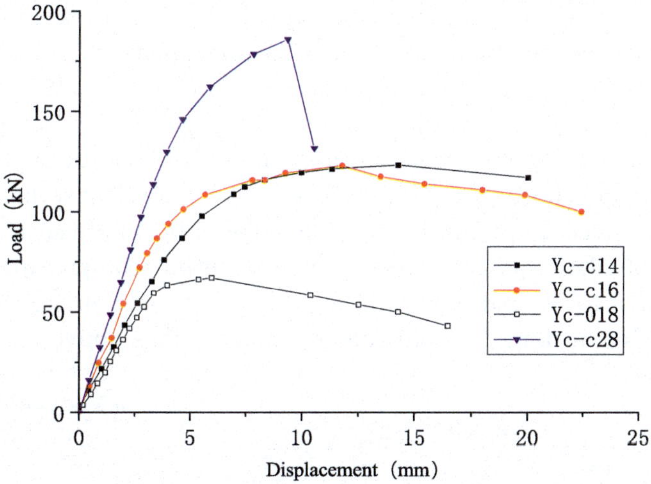

Figure 4.61 Load-displacement curve of the brace member for a Y-type joint.

Yc-c16, indicating that Y-type joints exhibit a trend similar to T-type joints—namely, as the branch-to-chord width ratio decreases, local deformation in the joint region becomes more pronounced. In Fig. 4.61, the load-displacement curves of specimens Yc-c28 and Yc-018 both show a distinct descending branch. For Yc-c28, a CFST joint, the failure load is very high, and before failure, no plastic hinge has formed in the chord section; the deformation is first chord bending, followed by local plastic buckling of the branch wall. For Yc-018, an RST, the failure load is very low, and at failure, excessive bending deformation occurs in the chord flange and brace within the joint region.

Fig. 4.62 helps us to understand the above test results and to further clarify the stress and deformation characteristics of CFRST T-type and Y-type compression joints. Fig. 4.62A corresponds to the local deformation of the joint region shown in Fig. 5.15B, which is further simplified into the model illustrated in Fig. 4.62B. In order to reflect the chord bending deformation of the joint shown in Fig. 4.59A, the simplified model uses equivalent springs to simulate the chord bending.

In Fig. 4.62B, the bearing capacity of the equivalent spring representing chord bending can be determined according to Eq. (4.14), while the bearing capacity of the equivalent strut representing local compression of the concrete inside the tube can be determined based on the transverse local bearing strength of CFRST. The bearing capacity of the equivalent strut in the chord compression zone depends on its cross-sectional area. If the direction of the branch axial force remains unchanged, the horizontal component H of the branch axial force in Fig. 4.62B produces a deformation u, and the vertical component V produces a deformation v, which should satisfy the relation: $v/u = \tan\theta$, where θ is the angle between the branch and chord. Since the stiffness of the equivalent spring for chord bending is inversely proportional to the cube of the chord span, while the stiffness of the equivalent strut in the chord compression zone is proportional to its cross-sectional area, their deformation capacities differ significantly, making it difficult to satisfy the above deformation

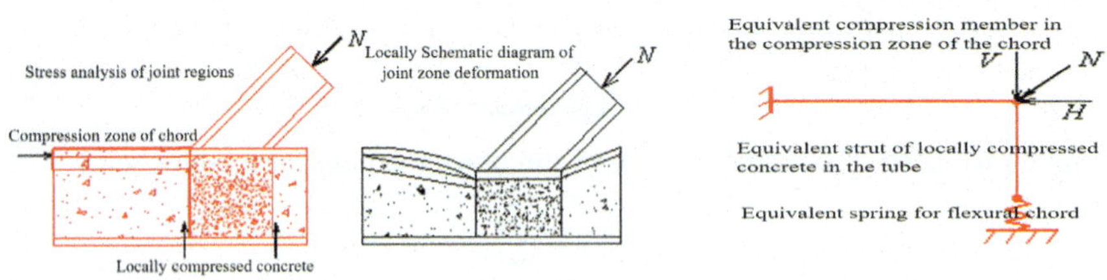

(A) (B)

Figure 4.62 Analysis model for force and deformation of a concrete-filled rectangular steel tubular Y-type joint under compression. (A) Analysis model for stress and deformation in joint regions. (B) Simplified model.

compatibility condition. When the chord span is relatively large, the joint deformation is dominated by chord bending. In this case, to maintain the direction of the branch axial force, the branch is subjected to eccentric compression, as shown in Fig. 4.63. When the branch-to-chord width ratio β is relatively small (specimens Yc-c14 and Yc-c16), the horizontal component of the branch axial force is mainly transferred from the chord top plate gradually to the chord sidewalls and the concrete inside the tube. At the front end of the branch, the local effect on the chord top plate is significant. Along with chord bending failure, the chord top plate exhibits local bulging deformation, as shown in Fig. 4.62A.

For joints with larger β values (e.g., Yc-c28), the distance between the branch wall and the chord sidewall is small, so the horizontal component of the branch axial force is transferred more directly to the chord sidewall. In addition, the chord wall is thicker, and the equivalent strut stiffness of the chord compression zone is larger. Therefore chord bending deformation becomes more pronounced, the branch eccentric effect increases, and the joint fails in the form of local buckling at the root of the branch.

4.4.3.2 Load-strain curves

Figs. 4.64 and 4.65 show the measured load-strain curves at the branch root section of the steel tube joint Yc-018 and the CFST joint Yc-c16, respectively. For the steel tube joint Yc-018, the branch flange is entirely under tensile strain, with strain values at measuring points 5 and 6 near the weld toe being greater than those at measuring points 9 and 10 near the weld root. The branch brace at measuring points 7 and 12 and at points 8 and 11 is under compressive strain, and the strain values at points 7 and 12

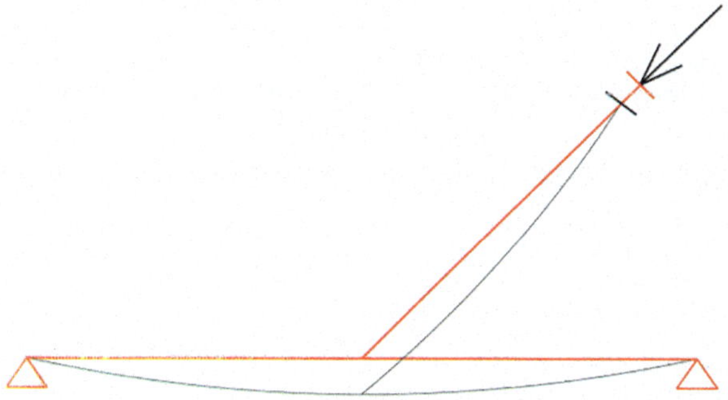

Figure 4.63 Eccentric compression of the brace member in a Y-type joint.

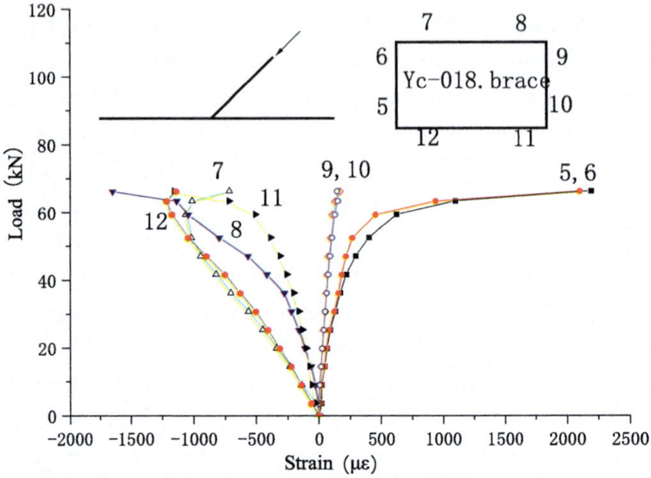

Figure 4.64 Load-strain relationship curve of the brace member for the steel Y-type joint Yc-018.

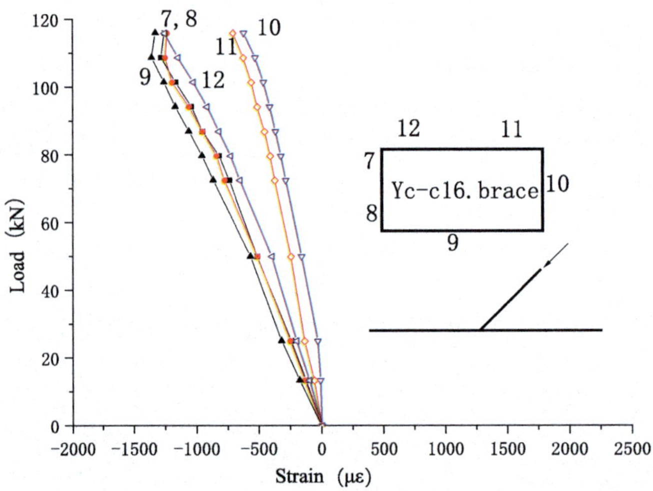

Figure 4.65 Load-strain relationship curve of the brace member for the CFST Y-type joint Yc-c16.

near the weld toe are greater than those at points 8 and 11 near the weld root, as shown in Fig. 4.64.

For the CFST joint Yc-c16, the branch is entirely under compressive strain. Among them, the strain values at measuring points 7, 8, and 12 near the weld toe are greater than those at measuring points 10 and 11 near the weld root, while the maximum strain occurs at measuring point 9 located in the branch brace, as shown in Fig. 4.66. Since the chord is filled with concrete, the bending of the chord flange plate is restrained; therefore, compared with the hollow steel tube joint, the branch of the CFST joint is entirely subjected to compressive stress. The stress state of the branch in the CFRST Y-type compression joint is illustrated in Fig. 4.66. As analyzed in Fig. 4.64, due to chord bending deformation, in order to maintain the direction of the branch axial force, the branch is subjected to a bending moment M. Meanwhile, since the horizontal component H of the branch axial force is mainly transmitted through the branch brace, the stress distribution in the branch is nonuniform.

Figs. 4.67 and 4.68 show the measured load-strain curves at measuring points on the compressed flange surface of the chord in the steel tube joint Yc-018 and the CFST joint Yc-c16, respectively. In the figure, sections B and C are symmetric and located 30 mm from the flange plate on either side of the branch; section A is 120 mm away from section B. Sections A and B are located on the weld toe side, that is, the front end of the branch, while section C is on the weld root side.

The strain at section C is mainly induced by chord bending, whereas section B not only experiences bending stress but also has to transfer the horizontal component of the branch axial force. Therefore the measured strains at section B are much larger than those at section C. For the hollow steel tube joint Yc-018, local bending strain of the chord flange plate is also present. Since the bending deformation at the midpoint of the chord top plate is larger than that at the edges, the strain at measuring point 2 in

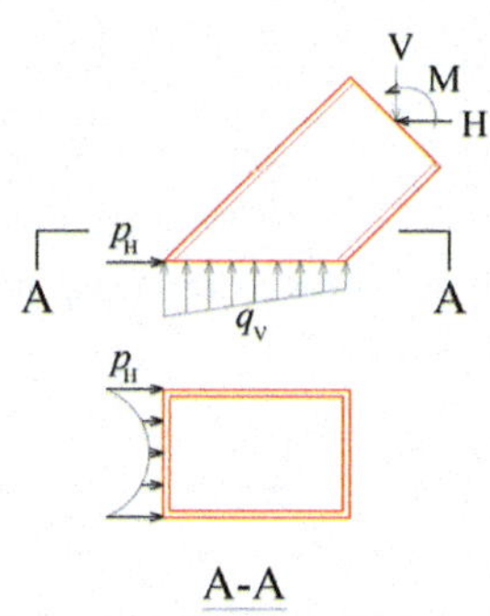

Figure 4.66 Force analysis of the brace member.

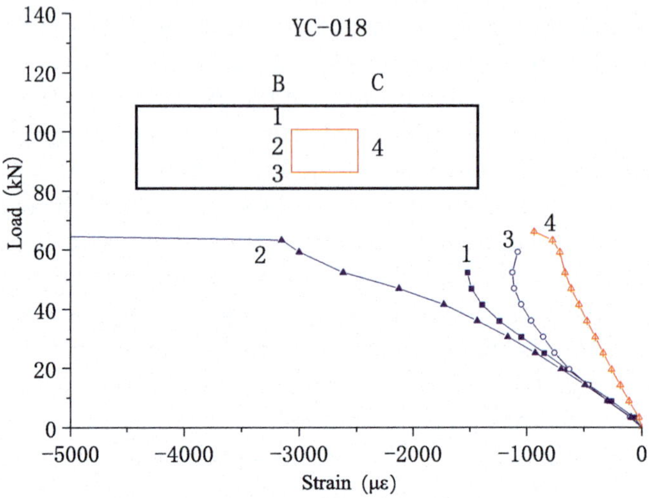

Figure 4.67 Load-strain relationship curve of chord flange measurement points for joint Yc-018.

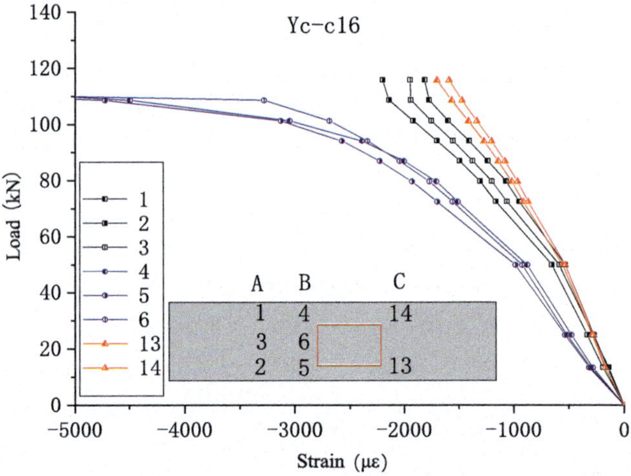

Figure 4.68 Load-strain relationship curve of chord flange measurement points for joint Yc-c16.

section B is greater than that at edge measuring points 1 and 3. A comparison of the load-strain curves at points 2, 1, and 3 in Fig. 4.67 shows that the local bending effect of the chord top plate in the hollow joint is quite pronounced.

For the CFST joint Yc-c16, due to the restraining effect of the infilled concrete, the bending of the chord top plate is limited, and the strain distribution at different measuring points within the same section is relatively uniform, as shown in Fig. 4.68. In Fig. 4.68, the measured strain at section A is much smaller than that at section B, indicating that the horizontal component of the branch axial force can be gradually transferred through the chord top plate to the entire chord section.

Fig. 4.69 shows the measured load-strain curves at the measuring points on the top plate surface of the chord section, located 30 mm from the branch side flange, for specimen Tc-c16. As seen in the figure, the strains at each measuring point are relatively uniform, with the measured strain at the central point 10 being slightly smaller.

Measuring points 7, 8 and 9, 10, 11 are located on the same cross section of the chord, 60 mm from the mid-span. Taking the average measured strain of top plate points 9, 10, and 11 as *E*, the longitudinal load–strain curve of this cross section is plotted as shown in Fig. 4.70.

From Fig. 4.68 it can be observed that, in the compression zone of the section, the strains at point 7 and point E maintain an approximately linear relationship with load up to about 70% of the failure load. Beyond this point, the strain at point E on the chord top plate increases rapidly, while the strain at brace point 7 remains nearly unchanged and shows a slight

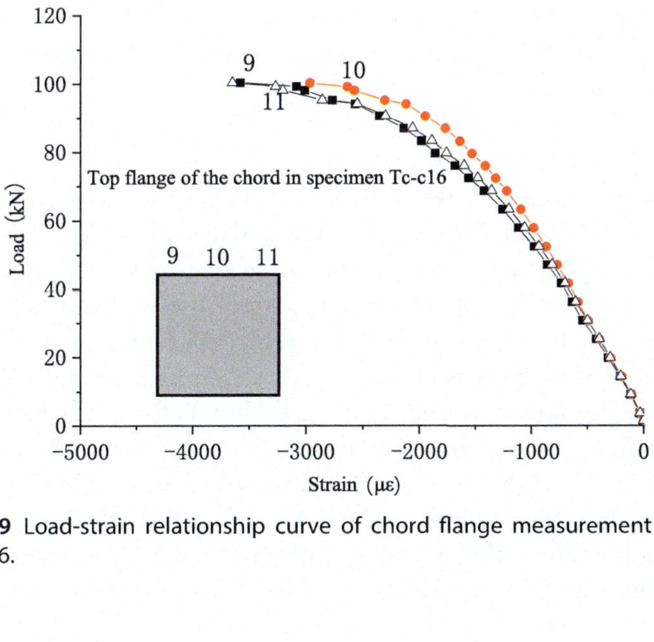

Figure 4.69 Load-strain relationship curve of chord flange measurement points for joint Tc-c16.

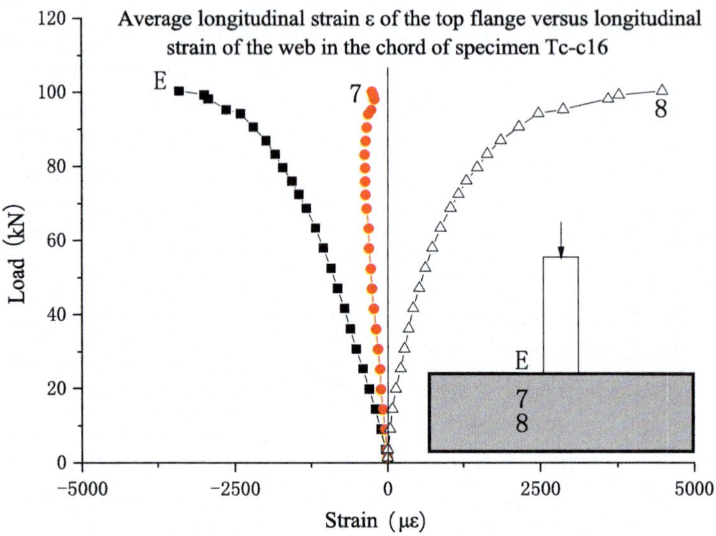

Figure 4.70 Load-longitudinal strain relationship curves of chord flange and brace measurement points for joint Tc-c16.

decrease near failure. In the tension zone of the section, the load-strain relationship at brace point 8 can be divided into three stages: when the load is less than about 20% of the failure load, the strain increases slowly; from this stage up to about 70% of the failure load, the strain increases rapidly; after about 70% of the failure load, the load-strain curve becomes distinctly nonlinear.

These load-strain relationships of the measuring points reveal the bending characteristics of the chord section in the joint: in the tension zone, before cracking of the concrete, the steel tube and concrete work together, and the load-strain relationship remains linear. After cracking, the stress in the steel tube of the tension zone increases, while the concrete gradually ceases to contribute; however, since the steel tube stress has not yet reached yield, the load-strain relationship still remains essentially linear. When the stress in the tension-zone steel tube reaches yield, the load-strain relationship becomes nonlinear until plastic hinges form, at which point the strain continues to increase while the load remains essentially unchanged.

Fig. 4.71 shows the measured longitudinal strain distribution along the section height under different load levels. It can be seen that the bending deformation of the CFRST chord conforms to the plane-section assumption. In the elastic stage, the neutral axis of the section nearly coincides

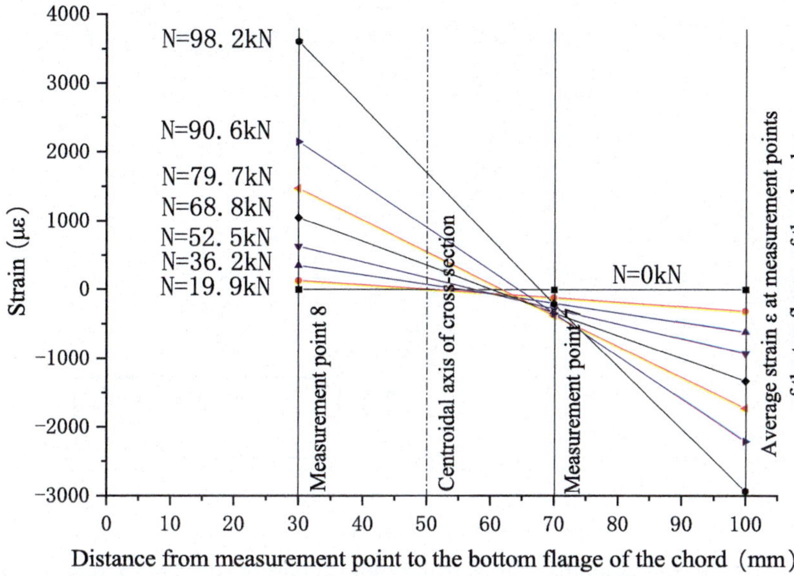

Figure 4.71 Distribution of longitudinal strain along the section height at chord measurement points for joint Tc-c16.

with the centroidal axis of the section. As the load increases, the neutral axis continuously shifts upward toward the upper flange. Near failure, the neutral axis is close to the location of measuring point 7, so the strain at point 7 hardly increases at this stage.

According to Eq. (4.14), the calculated height of the compression zone is $hx = 31.83$ mm, which is located near measuring point 7, in good agreement with Fig. 4.71. This demonstrates that the proposed formula for calculating the bending capacity of CFRST members can adequately reflect the actual mechanical behavior of the structure.

Figs. 4.72 and 4.73 show the load-strain relationships of the longitudinal and transverse strains at measuring points located 30 mm from the upper and lower flanges on the sidewall of the chord mid-span section. Here, the transverse strain refers to the strain measured along the section height direction. As shown in the figure, before reaching about 70% of the failure load, the transverse strain at the tension-zone measuring point (point 4) is nearly zero, while the longitudinal strain (point 6) increases with the load, indicating that the point is under uniaxial tensile stress. Beyond this stage, the point enters a biaxial stress state with opposite signs—longitudinal

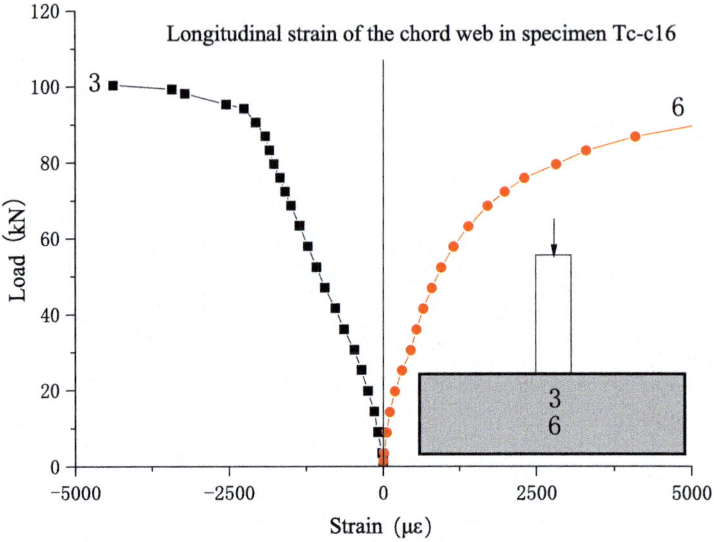

Figure 4.72 Load-longitudinal strain relationship curves of chord brace measurement points for joint Tc-c16.

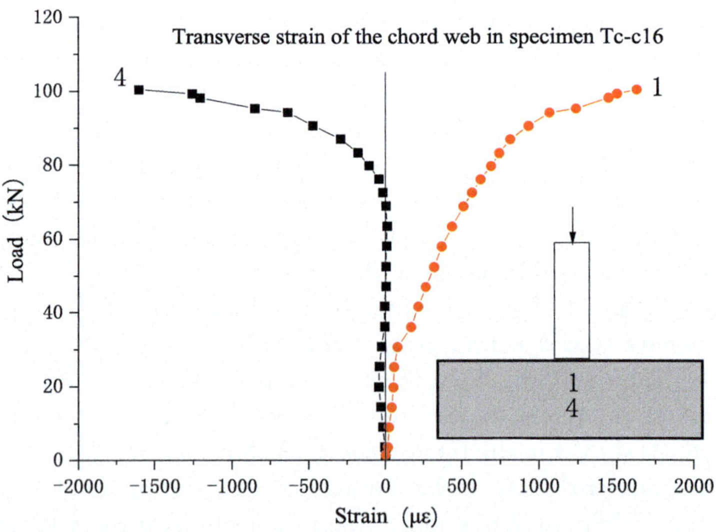

Figure 4.73 Load-transverse strain relationship curves of chord brace measurement points for joint Tc-c16.

tension accompanied by transverse compression—with the transverse compressive strain increasing rapidly. For the compression-zone measuring point, the state is also biaxial with opposite signs, that is, longitudinal compression (longitudinal strain at point 3) accompanied by transverse tension (transverse strain at point 1). After about 30% of the failure load, the transverse strain increases rapidly, and as failure approaches, both longitudinal and transverse strains grow quickly. At failure, the transverse strain reaches approximately 1700 $\mu\varepsilon$.

The occurrence of transverse tensile strain in the compression zone is related to the local stress behavior of the joint region. Under the action of the branch load, the concrete inside the chord is locally compressed. In the compression zone of the chord, longitudinal and transverse compression of the concrete produces lateral expansion, causing the chord sidewall to bend laterally; thus the transverse strain (point 1) on the sidewall surface appears as tensile strain. In the tension zone of the chord, the concrete inside the tube is under longitudinal tension, and due to the diffusion effect, the transverse compressive stress decreases, making it difficult to generate lateral expansion. At the same time, the longitudinal tension of the sidewall prevents it from bending laterally; therefore the brace in the tension zone does not develop transverse tensile strain.

As shown in Fig. 4.74, with increasing load, the concrete in the compression zone continues to expand laterally, causing the local bending deformation of the brace to extend downward. However, the longitudinal tension of the sidewall in the tension zone provides lateral restraint against

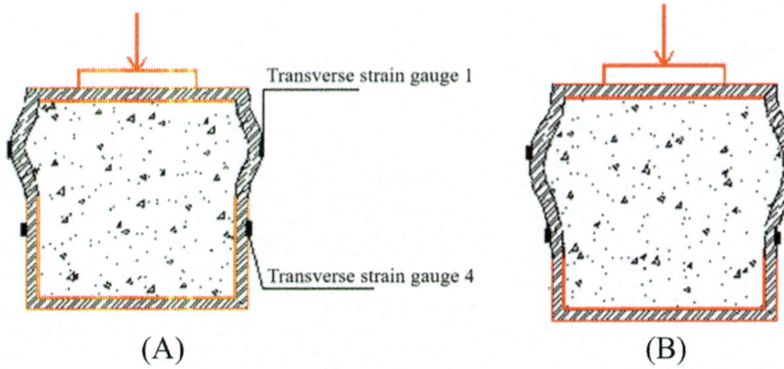

Figure 4.74 Schematic diagram of chord side plate deformation in the joint region. (A) Under small loads. (B) Under large loads

the sidewall bending, so that at the location of strain gauge 4 in the tension-zone brace, a negative bending moment develops, resulting in compressive strain on the sidewall surface. With the cracking of the concrete in the tension zone and the consequent reduction in the shear area of the chord, more of the branch load transferred through the concrete is redistributed to the chord sidewall. Together with the wedge effect shown in Fig. 4.74B, this leads to a rapid increase in the transverse compressive strain measured on the tension-zone brace surface.

Since the chord of the test joint is relatively long, the failure mode is chord bending failure, and the local deformation of the joint shown in Fig. 4.74 is not very pronounced. However, the observed transverse strain results confirm the characteristics of local joint deformation, indicating that T-type compression joints share a similar mechanism with the transverse local bearing tests of CFRST presented in Chapter 3. It can be expected that if the chord were sufficiently short, the joint failure load would be governed by local deformation, and the working mechanism of transverse local bearing would become more evident. Fig. 4.75 shows the relationship curves between the principal tensile strain at the brace measuring points and the load, calculated from the strain test results. It can be seen that, due to

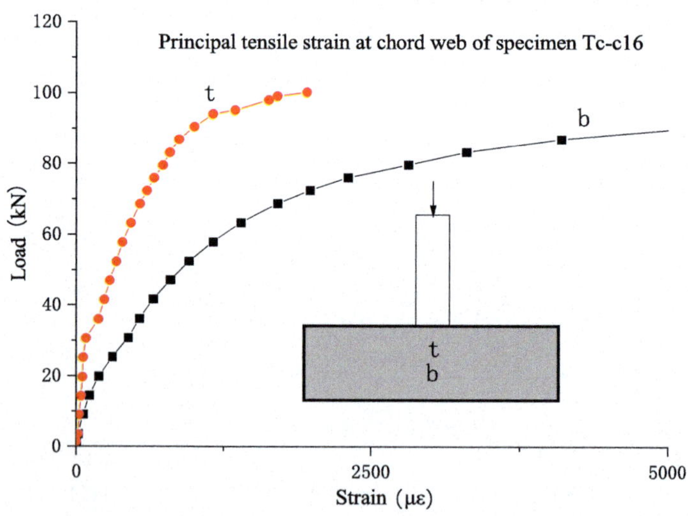

Figure 4.75 Load-principal tensile strain relationship curves of chord brace measurement points for joint Tc-c16.

lateral bending, the sidewall in the compression zone exhibits significant principal tensile strain on its surface.

4.4.4 Test results and analysis of K-type joints

4.4.4.1 Failure modes and failure process analysis

The possible failure modes of K-type steel tube joints can be classified into four categories: (1) excessive plastic deformation and failure of the chord top plate beneath the branch under branch tension or compression, (2) overall shear failure of the chord in the gap between two branches, (3) effective width failure or local instability failure of one branch, and (4) punching-shear failure of the chord top plate at the root of one branch. For K-type CFST joints, the presence of concrete inside the chord provides a highly rigid foundation beneath the compressed branch. For a compressed branch, the local buckling load of the branch and the transverse local bearing strength of the chord determine its load capacity, which is similar to the behavior of Y-type, T-type, and X-type CFST compression joints. For a tensioned branch, the punching-shear failure of the chord top plate beneath it, or the effective width failure of the branch itself, governs its capacity. In the test steel tube joint (K2-028), due to the relatively thick chord wall, the failure mode was local buckling at the root of the compressed branch, accompanied by bulging of the chord sidewall under the action of the compressed branch. Due to limited experimental experience, only in specimen K0-c14 of the K-type CFST joints did failure occur in the chord top plate beneath the tensioned branch, manifested as a protrusion on the chord top plate surface at the location of the tensioned branch. Specimens K0-c18, K2-c18, K4-c18, and K4-c28 all exhibited local buckling failure at the root of the compressed branch, as shown in Fig. 4.76A. Specimen K0-c16, due to eccentric loading, failed by overall bending of the compressed branch, which did not conform to the above failure modes. Fig. 4.76B shows the damage condition of the concrete inside the chord, where no distinct concrete failure mode was observed.

Table 4.11 summarizes the failure loads and failure modes of all K-type joints from the present tests and Packer's tests. In Packer's tests, the failure of K-type steel tube joints all occurred in the chord top plate beneath the compressed branch. In contrast, in the CFST K-type joints of the present study, failure consistently occurred in the chord top plate beneath the tensioned branch. Among them, specimens K1C, K2C, K3C, and K4C

(A) (B)

Figure 4.76 Failure modes of tested CFST K-type steel tubular joints. (A) Buckling at the root of the compression brace. (B) Condition of the concrete infill in the chord.

experienced punching-shear failure, while specimen K5C was expected to fail by effective width failure of the tensioned branch. However, due to premature yielding of the compressed branch, the actual failure mode of joint K5C was not obtained.

In Table 4.11, the values shown in parentheses represent the theoretical ultimate capacities of K-type steel tube joints, calculated using the yield-line mechanism. For the CFST K-type joints, the theoretical ultimate capacities in this column are calculated according to the formula for the transverse local bearing strength of CFRST members proposed in Chapter 3. Packer suggested that, for CFRST gap K-type joints, the capacity of a compressed branch should be calculated based on the local bearing strength of the infilled concrete, whereas the capacity of a tensioned branch should be checked against both punching-shear failure and effective width failure modes of the chord plate. The corresponding formulas for these bearing capacities are as follows:

Punching-shear failure:

$$N_{2u} = \frac{f_{y0}t_0}{\sqrt{3}}(2h_2/\sin\theta_2 + b_2 + b_{ep})/\sin\theta_2 \tag{4.15}$$

Effective width failure:

$$N_{2u} = f_{y2}\,t_2(2h_2 - 4t_2 + b_2 + b_e) \tag{4.16}$$

where: b_e—$b_e = \frac{10}{b_0/t_0}\frac{f_{y0}\,t_0}{f_{y2}\,t_2}b_2 \le b_2$, mm; b_{ep}—$b_{ep} = \frac{10}{b_0\,/\,t_0}b_2 \le b_2$, mm.

Table 4.11 Failure load and failure modes of K-type gap joints.

Joint number		Failure load	Punching shear		Effective width		Local bearing pressure (yield line)		Failure mode
		N_u^T (kN)	N_u (kN)	N_u^T/N_u	N_u (kN)	N_u^T/N_u	N_u (kN)	N_u^T/N_u	
K-type joints in this test	K0–c14	145	120	1.21	110	1.32	149	0.98	PS
	K0–c16	109	180	0.61	170	0.64	289	0.38	CW[a]
	K0–c18	205	240	0.86	231	0.89	469	0.44	CW
	K2–c18	185	240	0.77	231	0.80	469	0.39	CW
	K4–c18	172	240	0.72	231	0.74	469	0.37	CW
	K4–c28	179	354	0.51	238	0.75	578	0.31	CW
	K2–028	171	354	0.48	238	0.72	(165)	(1.04)	CW, CP
K-type joints in the reference test	K1	260	332	0.78	411	0.63	(166)	(1.57)	CP
	K2	224	335	0.67	412	0.54	(169)	(1.32)	CP
	K3	459	580	0.79	875	0.52	(290)	(1.58)	CP
	K4	529	580	0.91	875	0.60	(290)	(1.82)	CP
	K5	969	949	1.02	938	1.03	(569)	(1.70)	CP
	K1C	361	332	1.09	411	0.88	341	1.06	PS
	K1X	370	332	1.11	411	0.90	341	1.08	PS
	K2C	332	335	0.99	412	0.81	342	0.97	PS
	K3C	814	580	1.40	875	0.93	814	1.00	PS
	K4C	815	580	1.41	875	0.93	814	1.00	PS
	K5C	1 161	949	1.22	938	1.24	905	1.28	CW

Notes: In the column of joint failure modes:

CP stands for plastic failure of the main pipe top plate.

PS stands for punching-shear failure or tearing failure of the main pipe top plate under the tensioned branch pipe.

CW stands for failure of the compressed branch pipe.

[a] For this joint, the compressed branch pipe is loaded with eccentricity, which is the overall bending failure of the compressed branch pipe.

The effective width values b_e and b_{ep} can be applied to K-type CFST joints, since the concrete filled inside the chord has little influence on the tension brace.

Table 4.11 also lists the joint bearing capacities calculated according to Eqs. (4.15) and (4.16). As shown in Table 4.11, the failure modes of specimen K0-c14 in this study and the Packer test specimens are in good agreement with the corresponding bearing capacity calculation formulas. Although no transverse local bearing strength failure occurred in these specimens, the measured bearing capacities are still in good agreement with the calculated values of transverse local bearing strength. For the other K-type CFST joints tested in this study, the failure mode was compression brace failure; therefore their experimental ultimate loads were all lower than the theoretical values calculated by the above formulas. A comparison between the bearing capacities of hollow tube joints and CFST joints in the table shows that the infilled concrete in the chord prevents the yield-line failure mode and effectively improves the joint bearing capacity. Provided that the brace bearing capacity requirement is met, estimating the bearing capacity of gap K-type CFST joints by adopting the transverse local bearing failure mode for compression braces and the punching-shear and effective width failure modes for tension braces can reasonably reflect the joint's mechanical performance.

The tested CFST joints K0-c18, K2-c18, K4-c18, and K4-c28 did not experience joint failure, indicating that the joint bearing capacity is higher than the member bearing capacity. This shows that the sectional capacity of the brace members in the truss can be effectively utilized, with the brace member efficiency coefficient (brace member bearing capacity/joint bearing capacity) being greater than 1. From another perspective, this also demonstrates that the concrete infill in the chord enhances the joint bearing capacity. Since compression braces are more sensitive to end effects and imperfections, the observed failures of the test joints occurred in the compression braces rather than the tension braces. To prevent end buckling of compression braces, the CIDECT design guide specifies that the width-to-thickness ratio (or depth-to-thickness ratio) of the brace should satisfy:

$$\frac{b_1}{t_1} \leq 1.25\sqrt{\frac{E}{f_y}} \quad \text{or} \quad \frac{h_1}{t_1} \leq 1.25\sqrt{\frac{E}{f_y}} \tag{4.17}$$

Taking the elastic modulus $E = 2.06 \times 10^5$ and the yield strength of the steel tube material in the tested joints f_y=325 MPa, the right-hand limit value in the above equation is 31.5. In the present tests, the width-to-thickness (or depth-to-thickness) ratio of the braces for specimens K0-c18, K2-c18, K4-c18, and K4-c28 is 28.3, which is close to that limit. In view of the test results reported here, and considering the relatively high bearing capacity of CFRST compression joints, it is recommended that Eq. (4.17) for the compression braces of CFRST truss joints be revised to:

$$\frac{b_1}{t_1} \le \sqrt{\frac{E}{f_y}} \text{ or } \frac{h_1}{t_1} \le \sqrt{\frac{E}{f_y}} \tag{4.18}$$

The Canadian standard CAN/CSA-S16.1-M89 defines two categories of RST sections. A Category 1 section is defined as $\frac{b_1}{t_1} \le \frac{420}{\sqrt{f_y}}$ or $\frac{h_1}{t_1} \le \frac{420}{\sqrt{f_y}}$; a Category 2 section is defined as $\frac{b_1}{t_1} \le \frac{525}{\sqrt{f_y}}$ or $\frac{h_1}{t_1} \le \frac{525}{\sqrt{f_y}}$. Existing literature specifies that, for RST K-type overlapped joints, the minimum section of compression braces shall meet Category 1 requirements, while for K-type gap joints and T-type, Y-type, and X-type joints, the minimum section of compression braces shall meet Category 2 requirements. Therefore Eq. (4.18) falls between the Category 1 and Category 2 sections of the Canadian CSA-S16.1-M89 standard, being closer to Category 1, which is also more consistent with the literature requirements for the compression brace section size of K-type overlapped joints.

Boundary conditions and loading patterns have a certain influence on the joint bearing capacity. Table 4.12 presents the finite element analysis results of circular steel tube joints under various loading patterns. In loading patterns E and F, the tensile and compressive brace loads are equal. In other loading patterns, due to the influence of the chord's bending stiffness, the compressive brace load is, to varying degrees, greater than the tensile brace load. For CFST joints, the chord stiffness is higher than that of hollow tube joints, and local joint deformations are less pronounced; therefore, under loading patterns A–D, more of the compressive brace load is shared by the chord shear force, resulting in partial K-type and partial Y-type joints as described in Chapter 2.

The loading pattern E used by Packer requires a relatively complex loading device and synchronized control of tensile and compressive brace loading, making the test difficult to perform. Hence, for CFRST truss joints, loading pattern F is more suitable. Loading pattern F is consistent with the loading pattern adopted in this study.

A drawback of both loading pattern F and the one used in this study is that during loading, the braces undergo some bending deformation, as shown in Fig. 4.77. Since compressive braces are more sensitive to bending deformation, a P–δ effect will occur. Therefore, when the joint bearing capacity is relatively high, failure is more likely to occur in the compressive brace. This is also one of the main reasons why failures in the tested joints of this study occurred in the compressive braces.

4.4.4.2 Load-displacement curves

The measured load-displacement curves of the braces in K-type joints are shown in Fig. 4.78. Fig. 4.78A presents the load-displacement curves of specimens K0-c14, K0-c16, and K0-c18 with a gap of $g = 0$ between the two braces. As can be seen from Fig. 4.78A, the joint load–displacement curves differ due to the different failure modes: for K0-c14, the joint deformation is relatively large near failure, and the curve is relatively smooth; for K0-c18, local buckling occurred at the root of the compression brace, so the curve remained essentially linear before reaching the buckling load, but after buckling, the load dropped rapidly.

The initial stiffness of joint K0-c14 is greater than that of K0-c18, which is likely related to the distance between the axes of the two braces at the gap. Owing to the presence of concrete inside the chord, the compression brace is effectively supported on a stiffer foundation.

Table 4.12 The influence of boundary conditions and loading patterns on joint bearing capacity.

	Loading pattern	Compressive load	Tensile load		Loading pattern	Compressive load	Tensile load
A	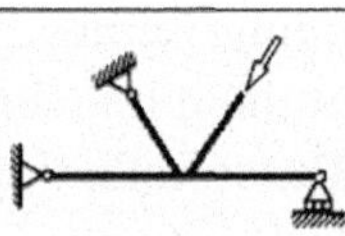	85.2kN	83.3kN	D	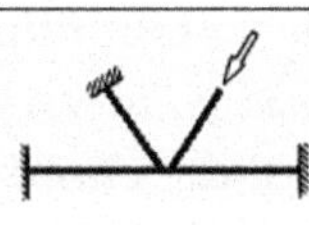	92.2kN	76.2kN
B	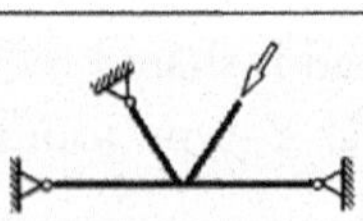	87.7kN	82.4kN	E	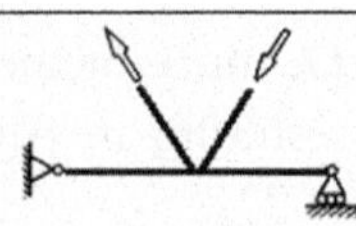	87.0kN	
C	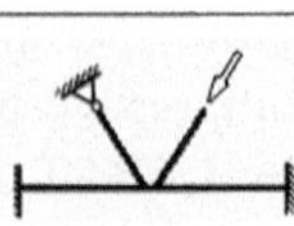	88.6kN	78.3kN	F	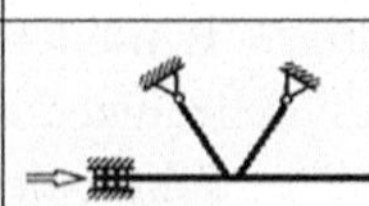	86.0kN	

Therefore the smaller the distance between the axes of the tension and compression braces on the supporting surface (the top plate of the chord), the smaller the joint deformation. In the case of joint K0-c16, premature failure occurred due to eccentric loading caused by testing operations; however, its relatively high initial stiffness also indicates a similar pattern—namely, as the distance between the axes of the two braces at the gap decreases, the stiffness of the joint region increases.

Fig. 4.78B shows the load-displacement curves of the braces in K-type joints with a branch-to-chord width ratio of $\beta = 0.8$. As seen in Fig. 4.78B, joint K0-c18 exhibits the highest bearing capacity and the greatest initial stiffness. Although the failures of the K-type joints with $\beta = 0.8$ all occurred at the root of the compression brace and their load-displacement curves follow a trend similar to that of K0-c18, the influence of the joint gap on the initial stiffness and bearing capacity can still be observed in Fig. 4.78B. Specifically, the smaller the joint gap, the higher the bearing capacity and the greater the initial stiffness.

4.4.4.3 Load-strain curves

In this test program, most K-type joints failed at the root of the compression brace. Except for K0-c14, no joint failure occurred; therefore the strain test results are particularly important for understanding the mechanical behavior of CFRST K-type joints.

First, the brace strains are analyzed. Fig. 4.79 shows the relationship between load and the average strain measured at points located 2 cm from the chord top surface on the weld root face, side face, and the weld toe face of the brace in the hollow tube joint K2-028. As seen from the figure, the tensile brace exhibits an approximately linear increase in strain, with the

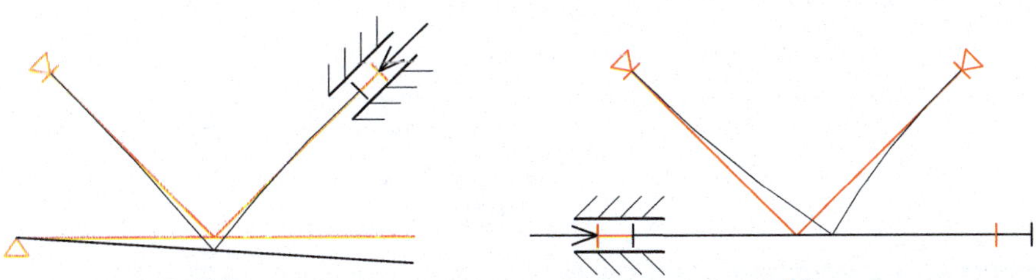

(A) (B)

Figure 4.77 Schematic diagram of brace bending deformation during the loading process. (A) Loading patterns in this study. (B) Loading pattern F in Table 4.12.

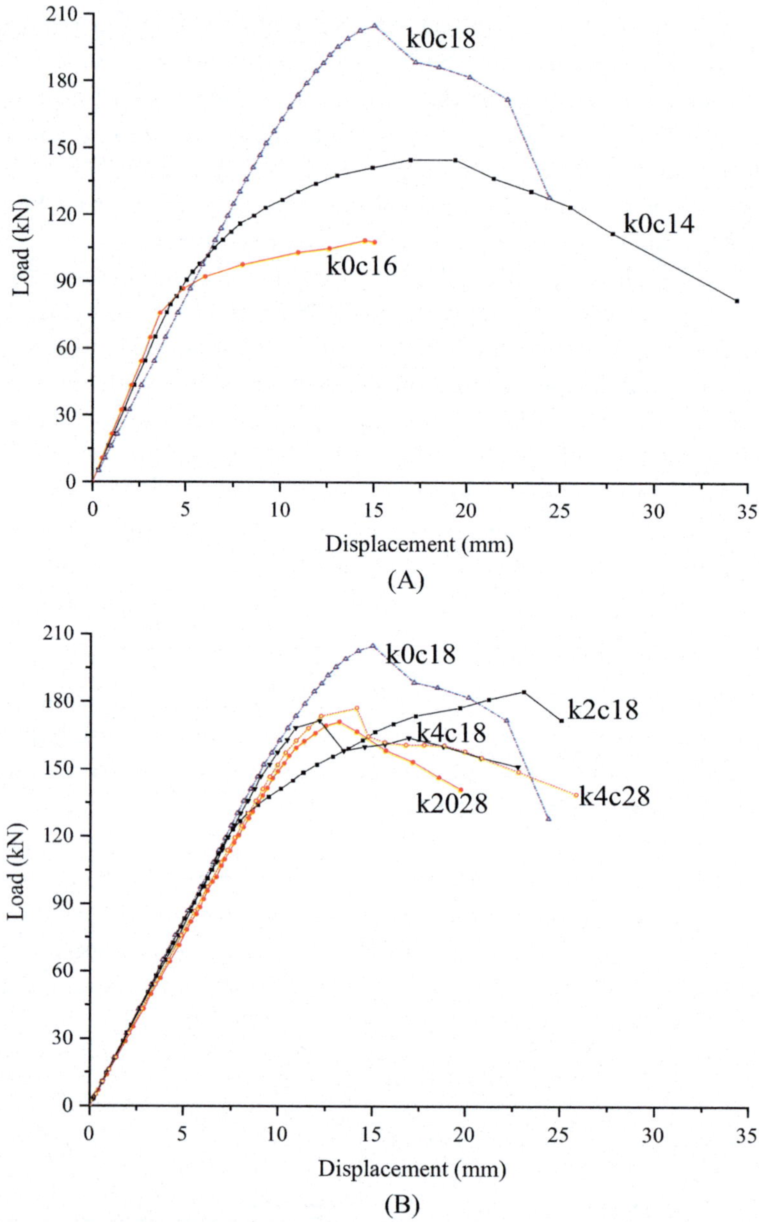

Figure 4.78 Load-displacement curves of brace members for K-type joints.

average strains at the different measurement faces being roughly equal. The strain at the weld toe face is slightly larger due to the constraint from the adjacent compression brace face, the weld root face strain is smaller, and the side face strain falls in between. For the compression brace, owing to the influence of local chord wall distortion, initial local bending of the brace wall, welding deformation, and other initial imperfections, the strain distribution is nonuniform: the weld toe face shows the largest strain, the side face is intermediate, and the weld root face exhibits tensile strain, though its magnitude is small compared with the compressive strain. Near failure, the compressive strain on the surface of the compression brace increases rapidly, indicating that local buckling occurred at the brace root.

The load-strain curves of the brace at the same measurement locations in the CFST K-type joint K2-c18 are shown in Fig. 4.80. It can be seen that the strain distribution on the three faces of the tension brace is nonuniform, which is likely related to the thinner chord wall. The chord wall thickness of K2-c18 is 2.75 mm (whereas that of K2-028 is 4 mm). The bending of the chord top plate induces bending in the brace wall; thus the surface strain of the tension brace includes bending strain, which is particularly evident at the weld root face. For the compression brace, the strains on the side face and the weld root face are essentially the same, while the strain at the weld toe face is smaller, likely due to the more concentrated welding deformation at the weld toe.

Figs. 4.81 and 4.82 present the load-strain curves of the average sectional strain of the braces in K2-028 and K2-c18, respectively. As shown, up to failure, the average sectional strains of the tension and compression braces are nearly equal, indicating that the loading method used in this study can effectively ensure force equilibrium between the tension and compression braces.

Next, the chord surface strains are analyzed. The test results of the longitudinal average strain at measurement points on the chord surface of K-type joints are shown in Figs. 4.83 and 4.84. In these figures, chord sections A–A, B–B, and C–C correspond to the midsection of the gap, the midsection where the tension brace intersects with the chord flange, and the front end section of the tension brace, respectively, as illustrated in Fig. 4.85.

In Fig. 4.83, for the hollow tube joint K2-028, during the early stage of loading, the front flange wall of the chord bulged outward under the action of the tension brace, producing tensile bending strain, and longitudinal tensile strain appeared at section A–A. As the load increased, under the

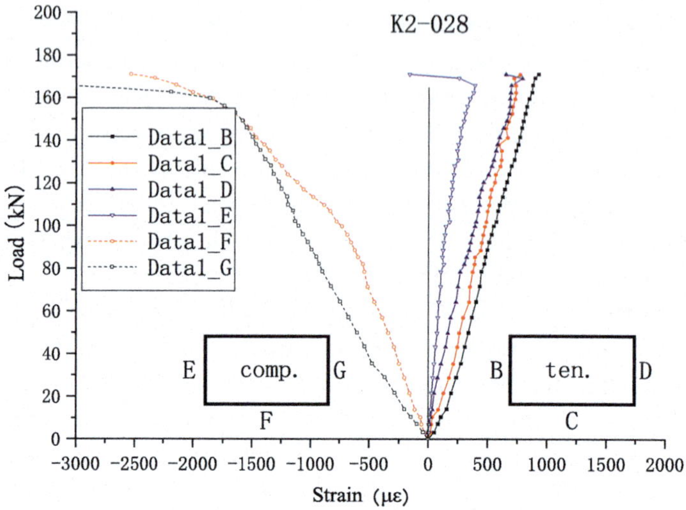

Figure 4.79 Load-strain curves of brace measurement points for K2-028.

same load level, the longitudinal average strain of the hollow joint K2-028 increased progressively from section C–C to section A–A, reflecting that the horizontal components of the axial forces in the compression brace and the tension brace were transmitted sequentially into the chord.

In contrast, as shown in Fig. 4.84, for the CFST joint K2-c18, the maximum longitudinal average strain occurred at section B–B. Although section A–A is located at the front end of the tension brace, the concrete infill inside the chord carried part of the horizontal force component from the brace, reducing the stress borne by the chord wall. Therefore the longitudinal strain at section A–A was smaller than that at section B–B.

The distributions of longitudinal strain along the section height under different load levels are shown in Figs. 4.86 and 4.87. In Fig. 4.86, for the hollow steel tube joint K2-028, the strain distribution along the section height at section C–C on the chord surface exhibits a triangular shape, with the lower flange strain being nearly zero. At sections A–A and B–B, the longitudinal strain decreases progressively from the upper flange to the lower flange. Because the tension brace causes bulging of the chord upper flange, bending strain appears on the chord surface, resulting in a nonlinear distribution. For hollow joints, the horizontal force component of the brace is gradually transferred to the chord brace through the connected chord flange. If the local bending of the chord wall is neglected, the

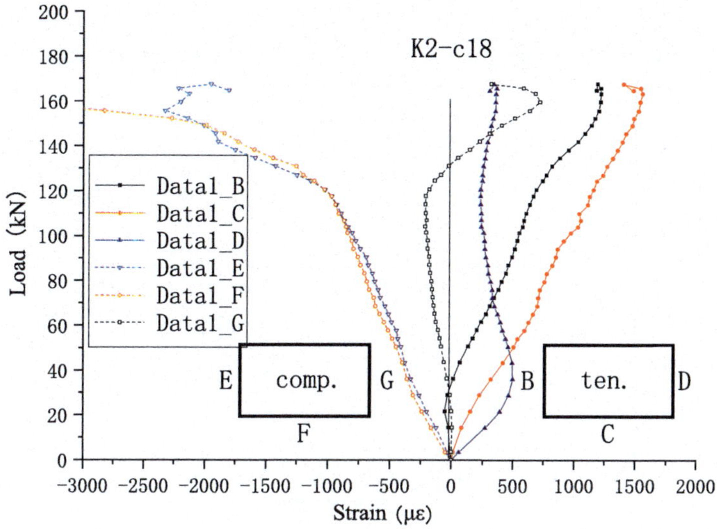

Figure 4.80 Load-strain curves of brace measurement points for K2-c18.

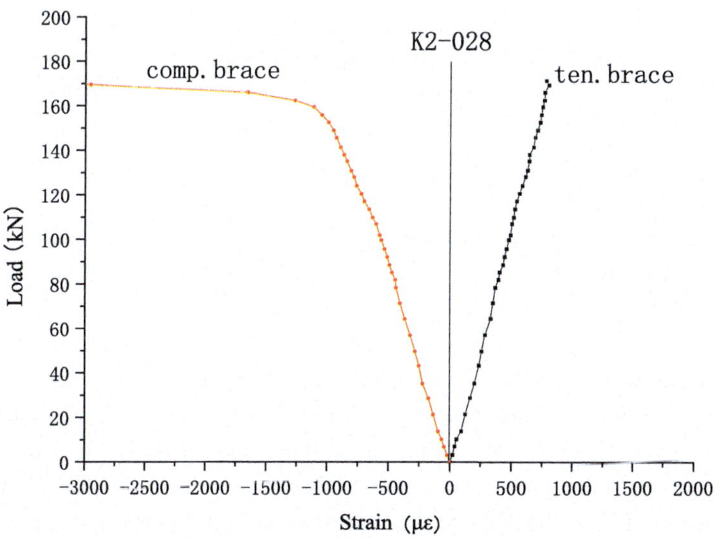

Figure 4.81 Load-average strain curves of brace measurement points for K2-028.

longitudinal stress distribution across the section can be assumed to be linear, as illustrated in Fig. 4.88A.

For CFST joints, the axial force of the compression brace is primarily transferred to the concrete inside the chord, which causes the point of action of the horizontal component of the compression brace force to

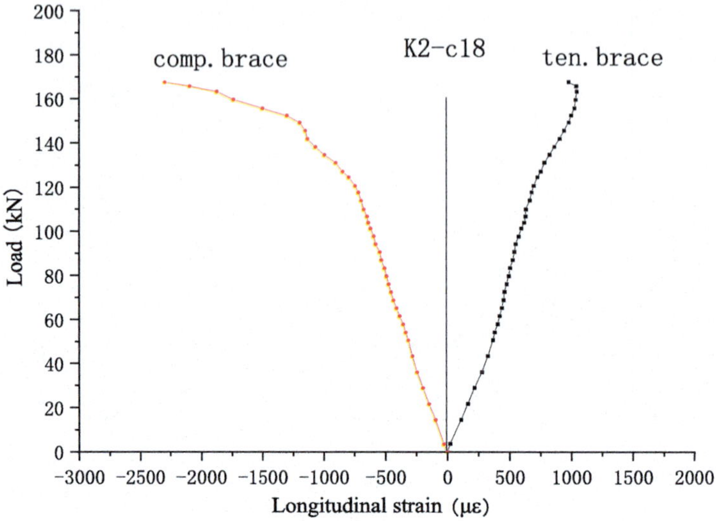

Figure 4.82 Load-average strain curves of brace measurement points for K2-c18.

move downward. If the point of action of the horizontal component of the tension brace force is still assumed to act on the surface of the chord top plate connected to the brace, then the stress distribution across the section is as shown in Fig. 4.88B. The test results of the longitudinal strains at each section of the CFST joint K2-c18 in Fig. 4.87 confirm this inference.

Fig. 4.89 illustrates the distributions of transverse strain along the section height at the brace measurement points of the chord in the tested K-type joints under different load levels.

For the hollow tube joint K2-028, the transverse strain on the chord sidewall surface is mainly generated by bending stress. As shown in Fig. 4.89, under the action of the compression brace, the chord top plate deflects inward while the sidewall bulges outward, producing positive bending stress on the sidewall surface. Therefore, along the direction of the compression brace axial force, the transverse tensile strain is larger at the sidewall measurement points near the top plate of section C–C and near the bottom plate of section B–B. In contrast, under the action of the tension brace, the chord top plate bulges outward while the sidewall deflects inward, producing negative bending stress on the sidewall surface. This reduces the tensile strain induced by the axial force of the tension brace, so the transverse strain at the sidewall measurement points near the top plate of section B–B is smaller. The sidewall measurement points near the

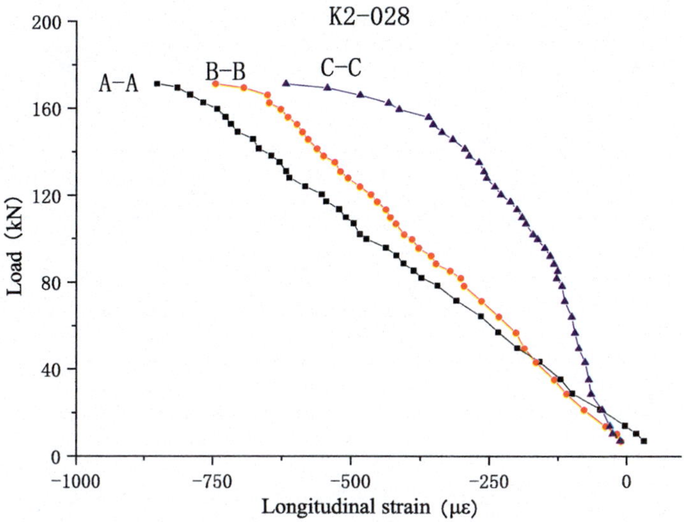

Figure 4.83 Relationship curves between load and average longitudinal strain at chord section measurement points for K2-028.

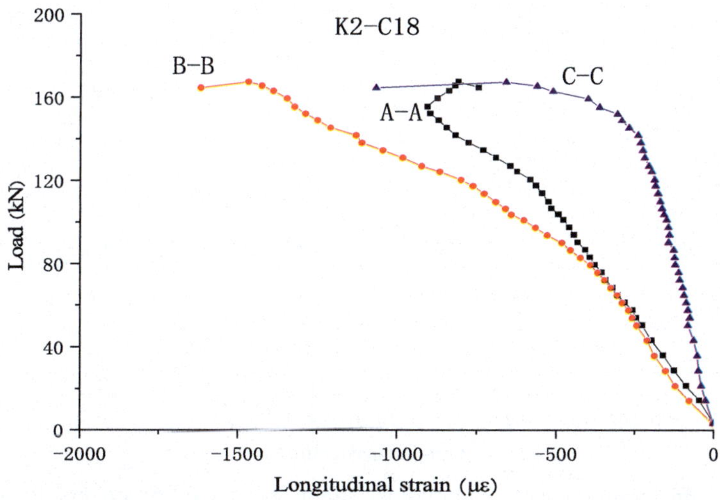

Figure 4.84 Relationship curves between load and average longitudinal strain at chord section measurement points for K2-c18.

bottom plate of sections C–C and A–A are less affected, and their transverse strains are relatively small.

For the CFST joint K2-c18, the infilled concrete prevents the inward deflection of the chord sidewall. The transverse strain on the sidewall

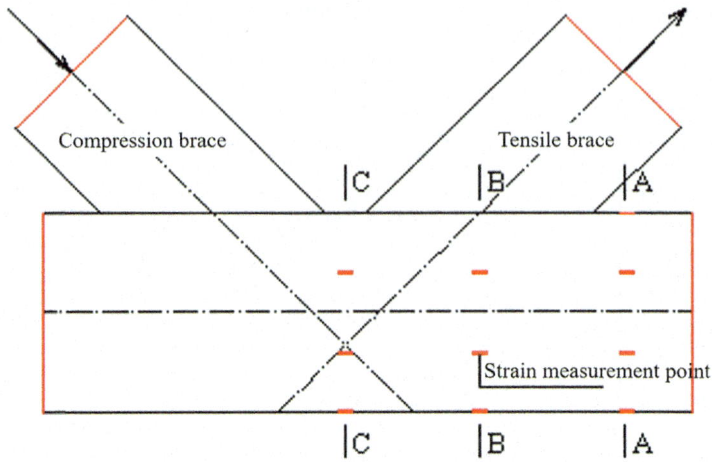

Figure 4.85 Schematic diagram of strain measurement sections and point locations for K-type joints.

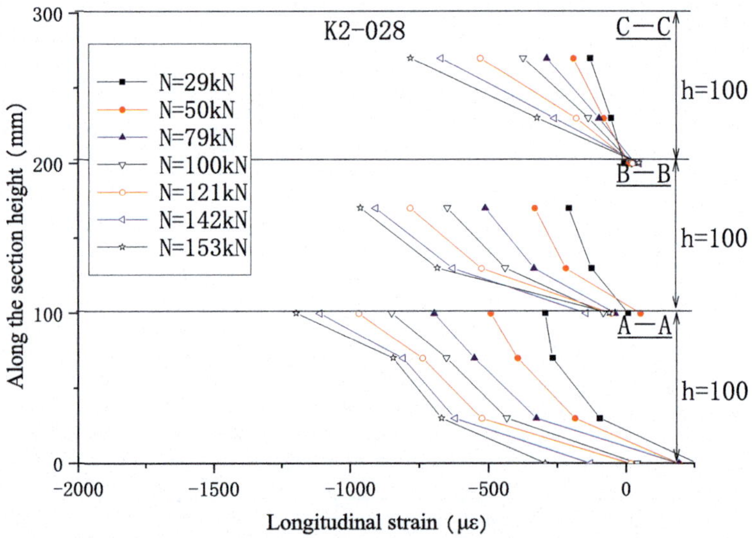

Figure 4.86 Distribution of longitudinal strain along the section height at chord measurement points for K2-028.

surface is mainly caused by the vertical component of the axial force from the tension brace. Due to stress diffusion, the transverse strain at the upper measurement points on the chord sidewall is greater than that at the lower measurement points, as shown in Fig. 5.46. Comparing the late-loading stage in Fig. 4.90, it can be observed that the transverse strain at the upper measurement points of section C–C is larger, which is likely due to the lateral expansion of the concrete inside the chord under the action of the compression brace.

4.5 Bearing capacity calculation formula for concrete-filled rectangular steel tubular truss joints

4.5.1 Gap determination formula for K-type joints

In CFRST truss K-type joints, the compressed brace member provides support to the top chord plate, while the tensioned brace member pulling on the top chord plate is restrained. Therefore, under the action of the tensioned brace member, only shear punching failure or effective width failure usually occurs. When the joint gap is relatively large, this restraining effect gradually diminishes, and the K-type joint tends to transform into a pair of Y-type joints. For joints with a larger β, since both K-type and Y-type joints experience either shear punching failure or effective width failure, and considering the effect of concrete infill inside the chord, the bearing capacities of K-type and Y-type joints are relatively close. Thus the influence of joint gap on the bearing capacity of CFRST truss K-type joints can be neglected. However, for CFRST truss K-type joints with relatively small tensioned brace member width, when the gap is large, the top chord

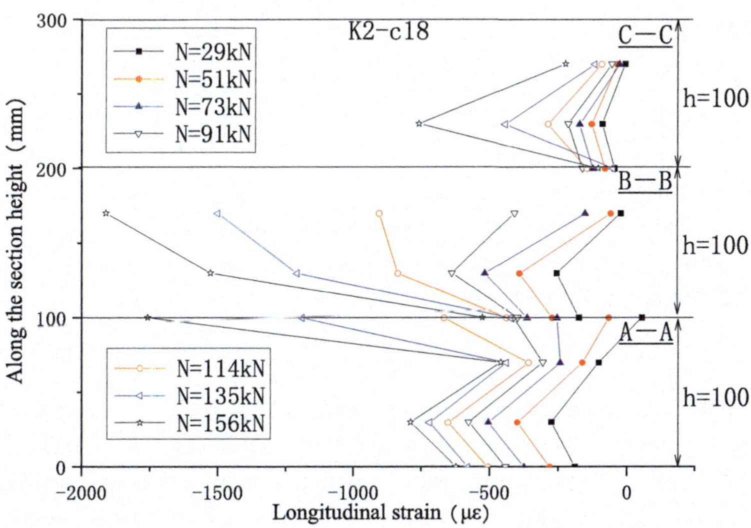

Figure 4.87 Distribution of longitudinal strain along the section height at chord measurement points for K2-c18.

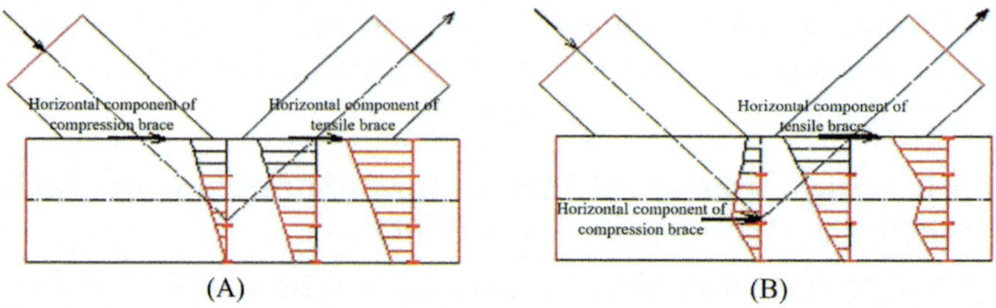

Figure 4.88 Assumed distribution of longitudinal strain in the chord member of K-type joints. (A) Hollow pipe joint. (B) Concrete-filled steel tubular joint.

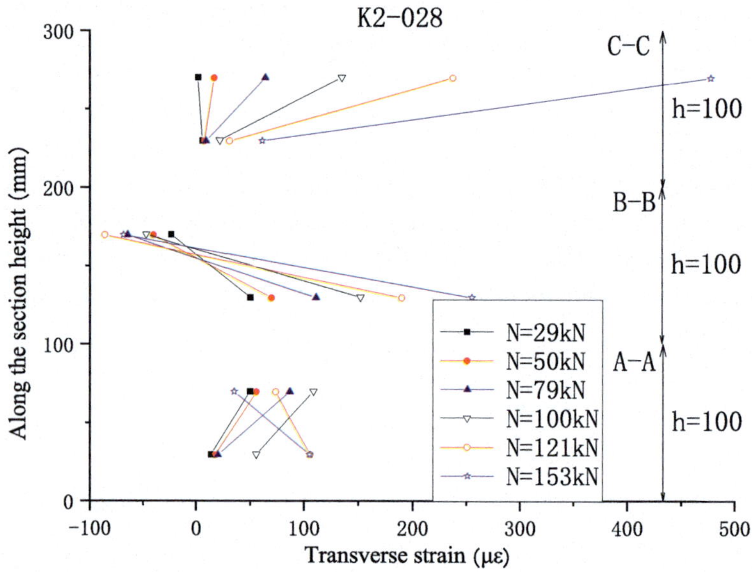

Figure 4.89 Distribution of transverse strain along the section height at chord measurement points for K2-028.

plate at the tensioned brace member will develop yield-line failure. In this case, the effect of the joint gap becomes significant. Accordingly, for CFRST truss K-type joints with smaller β, the criterion for judging whether they should be regarded as a pair of Y-type joints is whether the compressed brace member lies outside the yield-line region.

As shown in Fig. 4.91, the tensioned brace member of a K-type joint is simplified as a branch member oriented perpendicular to the chord, with a

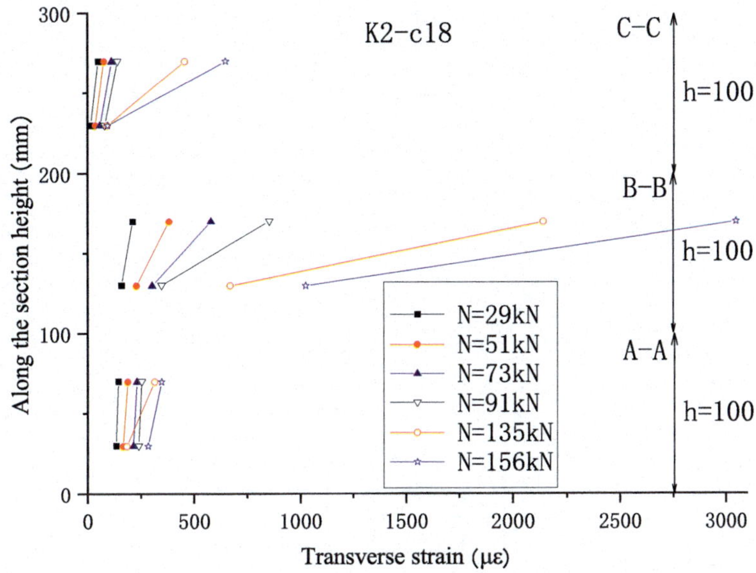

Figure 4.90 Distribution of transverse strain along the section height at chord measurement points for K2-c18.

cross-sectional height of $h_2/\sin\theta_2$. When the width ratio of the tensioned brace member to the chord, $\beta = b_2/b_0$, is relatively small and the joint gap is large, yield-line failure occurs in the chord flange plate at the location of the tensioned brace member, as illustrated in Fig. 4.91. From Fig. 4.91, it can be seen that the condition for judging whether a K-type joint becomes a pair of Y-type joints is:

$$g > \frac{b_0 - b_2}{2\tan\alpha} \tag{4.19}$$

Assuming the pullout displacement of the tensioned brace member is δ, the work done by the external load is:

$$W_e = N_2 \sin\theta_2 \delta \tag{4.20}$$

The internal work produced along yield lines 1, 2, 3, 4, and 5 in Fig. 4.91 is

1. Yield line 1: $b_0 \frac{2\delta tg\alpha}{b_0 - b_2} m_p = \frac{2tg\alpha}{1-\beta}\delta m_p$

2. Yield line 2: $b_2 \frac{2\delta tg\alpha}{b_0 - b_2} m_p = \frac{2\beta tg\alpha}{1-\beta}\delta m_p$

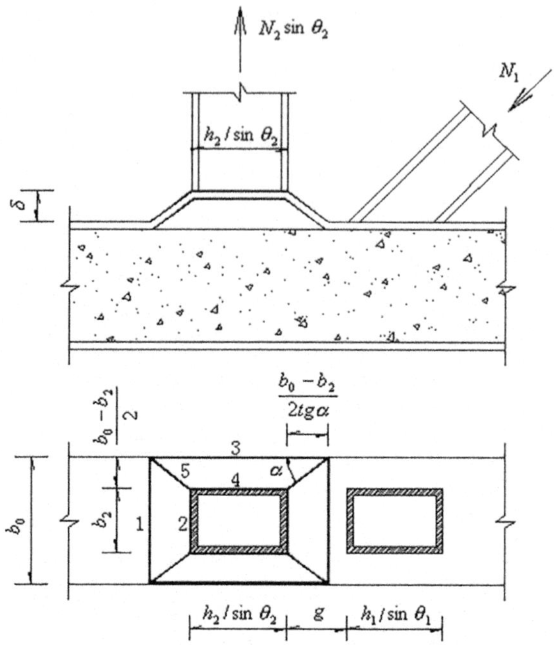

Figure 4.91 Yield-line failure model of tension brace in K-type joints.

3. Yield line 3: $\left(\frac{h_2}{\sin\theta_2}+2\frac{b_0-b_2}{2tg\alpha}\right)\frac{2\delta}{b_0-b_1}m_p=\left(\frac{\eta}{\sin\theta_2}+\frac{1-\beta}{tg\alpha}\right)\frac{2}{1-\beta}\delta m_p$

4. Yield line 4: $\frac{h_2}{\sin\theta_2}\frac{2\delta}{b_0-b_1}m_p=\frac{2\eta}{(1-\beta)\sin\theta_2}\delta m_p$

5. Yield line 5: $l_5\left(\frac{\delta}{l_5 tg\alpha}+\frac{\delta}{l_5 ctg\alpha}\right)m_p=(tg\alpha+ctg\alpha)\delta m_p$

In the equation, $m_p=\frac{1}{4}f_{y0}\,t_0^2$ is the plastic yield moment, and $\eta=\frac{h_2}{b_0}$. Summing the internal work of the yield lines yields the total internal work:

$$W_i=\frac{8m_p}{1-\beta}\left(tg\alpha+\frac{1-\beta}{tg\alpha}+\frac{\eta}{\sin\theta_2}\right)\delta \tag{4.21}$$

By equating the internal work to the external work, we obtain:

$$N_2\sin\theta_2=\frac{8m_p}{1-\beta}\left(tg\alpha+\frac{1-\beta}{tg\alpha}+\frac{\eta}{\sin\theta_2}\right) \tag{4.22}$$

To find the extremum of N_2, setting $\frac{dN_2}{d\alpha}=0$ yields:

$$tg\alpha = \sqrt{1-\beta} \tag{4.23}$$

Substituting Eqs. (4.23) into (4.19) yields:

$$\frac{g}{b_0} > \frac{\sqrt{1-\beta}}{2} \tag{4.24}$$

Therefore, when β is relatively small and satisfies Eq. (4.24), the K-type joint can be regarded as a pair of Y-type joints, in which the tension brace induces yield-line failure at its connection to the chord. The corresponding bearing capacity is obtained by substituting Eqs. (4.23) into (4.22).

$$N_{2u} = \frac{f_{y0}\, t_0^2}{(1-\beta)\sin\theta_2}\left[\frac{2\eta}{\sin\theta_2} + 4\sqrt{1-\beta}\right] \tag{4.25}$$

4.5.2 Bearing capacity calculation formula for concrete-filled rectangular steel tubular truss joints

Experimental studies and finite element numerical simulation results show that the bearing capacity of CFRST truss joints should be calculated based on the corresponding failure mode. Specifically: For compression joints, the calculation can be carried out according to the transverse local bearing failure mode. For tension joints, the calculation method depends on the chord-to-brace width ratio β and the joint gap:

1. When β is relatively large, the capacity is calculated based on the chord flange punching-shear failure mode or the brace effective width failure mode.
2. When β is relatively small but the gap g is also small, K-type joints are still calculated according to the punching-shear or effective width failure mode.
3. For T, Y, and X joints with relatively small β, as well as K-type joints with small β and large gap g, the calculation should be based on the yield-line failure mode.

4.5.2.1 Bearing capacity calculation formula for tension joints

Based on the experimental results of CFRST truss T-type, Y-type, and X-type tension joints, it is recommended that when $\beta < 0.6$, the joint bearing capacity should be calculated according to the yield-line failure mode. For joints with $\beta \geq 0.6$, the bearing capacity should still be determined using the corresponding formulas for punching-shear failure or effective width failure of T-, Y-, and X-type steel tube joints. For practical application, it is reasonable to adopt the same punching-shear failure and effective width failure formulas for CFRST T-type, Y-type, X-type, and K-type joints. This approach provides two main advantages: on the one hand, it indirectly accounts for the improvement in joint performance contributed by the concrete infill in the chord; on the other hand, for K-type joints, it only requires using Eq. (4.24) to determine whether the K-type joint should be considered as a Y-type joint.

4.5.2.2 Bearing capacity calculation formula for compression joints

The bearing capacity of compression joints in CFRST trusses can be calculated based on the transverse local bearing strength of the chord member. The calculation formula for the transverse local bearing strength of CFRST members is:

$$N_{\text{ucs}} = k_{\text{cs}} f_c A_b \tag{4.26}$$

In the formula, k_{cs} is the enhancement factor for the transverse local bearing strength of CFRST, expressed as $k_{\text{cs}} = (1 + k_s)(1 + k_L)k_c$. Here, k_L is the length influence coefficient of the infilled concrete, k_s is the participation coefficient of the steel tube in transverse local bearing, and ks is the enhancement factor for the local bearing strength of plain concrete.

When the infilled concrete length inside the tube satisfies $L_c > h_1 + 2h_0$, the coefficient k_L falls within $0 < k_L \leq 0.1$. For engineering applications, the influence of the concrete infill length can thus be neglected. Therefore the bearing capacity of compression joints in CFRST trusses can be conservatively calculated using the following formula:

$$N_{1u} = (1 + k_s) k_c f_{\text{ck}} A_b \tag{4.27}$$

The participation coefficient of the steel tube in transverse local bearing is rewritten as:

$$k_s = \frac{3 f_{y0}}{\nu_c f_{\text{ck}}} \frac{(k_c h_1 / h_0)^2}{(h_0 / t_0)^2} \tag{4.28}$$

where $\nu_c = 0.23$, f_{ck} is the standard axial compressive strength of concrete, and f_{y0} is the standard yield strength of the chord steel. Eq. (4.28) effectively reflects the confinement effect of the chord side plate on the infilled concrete. As the side plate height-to-thickness ratio h_0/t_0 decreases, the confinement effect becomes stronger. The term $k_c h_1/h_0$ essentially represents the effective working range of the side plate; the larger this value, the stronger the confinement effect of the side plate on the infilled concrete.

Enhancement factor for the local bearing strength of plain concrete:

$$k_c = \sqrt{A_d/A_b} \tag{4.29}$$

When the concrete grade inside the tube exceeds C40:

$$k_c = \frac{190 - f_{cu}}{150}\sqrt{\frac{A_d}{A_b}} \tag{4.30}$$

In the formula, f_{cu} is the strength of a 150 mm cube specimen, expressed in MPa. $A_b = b_1 h_1/\sin\theta_1$ is the local bearing area. A_d is the design-bearing base area, determined according to the "principle of concentric symmetry" specified in the Chinese concrete structure design code, as shown in Fig. 4.92. The code restriction that the local bearing height should be greater than the local bearing width is neglected. From Fig. 4.92, the calculation formula for the design-bearing base area A_d is obtained.

When $3b_1 < b_0$:

$$A_d = 3b_1(h_1/\sin\theta_1 + 2b_1) \tag{4.31}$$

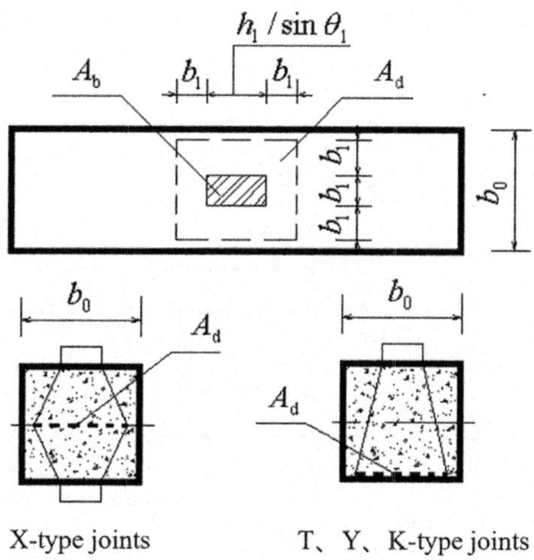

Figure 4.92 Calculation of base area A_d.

When $3b_1 \geq b_0$:

$$A_d = b_0 h_1 (1 + 2\alpha)/ \sin\theta_1,\ \alpha = b_1 \sin\theta_1/h_1 \leq 1.0 \tag{4.32}$$

4.5.3 Recommended code formulas for the bearing capacity of concrete-filled rectangular steel tubular truss joints

4.5.3.1 Design formula for the tensile bearing capacity of concrete-filled rectangular steel tubular truss joints

Taking the resistance partial factor into account, the design formulas for the tensile bearing capacity of CFRST truss joints are summarized as follows:

1. T-type, Y-type, and X-type joints

When $\beta < 0.7$, calculate according to the yield-line failure mode (Fig. 4.93).

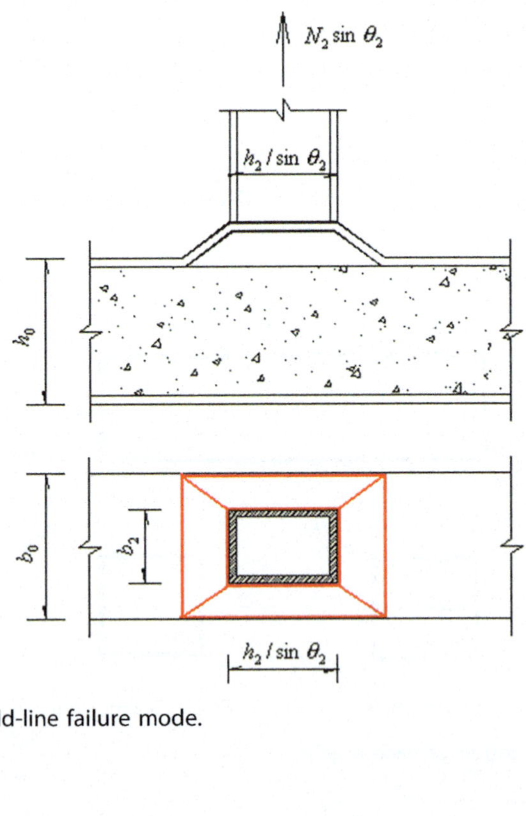

Figure 4.93 Yield-line failure mode.

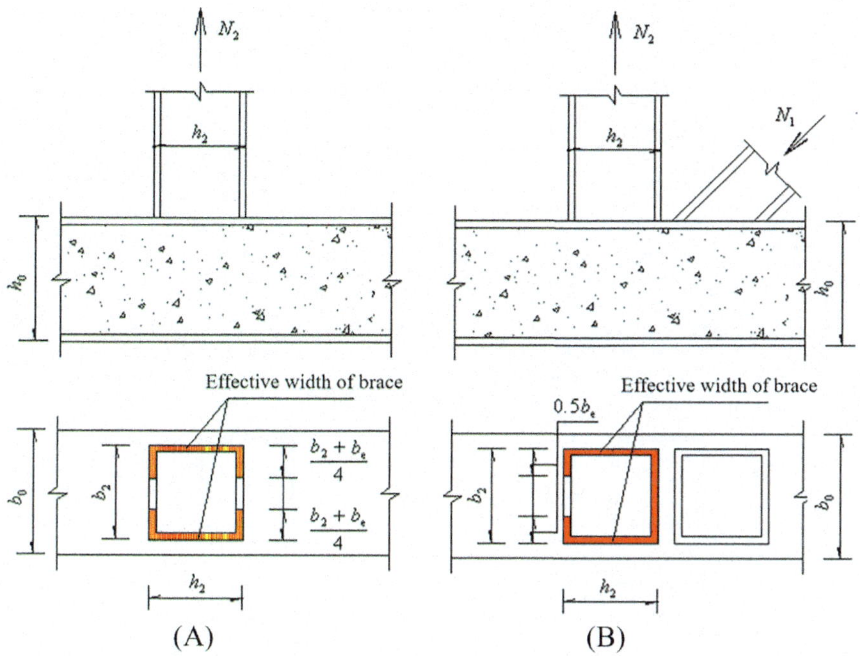

Figure 4.94 Effective width failure mode of the brace member. (A) T-type, Y-type, and X-type nodes. (B) K-type nodes.

$$N_2^u = \frac{ft_0^2}{(1-\beta)\sin\theta_2}\left[\frac{2h_2}{b_0\sin\theta_2} + 4(1-\beta)^{0.5}\right] \quad (4.33)$$

When $1 \le \beta < 0.7$, the design value of the joint bearing capacity N_2^u is taken as the lesser of the following expressions:

Effective width failure (Fig. 4.94A):

$$N_2^u = f_2 t_2 (2h_2 - 4t_2 + b_2 + b_e) \quad (4.34)$$

Punching-shear failure (Fig. 4.95A):

$$N_2^u = \frac{\sqrt{3}}{3} ft_0 (2h_2/\sin\theta_2 + b_2 + b_{ep})/\sin\theta_2 \quad (4.35)$$

In the formula: $b_e = \frac{10}{b_0/t_0}\frac{f_{y0}t_0}{f_{y2}t_2}b_2 \le b_2, b_{ep} = \frac{10}{b_0/t_0}b_2 \le b_2$. Here, f_{y2} and f_2 are the standard and design values of the yield strength of the branch tube steel; f_{y0} and f are the standard and design values of the yield strength of the chord steel.

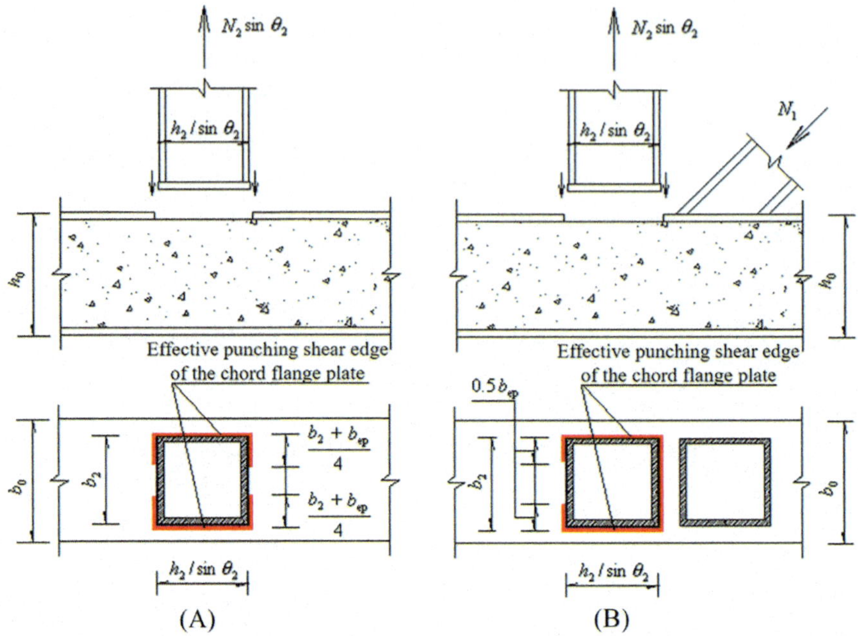

Figure 4.95 Punching-shear failure mode of the chord flange plate. (A) T-type, Y-type, and X-type nodes. (B) K-type nodes.

2. K-type joints:

When $\beta < 0.7$ and $g \geq 0.5b_0\sqrt{1-\beta}$, the design value of the tensile branch joint bearing capacity N_2^u of the K-type joint is calculated according to Eq. (4.33).

When $\beta < 0.7$ but $g < 0.5b_0\sqrt{1-\beta}$, or when $1 \leq \beta \leq 0.7$, the design value of the tensile branch joint bearing capacity N_2^u of the K-type joint is determined by taking the lesser value between the effective width failure mode of the tensile branch (Fig. 4.94B) and the punching-shear failure mode of the chord flange plate (Fig. 4.95B). The corresponding calculation formulas are the same as those given in Eq. (4.34) or Eq. (4.35).

4.5.3.2 Design formula for the compressive bearing capacity of concrete-filled rectangular steel tubular truss joints

After considering the resistance partial factor, the design value of the compressive joint bearing capacity N_1^u is:

$$N_1^u = (1 + k_s)k_c f_c A_b \tag{4.36}$$

In the formula, f_c is the design value of the axial compressive strength of concrete, and the meanings of the other symbols are the same as those in the above calculation formulas. According to the Chinese concrete structure design code, $f_c = f_{ck}/1.4$. Therefore the ratio of the bearing capacity calculated by Eq. (4.36) to that calculated by Eq. (4.27) is $N_1^u/N_{1u} = 1/1.4$.

The bearing capacity of strengthened compressive joints in RST trusses, where concrete is locally infilled near the joints, can also be calculated using Eq. (4.36). However, the infill length of the concrete inside the tube should meet certain requirements, which are recommended as follows:

$$L_c \geq h_1/\sin\theta_1 + 4h_0 \tag{4.37}$$

Considering that the chord height-to-width ratio of the test joints in this study and in Packer's tests ranges between 0.7 and 1.4, it is therefore recommended that Eq. (4.36) should be applied within $0.7 \leq h_0/b_0 \leq 1.4$.

4.5.3.3 Design formula for the flexural bearing capacity of concrete-filled rectangular steel tubular truss joints

The infilled concrete in the chord can enhance the flexural bearing capacity of the joint. When the branch member of a CFRST truss is subjected to bending moment, the effective width method (Fig. 4.96) can be applied to the tension flange of the branch member to obtain the reduced flexural capacity of the branch. The joint flexural bearing capacity is then taken as equal to the reduced branch flexural capacity, M_u:

$$M_u = f_i\left(Z_i - (1 - b_e/b_i)\,b_i t_i (h_i - t_i)\right) \tag{4.38}$$

In Eq. (4.38), Z_i is the plastic section modulus of the branch member, which can be approximately taken as:

$$Z_i = b_i h_i t_i + \frac{t_i (h_i - 2t_i)^2}{2} \tag{4.39}$$

b_e is the effective width of the branch flange, $b_e = \frac{10}{b_0/t_0}\frac{f_{y0}\,t_0}{f_{yi}\,t_i} b_i \leq b_i$.

4.6 Summary

In this chapter, a total of 37 joint tests were conducted, including 30 tests on CFRST truss joints of types T, Y, X, and K, as well as 7 bearing capacity tests on RST of types Y, X, and K. Through comparative testing, the influence patterns of parameters such as concrete infill and joint geometric dimensions on the mechanical performance of joints were

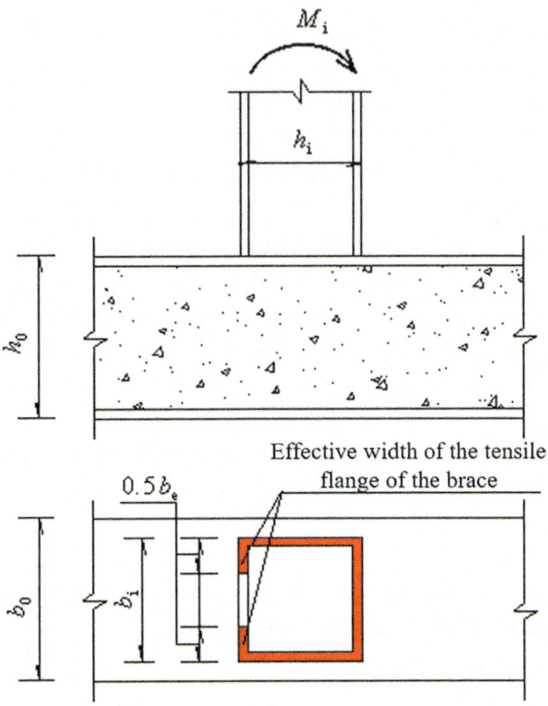

Figure 4.96 Effective width failure mode of the brace tension flange.

obtained. Based on the test results, calculation formulas applicable to the bearing capacity of CFRST truss joints were proposed.

By combining the research findings on the static performance of CFRST joints in this chapter with the interaction relationship between truss flexural capacity and joint bearing capacity established in Chapter 3, guidance for truss design can be provided. To ensure the full utilization of the mechanical performance of CFRST trusses, premature failure of certain joints should be avoided, as it may lead to a reduction in the flexural capacity of the truss.

Extended reading

CECS: 280-2010. Technical specification for structures with steel hollow sections [S]. Beijing: Beijing China Planning Publishing House; 2010.

Cheng G, Liu YJ, Tian ZJ, et al. Tensile behavior of PBL stiffened CFST unequal T-connections[J]. Journal of Chang'an University(Natural Science Edition) 2015;35(03):83–90.

CIDECT Design Guide No.3. Design guide for rectangular hollow section (RHS) joints under predominantly static loading[S]. Cologne: Verlag Tuv Rheinland Press; 2009.
Liu YJ. Experiments on ultimate bearing capacity and research on design method of joints of CFST truss[D]. Changsha: Hunan University; 2003.
Liu YJ, Jiang L, Xiong ZH. Comprehensive failure model of yield line and punching shear for RST[J]. Journal of Architecture and Civil Engineering 2016;33(6):6–13.
Liu YJ, Zhou XH, Liu JP. Experiment on force performance of CFST K-type joints[J]. Journal of Architecture and Civil Engineering 2007(02):36–42.
Liu YJ, Zhou XH, Liu JP. Behavior of CFST T-joints and Y-type joints under compression [J]. Journal of Chang'an University(Natural Science Edition) 2008(05):48–52.
Liu YJ, Zhou XH, Liu JP. Experimental research on RST X-joints with chord concrete-inside subjected to tension and bending[J]. Journal of Building Structures 2009;30(1):82–94.
Liu YJ, Zhou XH, Zou YS, et al. Experimental research on local bearing strength of CFST under transverse load[J]. Journal of Building Structures 2003;24(2):42–8.
MA YP, LIU YJ, KUN W, et al. Axial stiffness of concrete filled rectangular steel tubular (CFRST) truss joints [J]. Journal of Constructional Steel Research 2021;184:106820.
Ma Y.P., Liu Y.J., Wang K., Ma T.Y. Study on joint efficiency coefficient of steel tubular truss[J/OL]. Journal of Building Structures: 1-18[2020-12-22]. https://doi.org/10.14006/j.jzjgxb.2020.0570.
Packer JA. Concrete-filled HSS connections [J]. Journal of Structural Engineering 1995;121(3):458–67.
Packer J.A., Fear C.E. Concrete-filled rectangular hollow section X and T connections[C]. In: Proceeding of the 4th international symposium on tubular structures, Delft. The Netherlands, 1991: 382-391.

CHAPTER 5

Fatigue performance of joints in concrete-filled rectangular steel tubular trusses

5.1 Overview

To prevent fatigue failure of concrete-filled rectangular steel tubular (CFRST) trusses during their service life, research on their fatigue performance is essential. Welds in trusses typically converge at the joints, resulting in a high degree of stress concentration at these locations, making them more susceptible to fatigue failure. This chapter experimentally analyzes the influence of joint geometric parameters on the Stress Concentration Factor (SCF). Subsequently, fatigue tests are conducted to investigate the fatigue life and crack propagation mechanisms of the truss joints, leading to the proposal of a hot-spot stress range *S-N* curve for the joints. Finally, through finite element analysis, fitted formulas for the joint SCF parameters are derived, and a hot-spot stress method suitable for fatigue assessment of CFRST truss joints is proposed.

5.2 Experimental study on stress concentration in joints of CFRST trusses

At the intersection lines of CFRST joints, significant concentration of stress occurs due to geometric discontinuities and the presence of welds. Fatigue initial cracks often originate at the location with the highest degree of stress concentration. The hot-spot stress method uses SCF to characterize the severity of stress concentration at the weld toe. SCF is independent of the absolute geometric dimensions of the joint but is related to the dimensionless geometric parameters (θ, β, 2γ, and τ). This chapter aims to investigate the influence of these dimensionless geometric parameters on the SCF of CFRST joints through experimental tests. Additionally, tests on CFRST truss segments will be conducted to study the distribution of SCF in the joints within a truss system.

5.2.1 Overview of node trials

The typical configuration of a CFRST Y-type joint is shown in Fig. 5.1. A total of 11 joint specimens were designed, and their geometric

Composite Truss Bridges With Concrete-filled Rectangular Steel Tube Members
DOI: https://doi.org/10.1016/B978-0-443-44011-3.00006-8

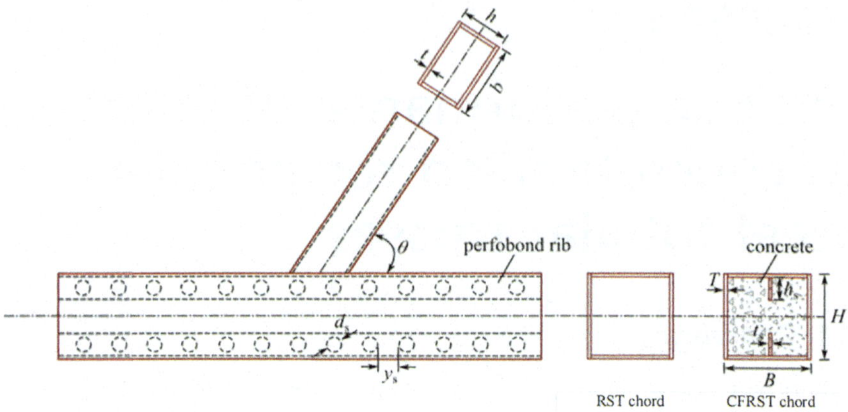

Figure 5.1 Typical joint configuration.

parameters are summarized in Table 5.1. Among them, 9 specimens had concrete-filled chords with PBL stiffeners and were designated as "YCP". To investigate the influence of dimensionless geometric parameters on joint stress concentration, the following parameter values were selected: brace-to-chord angle θ at 45°, 60°, and 90°, brace-to-chord width ratio β at 0.6, 0.8, and 1.0, chord width-to-thickness ratio 2γ at 18.75, 25, and 37.5, and brace-to-chord thickness ratio τ at 0.5, 0.75, and 1.0.

Simultaneously, to study the influence of concrete filling in the chord and the presence of PBL stiffeners on joint stress concentration, two control specimens were designed: one specimen with only concrete filling in the chord (designated "YC") and one empty hollow section joint specimen (designated "YH").

The specimen fabrication process is shown in Fig. 5.2. Both the brace and chord were fabricated by welding together four steel plates. First, the PBL stiffeners were welded along the entire length at the center of the top and bottom plates of the chord, as shown in Fig. 5.2(A). Then, the four plates were welded to form the brace and the chord. The brace was connected to the chord using full-penetration welds to create the hollow section joint, and one end plate of the chord was welded on, as shown in Fig. 5.2(B). Subsequently, the chord of the joint was placed upright, and concrete was poured to fill it, as shown in Fig. 5.2(C). Finally, the end plate on the other side of the chord was welded on to complete the CFRST joint specimen, as shown in Fig. 5.2(D). After processing, all welds were inspected to ensure welding quality.

Table 5.1 Geometric parameters of the specimens.

Identification number	Chord member (mm)			Brace member (mm)			PBL (mm)				Dimensionless parameter			
	b_0	h_0	t_0	b_1	h_1	t_1	h_p	d_p	y_p	t_p	θ	β	2γ	τ
YCP-1	100	100	4	100	60	4	30	20	40	4	60°	1	25	1
YCP-2	150	150	6	150	90	6	50	30	60	4	45°	1	25	1
YCP-3	150	150	6	150	90	6	50	30	60	4	90°	1	25	1
YCP-4	150	150	6	120	90	6	50	30	60	4	60°	0.8	25	1
YCP-5	150	150	6	90	90	6	50	30	60	4	60°	0.6	25	1
YCP-6	150	150	4	150	90	4	50	30	60	4	60°	1	37.5	1
YCP-7	150	150	8	150	90	8	50	30	60	4	60°	1	18.75	1
YCP-8	150	150	8	150	90	6	50	30	60	4	60°	1	18.75	0.75
YCP-9	150	150	8	150	90	4	50	30	60	4	60°	1	18.75	0.5
YH	150	150	6	150	90	6	–				60°	1	25	1
YC	150	150	6	150	90	6	–				60°	1	25	1

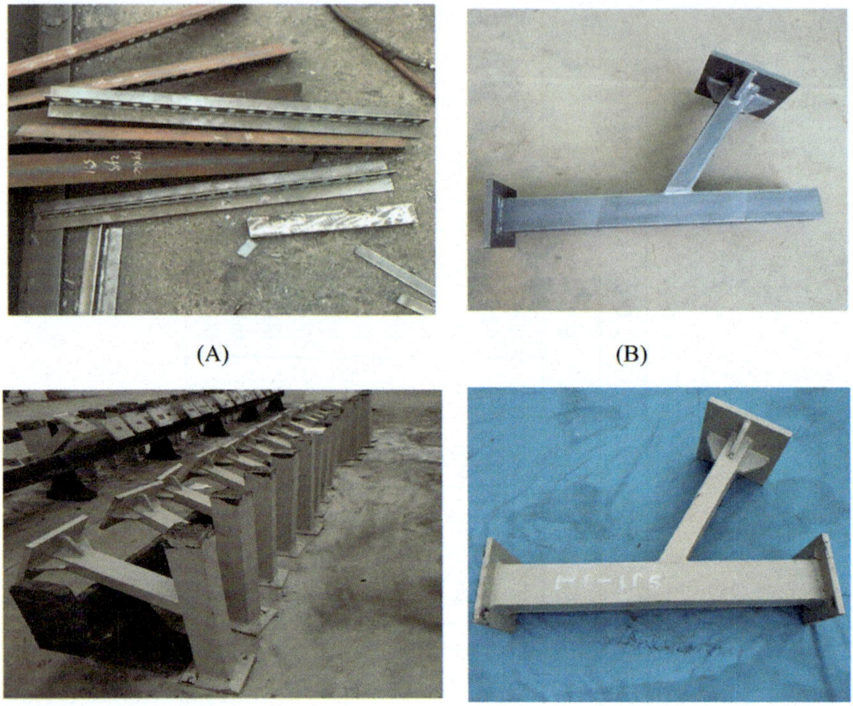

(C) (D)

Figure 5.2 Specimen fabrication. (A) Welded PBL stiffener. (B) Connection between brace and chord members. (C) Concrete-filled chord. (D) Joint specimen fabrication completed.

The weld details in the specimens conformed to the *Code for Welding of Steel Structures* (GB 50661-2011). The PBL stiffener, brace and chord were all made of Q235 steel. In accordance with *Metallic materials-Tensile testing-Part 1: Method of test at room temperature* (GB/T 228–2010), steel samples were tested, and the measured material properties are listed in Fig. 5.2 Specimen fabrication

Table 5.2. C50 grade concrete was used for the infill. Following *Standard for test method of mechanical properties on ordinary concrete* (GB/T 50081-2002), material property tests were conducted on 150 mm × 150 mm × 150 mm standard concrete cubes. The concrete mix proportions and material properties are listed in Table 5.3.

The test was conducted on a 100kN MTS fatigue testing machine. The test setup is shown in Fig. 5.3. Steel end plates with a thickness of 15 mm

Table 5.2 Steel properties.

Steel grade	Thickness (mm)	E_s (GPa)	f_y (MPa)	f_u (MPa)	ε_f (%)
Q235B	4	205	290	415	28
	6	205	280	390	29
	8	205	305	435	28.5

were installed at both ends of the chord, and a steel end plate with a thickness of 20 mm was installed at the end of the brace. Additionally, 15 mm thick stiffeners were added at the ends of both the brace and the chord to ensure that the specimen would not fail prematurely at the ends during the test. Both ends of the chord were connected via pins to achieve hinged support, while the end of the brace was directly connected to the loading device.

The assessment of stress concentration in the joint required only static testing. By measuring the hot-spot stress levels along the brace-chord intersection line under a specific load, the degree of stress concentration at the weld could be determined. Prior to formal testing, the specimen underwent 9 pre-loading cycles to eliminate joint eccentricity and ensure the proper functioning of the strain gauges. The test was conducted under load control, with each load step inducing a nominal stress of 5 MPa in the brace, until reaching the target load of 50 MPa. The entire process ensured the steel remained within its elastic range. This test program included stress concentration measurements under both brace tension and brace compression load cases to identify the most critical fatigue loading condition for this joint type. Three tests were conducted for each load case.

The measurement point arrangement in the test was divided into two categories: one for measuring the hot-spot stress at the weld of the brace-chord intersection line, and the other for measuring the nominal stress of the brace and chord.

5.2.1.1 Hot-spot stress measurement scheme

Due to the notch effect of the weld, the hot-spot stress cannot be obtained directly by measuring the stress at the weld toe, it must be determined using an extrapolation method. Extrapolation methods are divided into linear extrapolation and quadratic extrapolation, as shown in Fig. 5.4. The

Table 5.3 Concrete constituents and properties.

Concrete strength grade	Mix proportion (kg/m^3)						Cube compressive strength f_{cu} (MPa)	Slump(mm)
	Water	Cement	Fly ash	Fine aggregate	Coarse aggregate	Water reducer		
C50	167	471	52	624	1109	10.46	56.3	170

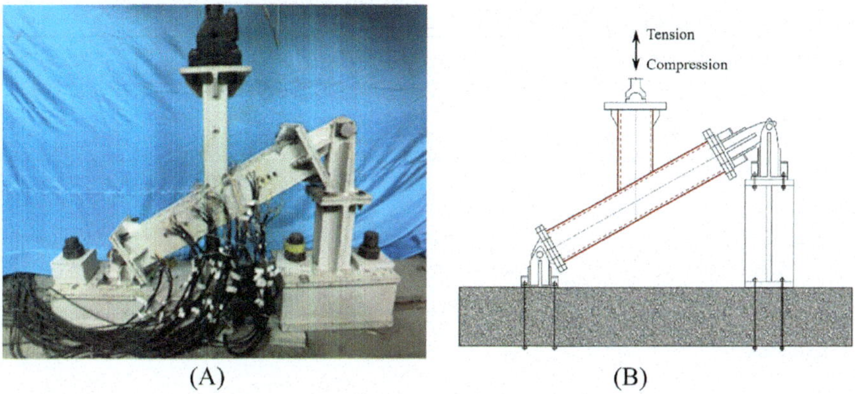

Figure 5.3 Test setup (A) Actual site diagram (B) Schematic diagram.

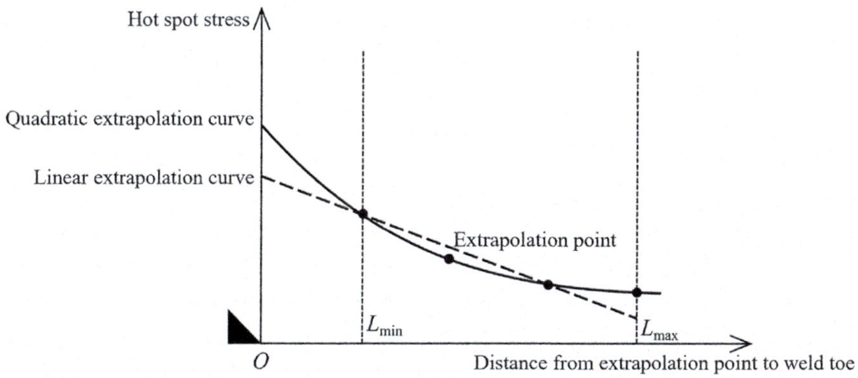

Figure 5.4 Hot-spot stress extrapolation method.

CIDECT code specifies that linear extrapolation is suitable for circular hollow section joints, while quadratic extrapolation should be used for RST joints. Based on experimental measurements, the stress distribution in the extrapolation region of CFRST joints is nonlinear. Therefore, this test employed the quadratic extrapolation method for hot-spot stress measurement. A minimum of three strain gauges must be installed within the extrapolation region. The selection of the extrapolation region follows the specifications outlined in the CIDECT code, as detailed in Table 5.4.

The hot-spot strain at the weld toe is obtained by extrapolating the measured strains within the extrapolation region. According to Hooke's

Table 5.4 Extrapolation region specified by CIDECT.

Distance from weld toe	Chord	Brace
L_{min} [a]	$0.4\ T$	$0.4t$
L_{max}	$L_{min} + T$	$L_{min} + t$

[a] Minimum value for L_{min} is 4 mm.

Law, the relationship between the hot-spot strain and the hot-spot stress is as follows:

$$\sigma_{hs} = \frac{E_s}{1 - \nu_s^2}\left[\varepsilon_{hs} + \nu_s \cdot (\varepsilon_y + \varepsilon_z)\right] \tag{5.1}$$

Where $\nu_s = 0.3$ is the Poisson's ratio of concrete, ε_{hs} is the strain perpendicular to the weld toe, i.e., the hot-spot strain, and ε_Y and ε_z are the strains in the other two directions. In Eq. (5.1), ε_Y and ε_z are very small and can be neglected. Therefore, Eq. (5.2) is simplified as:

$$\sigma_{hs} = 1.1 E_s \varepsilon_{hs} \tag{5.2}$$

After obtaining the hot-spot stress value using Eq. (5.2), the ratio of this value to the corresponding nominal stress is the SCF. Fig. 5.5 shows the arrangement of the hot-spot stress measurement points. Since the angle between the brace and the chord is not necessarily 90°, the degree of stress concentration on the weld toe side and the weld heel side is not the same. Therefore, hot-spot stress measurements were taken at a total of 14 locations on both the weld toe side and the weld heel side. The selection of points A to E was based on the provisions for RST joints in the CIDECT code, as these points are considered to have a high degree of stress concentration. Points F and G were added considering that the presence of PBL stiffeners could also lead to an increase in stress concentration at these locations.

5.2.1.2 Nominal stress measurement scheme

In addition to measuring the hot-spot stress at the weld toe of the intersection line, it is also necessary to measure the corresponding nominal stresses in the brace and chord. Fig. 5.6 shows the arrangement of the nominal stress measurement points. Strain gauges need to be placed on the

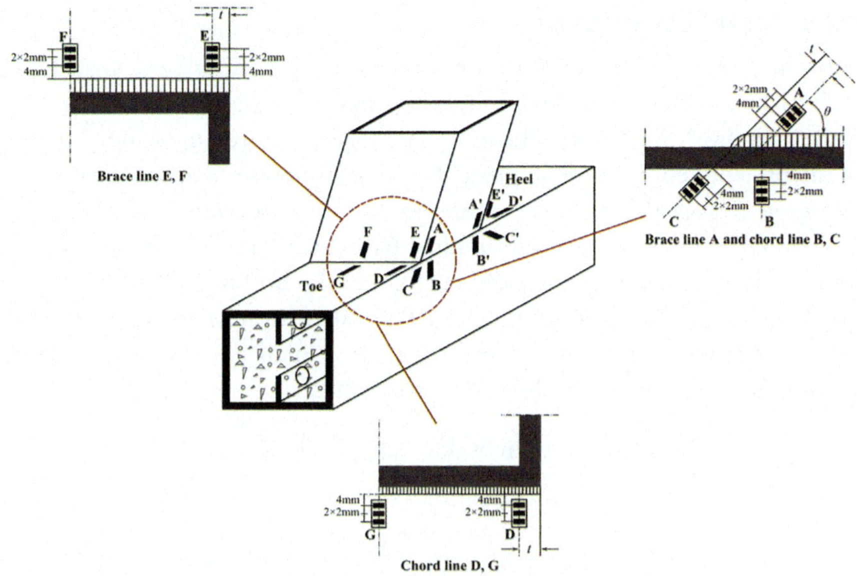

Figure 5.5 Arrangement of hot-spot stress measurement points.

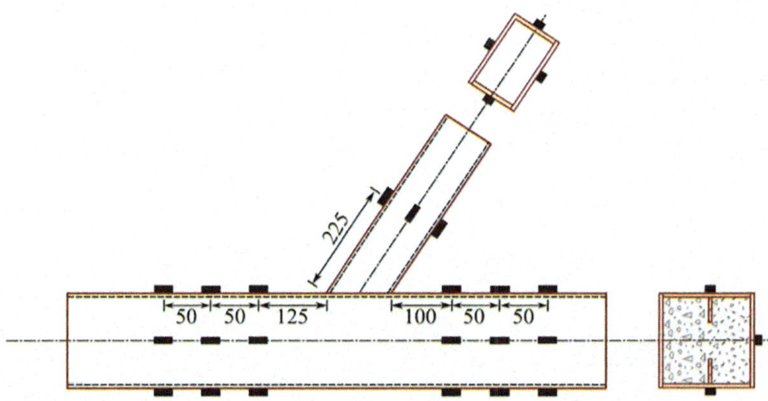

Figure 5.6 Arrangement of nominal stress measurement points (unit: mm).

surfaces of the brace and chord, away from the extrapolation region. Specifically, strain gauges are installed on all four sides of the mid-span cross-section of the brace to measure its axial nominal stress. Three rows of strain gauges are arranged on both the weld toe side and the weld heel side of the chord to measure its axial and bending stresses.

5.2.2 Joint test results

Joints are typically subjected to a combination of various basic load types, such as brace axial force, brace in-plane bending moment, brace out-of-plane bending moment, chord axial force, chord in-plane bending moment, and chord out-of-plane bending moment. Consequently, the hot-spot stress at the weld toe of the joint is the superposition of the hot-spot stress generated by each individual basic load type. The hot-spot stress produced by each basic load is the product of the nominal stress induced by that load and its corresponding SCF. The calculation of the total hot-spot stress at the weld toe is given by Eq. (5.3), which is consistent with the method for calculating the hot-spot stress range.

$$\sigma_{hs} = \sigma_{AB}\cdot SCF_{AB} + \sigma_{B}\cdot SCF_{B} + \sigma_{OB}SCF_{OB} + \sigma_{AC}\cdot SCF_{AC} + \sigma_{IC}\cdot SCF_{IC} + \sigma_{OC}\cdot SCF_{OC} \tag{5.3}$$

In the equation, σ_{AB}, σ_{IB}, σ_{OB}, σ_{AC}, σ_{IC}, and σ_{OC} represent the nominal stresses generated by brace axial force, brace in-plane bending moment, brace out-of-plane bending moment, chord axial force, chord in-plane bending moment, and chord out-of-plane bending moment, respectively. SCF_{AB}, SCF_{IB}, SCF_{OB}, SCF_{AC}, SCF_{IC}, and SCF_{OC} are the corresponding stress concentration factors.

Fig. 5.7(A) shows the measured nominal stresses in the chord of the typical joint YCP-1 when the brace axial tensile stress is 20 MPa. The

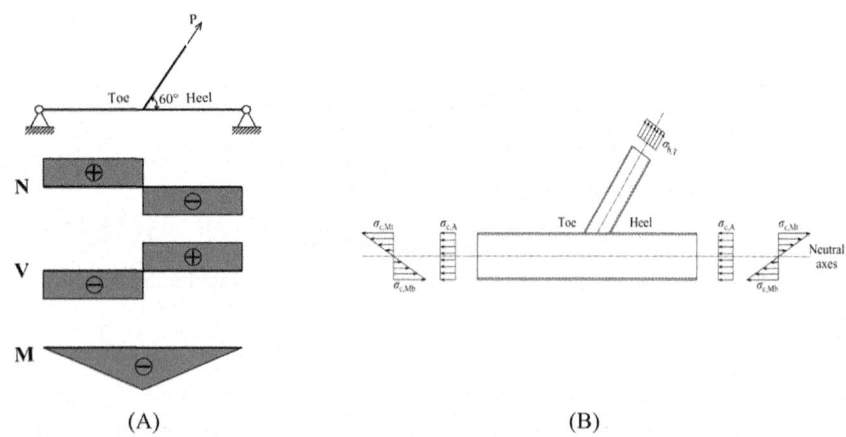

Figure 5.7 Joint force analysis. (A) Measured nominal stresses. (B) Joint internal force diagram.

Figure indicates that under brace axial force, the chord develops additional axial force and in-plane bending moment. This result is consistent with the joint internal force diagram shown in Fig. 5.7(B). According to Eq. (5.3), the hot-spot stress values at points A, E, and F on the brace surface are expressed by Eq. (5.4). Their SCF can be calculated based on the measured brace surface hot-spot stresses and nominal stresses.

$$\sigma_{\mathrm{hs,b}} = \sigma_{AB} \cdot \mathrm{SCF}_{\mathrm{AB,b}} \tag{5.4}$$

Where $\sigma_{\mathrm{hs,b}}$ represents the hot-spot stress value at the weld toe on the brace surface, and $\mathrm{SCF}_{\mathrm{AB,b}}$ is the stress concentration factor generated by the brace axial force at the weld toe on the brace surface.

The hot-spot stress values at points B, C, D, and G on the chord surface can be expressed by the following equation:

$$\sigma_{\mathrm{hs,c}} = \sigma_{\mathrm{AB}} \cdot \mathrm{SCF}_{\mathrm{AB,c}} + \sigma_{\mathrm{AC}} \cdot \mathrm{SCF}_{\mathrm{AC}} + \sigma_{\mathrm{IC}} \cdot \mathrm{SCF}_{\mathrm{IC}} \tag{5.5}$$

In the equation, $\sigma_{\mathrm{hs,c}}$ represents the hot-spot stress value at the weld toe on the chord surface, and $\mathrm{SCF}_{\mathrm{AB,c}}$ is the stress concentration factor generated by the brace axial force at the weld toe on the chord surface, which is distinct from $\mathrm{SCF}_{\mathrm{AB,b}}$.

Eq. (5.5) contains three unknowns: $\mathrm{SCF}_{\mathrm{AB,c}}$, $\mathrm{SCF}_{\mathrm{AC}}$, and $\mathrm{SCF}_{\mathrm{IC}}$, which cannot be solved directly from a single equation. The stress ratios from Fig. 5.7(a) are σ_{AB}: σ_{AC}: σ_{IC} = 5.1: 1: 21.1, indicating that the chord bending stress is the largest. Furthermore, fatigue test results show that the joints ultimately failed due to fatigue cracking in the chord caused by bending. The proportion of chord axial stress is minimal and can be neglected. Although the load proportion from the brace axial force cannot be ignored, considering that the chord is filled with concrete and equipped with PBL stiffeners, the stress concentration effect it induces in the chord is very small and can also be neglected. Based on this analysis, assuming $\mathrm{SCF}_{\mathrm{AB,\ c}} = \mathrm{SCF}_{\mathrm{AC}} = 0$, the stress concentration factor at the weld toe on the chord surface can be calculated based on the chord bending stress:

$$\mathrm{SCF} = \frac{\sigma_{hs,c}}{\sigma_{IC}} \tag{5.6}$$

Table 5.5 lists the comparison of SCF on the weld toe side and the weld heel side under brace tension. The data in the Table shows that the SCF values on the weld toe side are generally higher than those on the weld heel side, particularly for the SCF on the chord, where this phenomenon is more pronounced. This observation is largely consistent with findings from

Table 5.5 Comparison of SCF on the weld toe side and weld heel side under brace tension.

Number	θ	β	2γ	τ	Brace SCF			Chord SCF			
					A	E	F	B	C	D	G
YCP-1 Weld toe side	60°	1	25	1	2.38	2.27	1.25	0.49	0.44	**2.10**	1.83
YCP-1 Weld root side					2.18	2.44	1.79	-	-	1.84	1.81
YCP-1 Relatively large value					2.38	2.44	1.79	0.49	0.44	2.10	1.83
YCP-2 Weld toe side	45°	1	25	1	1.82	2.38	1.72	0.10	0.37	1.89	2.10
YCP-2 Weld root side					1.69	2.17	1.57	-	0.20	1.50	1.64
YCP-2 Relatively large value					1.82	2.38	1.72	0.10	0.37	1.89	2.10
YCP-3 Weld toe side	90°	1	25	1	1.83	2.28	0.95	0.55	0.23	1.78	2.18
YCP-3 Weld root side					2.47	2.57	1.91	0.53	0.53	2.29	1.65
YCP-3 Relatively large value					2.47	2.57	1.91	0.55	0.53	2.29	2.18
YCP-4 Weld toe side	60°	0.8	25	1	2.73	2.65	1.92	1.30	1.13	2.44	2.33
YCP-4 Weld root side					2.39	1.99	1.76	0.85	1.09	1.95	1.70
YCP-4 Relatively large value					2.73	2.65	1.92	1.30	1.13	2.44	2.33
YCP-5 Weld toe side	60°	0.6	25	1	2.15	2.37	1.06	3.46	1.78	2.11	2.02
YCP-5 Weld root side					2.07	2.41	1.03	2.67	1.53	1.93	1.77
YCP-5 Relatively large value					2.15	2.41	1.06	3.46	1.78	2.11	2.02
YCP-6 Weld toe side	60°	1	37.5	1	2.41	2.69	1.41	0.71	0.63	2.46	2.38
YCP-6 Weld root side					2.06	2.16	1.90	-	-	2.05	1.47
YCP-6 Relatively large value					2.41	2.69	1.90	0.71	0.63	2.46	2.38
YCP-7 Weld toe side	60°	1	18.75	1	2.10	2.22	0.93	0.46	0.41	1.72	1.61
YCP-7 Weld root side					1.78	1.90	1.62	0.37	0.29	2.05	1.77
YCP-7 Relatively large value					2.10	2.22	1.62	0.46	0.41	2.05	1.77
YCP-8 Weld toe side	60°	1	18.75	0.75	1.68	2.05	1.10	0.15	0.19	2.02	1.74
YCP-8 Weld root side					1.59	1.58	1.33	0.40	0.29	1.77	1.32

YCP-8 Relatively large value					1.68	2.05	1.33	0.40	0.29	2.02	1.74
YCP-9 Weld toe side	60°	1	18.75	0.5	1.63	2.03	0.81	-	-	1.90	1.26
YCP-9 Weld root side					1.65	1.91	1.22	0.25	0.23	1.54	0.81
YCP-9 Relatively large value					1.65	2.03	1.22	0.25	0.23	1.90	1.26
YH Weld toe side	60°	1	25	1	4.16	3.20	1.04	0.78	1.36	3.09	1.44
YH Weld root side					3.68	1.53	1.17	0.56	1.33	1.67	1.43
YH Relatively large value					4.16	3.20	1.17	0.78	1.36	3.09	1.44
YC Weld toe side	60°	1	25	1	3.86	2.87	1.16	0.68	0.95	2.51	1.63
YC Weld root side					2.51	2.64	1.40	0.41	1.10	1.56	1.01
YC Relatively large value					3.86	2.87	1.40	0.68	1.10	2.51	1.63

Note: bold font indicates the maximum SCF on the branch pipe and main pipe of each test piece; "-" indicates that the SCF is very small and can be ignored.

stress concentration tests on non-90° RST joints. On the brace, the maximum SCF may occur at point A or point E. On the chord, except for specimens YCP-2 and YCP-5, the maximum SCF consistently occurs at point D. For specimen YCP-2, the SCF at point G on the chord is greater than that at point D, which may be attributed to the stronger effect of the PBL stiffeners in YCP-2 compared to other specimens. For specimen YCP-5, which has the smallest brace-to-chord width ratio (β), point B in the width direction of the chord may experience higher bending stress, resulting in a larger SCF at point B that exceeds the SCF at point D.

Table 5.6 lists the comparison of SCF on the weld toe side under brace tension and compression, respectively. The Table shows that for the hollow section joint YH, under either brace tension or compression, the top plate of the chord undergoes outward bulging or inward buckling deformation, and the SCF levels are generally comparable. Therefore, both brace tension and compression load cases should be considered in the fatigue calculation of RST joints. After the chord is filled with concrete, it can be observed that the joint does not undergo inward buckling deformation. The SCF values under the brace compression load case are significantly lower than those under the brace tension load case. Consequently, brace tension loading is more critical for the fatigue performance of CFRST joints.

The following sections will select the maximum SCF values at points A to G on both the weld toe side and weld heel side of each joint to analyze the influence of concrete filling in the chord and the presence of PBL stiffeners on the SCF, as well as the effects of variations in the dimensionless geometric parameters θ, β, 2γ, and τ on the SCF.

5.2.2.1 The influence of concrete filling in the chord and PBL stiffeners on SCF

Fig. 5.8 shows a comparison of SCF among three types of joints: the RST joint (YH), the CFRST joint (YC), and the PBL-stiffened CFRST joint (YCP-1). To more intuitively depict the influence of concrete filling in the chord and the addition of PBL stiffeners on the SCF, the hot-spot stress distribution along the intersection line for these three joint types is plotted based on the data in Fig. 5.8, as shown in Fig. 5.9.

The Figure reveal that the hot-spot stress distribution along the intersection line of the three joint types is related to the stiffness distribution. For the RST joint (YH), under brace tension, the brace and chord maintain deformation compatibility at the intersection line. Consequently, the stress distribution on the surfaces of the brace and

Table 5.6 Comparison of SCF on the weld toe side under brace tension and compression.

Number	θ	β	2γ	τ	Brace SCF			Chord SCF			
					A	E	F	B	C	D	G
YCP-1 Tensile force	60°	1	25	1	2.38	2.27	1.25	0.49	0.44	2.10	1.83
YCP-1 Compressive force					1.16	1.81	1.78	0.32	0.76	1.82	1.43
YCP-1 Relatively large value					2.38	2.27	1.78	0.49	0.76	2.10	1.83
YCP-2 Tensile force	45°	1	25	1	1.82	2.38	1.72	0.10	0.37	1.89	2.10
YCP-2 Compressive force					1.57	2.08	1.64	0.10	0.25	1.76	1.98
YCP-2 Relatively large value					1.82	2.38	1.72	0.10	0.37	1.89	2.10
YCP-3 Tensile force	90°	1	25	1	1.83	2.28	0.95	0.55	0.23	1.78	2.18
YCP-3 Compressive force					1.39	1.24	1.67	0.47	0.19	1.71	1.97
YCP-3 Relatively large value					1.83	2.28	1.67	0.55	0.23	1.78	2.18
YCP-4 Tensile force	60°	0.8	25	1	2.73	2.65	1.92	1.30	1.13	2.44	2.33
YCP-4 Compressive force					1.98	1.92	1.90	0.75	0.86	1.78	2.07
YCP-4 Relatively large value					2.73	2.65	1.92	1.30	1.13	2.44	2.33
YCP-5 Tensile force	60°	0.6	25	1	2.15	2.37	1.06	3.46	1.78	2.11	2.02
YCP-5 Compressive force					1.22	1.54	1.31	1.89	1.11	1.71	1.94
YCP-5 Relatively large value					2.15	2.37	1.31	3.46	1.78	2.11	2.02
YCP-6 Tensile force	60°	1	37.5	1	2.41	2.69	1.41	0.71	0.63	2.46	2.38
YCP-6 Compressive force					1.58	2.42	1.14	0.42	0.38	1.31	1.88
YCP-6 Relatively large value					2.41	2.69	1.41	0.71	0.63	2.46	2.38
YCP-7 Tensile force	60°	1	18.75	1	2.10	2.22	0.93	0.46	0.41	1.72	1.61
YCP-7 Compressive force					1.60	1.34	1.20	0.14	0.11	1.58	1.58
YCP-7 Relatively large value					2.10	2.22	1.20	0.46	0.41	1.72	1.61
YCP-8 Tensile force	60°	1	18.75	0.75	1.68	2.05	1.10	0.15	0.19	2.02	1.74
YCP-8 Compressive force					1.56	1.05	0.66	/	/	1.79	1.84

(*Continued*)

Table 5.6 (Continued)

Number	θ	β	2γ	τ	Brace SCF			Chord SCF			
					A	E	F	B	C	D	G
YCP-8 Relatively large value					1.68	2.05	1.10	0.15	0.19	2.02	1.84
YCP-9 Tensile force	60°	1	18.75	0.5	1.63	2.03	0.81	/	/	1.90	1.26
YCP-9 Compressive force					1.11	1.15	0.51	/	/	1.02	0.83
YCP-9 Relatively large value					1.63	2.03	0.81	/	/	1.90	1.26
YH Tensile force	60°	1	25	1	4.16	3.20	1.04	0.78	1.36	3.09	1.44
YH Compressive force					4.31	3.21	1.05	0.43	0.87	2.73	1.43
YH Relatively large value					4.31	3.21	1.05	0.78	1.36	3.09	1.44
YC Tensile force	60°	1	25	1	3.86	2.87	1.16	0.68	0.95	2.51	1.63
YC Compressive force					3.43	2.35	1.05	0.35	0.48	2.14	1.34
YC Relatively large value					3.86	2.87	1.16	0.68	0.95	2.51	1.63

Note: "-" indicates that the SCF is very small and can be ignored.

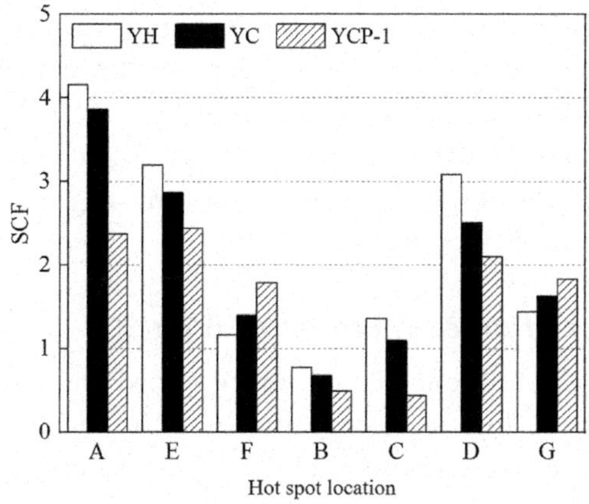

Figure 5.8 SCF comparison of specimens YH, YC and YCP-1.

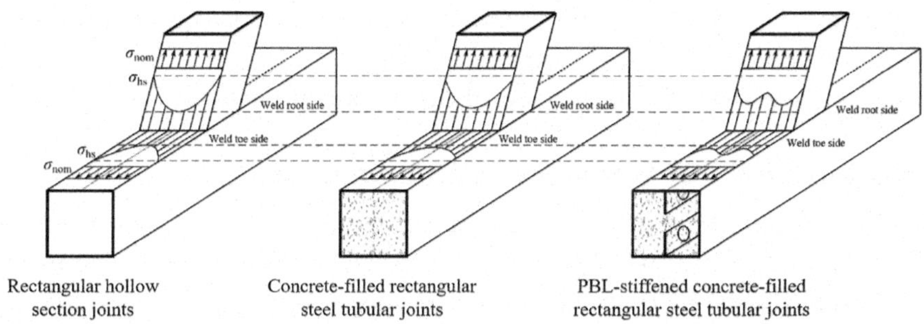

Figure 5.9 Hot-spot stress distribution along the intersection line for the three types of joints.

chord is non-uniform. The corner regions of the brace, which have higher stiffness, exhibit higher hot-spot stress levels, while the central regions, with lower stiffness, show lower hot-spot stress levels.

After the chord is filled with concrete (YC), the overall stress distribution trend remains similar to that of the RST joint. However, the hot-spot stress levels at the corner regions decrease, while the levels in the central regions increase. This indicates that filling the chord with concrete leads to a more uniform stiffness distribution along the intersection line, thereby reducing the peak stress concentration.

When the chord is both filled with concrete and equipped with PBL stiffeners (YCP-1), the stiffness at the PBL-stiffened locations increases.

New peaks in hot-spot stress appear in the central regions of both the brace and chord. However, the stress levels at these new peaks remain lower than those at the corner regions. It can be concluded that filling the chord with concrete and adding PBL stiffeners further homogenizes the stiffness distribution along the intersection line, resulting in a further reduction of the peak stress concentration.

5.2.2.2 Influence of brace-chord angle θ on SCF

Fig. 5.10 shows the influence of the brace-chord angle θ on the SCF. As can be seen from the Figure, except for point G, the SCF values at various points on the brace and chord increase with the increase of θ. This is because a larger θ means a larger vertical component force, leading to a higher degree of stress concentration at the intersection line. The stress concentration level at point G is closely related to the stiffening and shear-resisting effects of the PBL. The effectiveness of these effects depends on the material properties of the concrete dowels passing through the PBL holes. Therefore, the uncertainty of the concrete dowels within the holes leads to uncertainty in the degree of stress concentration at this point.

5.2.2.3 Influence of brace-chord width ratio β on SCF

Fig. 5.11 shows the influence of the brace-chord width ratio β on the SCF. As can be seen from the Figure, as β increases, the SCF values at points on the brace (points A, E, and F) and at points C and D on the chord first increase and then decrease, reaching their maximum values when $\beta = 0.8$. In contrast, for points B and G on the chord, smaller β values correspond to

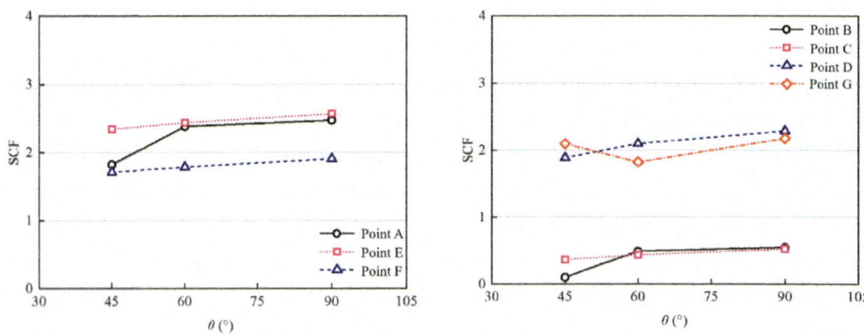

(A) (B)

Figure 5.10 θ effects on SCF (YCP-1 ~ YCP-3). (A) Brace (B) Chord.

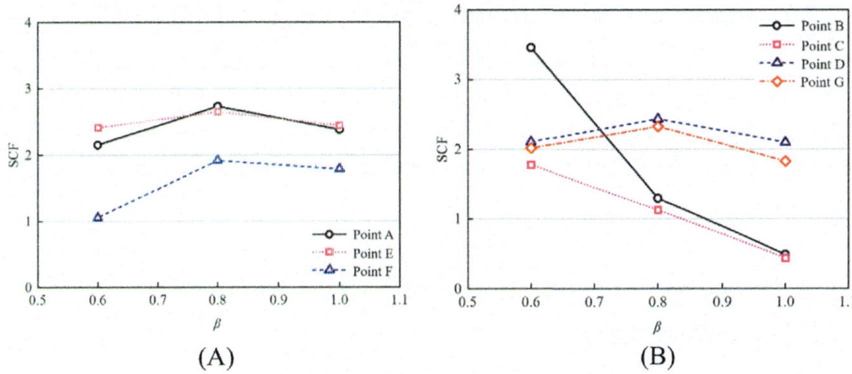

Figure 5.11 β effects on SCF (YCP-1 ~ YCP-3). (A) Brace. (B) Chord.

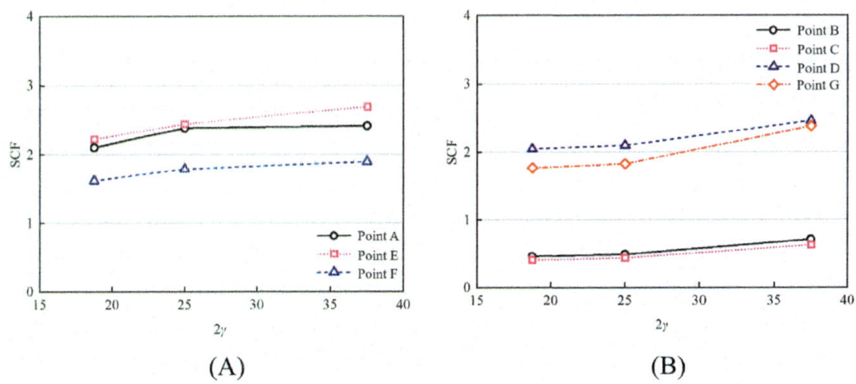

Figure 5.12 2γ effects on SCF (YCP-1, YCP-6, and YCP-7). (A) Brace. (B) Chord.

larger SCF values. This is because a smaller brace-chord width ratio induces greater bending stress at points B and G on the chord.

5.2.2.4 Influence of chord width-to-thickness ratio 2γ on SCF

Fig. 5.12 shows the influence of the chord width-to-thickness ratio 2γ on the SCF. As can be seen from the Figure, the SCF at all points on the surfaces of both the brace and the chord is positively correlated with 2γ. This is because a larger 2γ indicates a thinner chord wall, leading to greater bending stress in the chord top plate under the same load, which consequently results in a higher SCF value.

5.2.2.5 Influence of brace-chord thickness ratio τ on SCF

Fig. 5.13 shows the influence of the brace-chord thickness ratio τ on the SCF. As can be seen from the Figure, a larger τ value corresponds to higher SCF values at various points on the surfaces of both the brace and the chord. This conclusion is consistent with experimental observations for RST joints.

5.2.3 Overview of truss segment test

5.2.3.1 Specimen design

Simultaneously with the test on the negative moment region of the CFRST composite truss girder described in Section 6.2, an experiment was conducted to investigate the degree of stress concentration in the joints. The specimen is introduced in Section 6.2.1, and the tested joint is shown in Fig. 5.14.

5.2.3.2 Arrangement of measurement points

During the test, the hot-spot stress at the brace-chord intersection line of the truss joint was measured, along with the nominal stress on the surfaces of the truss brace and chord.

Fig. 5.15 shows the arrangement of hot-spot stress measurement points for the K-type joint within the truss. The potential locations for hot-spot stress at the brace-chord intersection line are similar to those for the Y-type joint shown in Fig. 5.5. Seven measurement points (A to G) were arranged on both the weld toe side and the weld heel side. However, unlike the

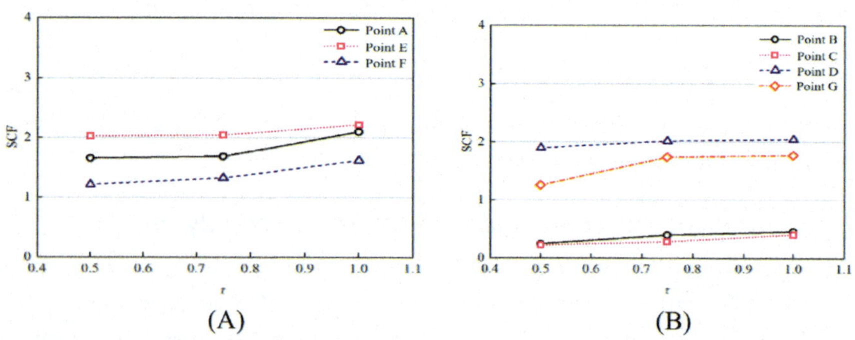

Figure 5.13 τ effects on SCF (YCP-7 to YCP-9). (A) Brace. (B) Chord.

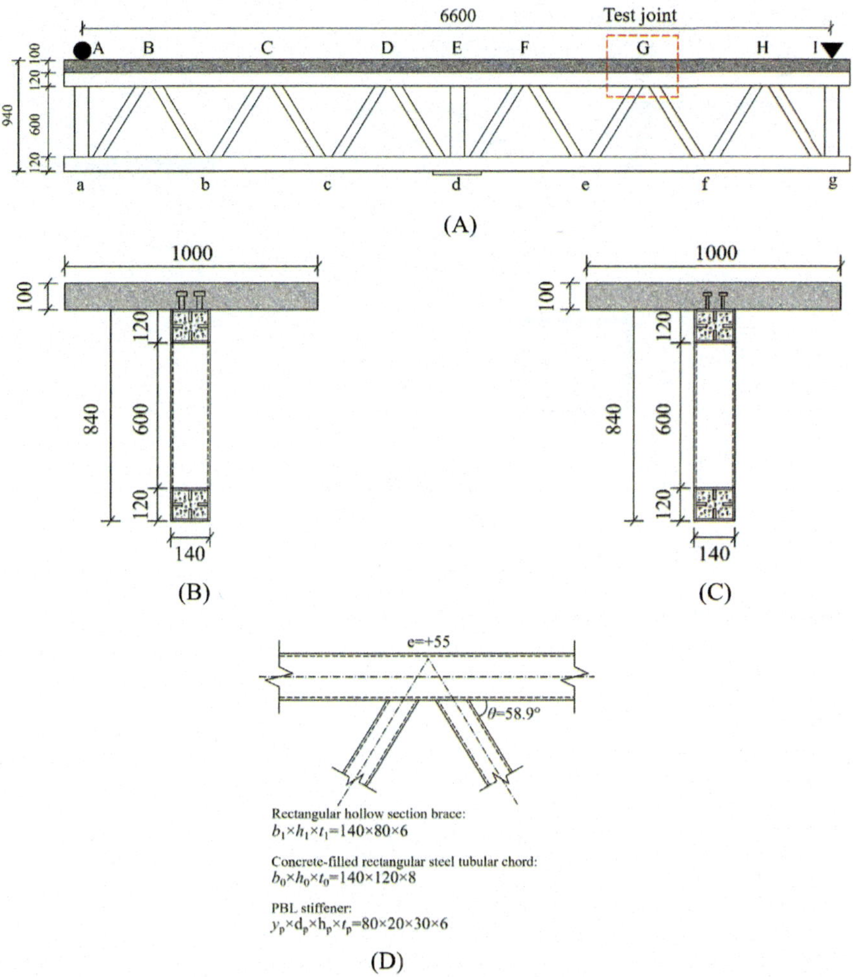

Figure 5.14 Specimen configuration and dimensions (unit: mm). (A) Truss elevation drawing. (B) Cross-section view of CTB1. (C) Cross-section view of CTB2. (D) Test nodes.

Y-type joint, the K-type joint has a tension brace and a compression brace. The distribution of hot-spot stress under different loads varies between the two braces. Therefore, hot-spot stress measurement points should be installed on both the tension brace and the compression brace, totaling 28 locations. The hot-spot stress at the weld toe was obtained using quadratic extrapolation.

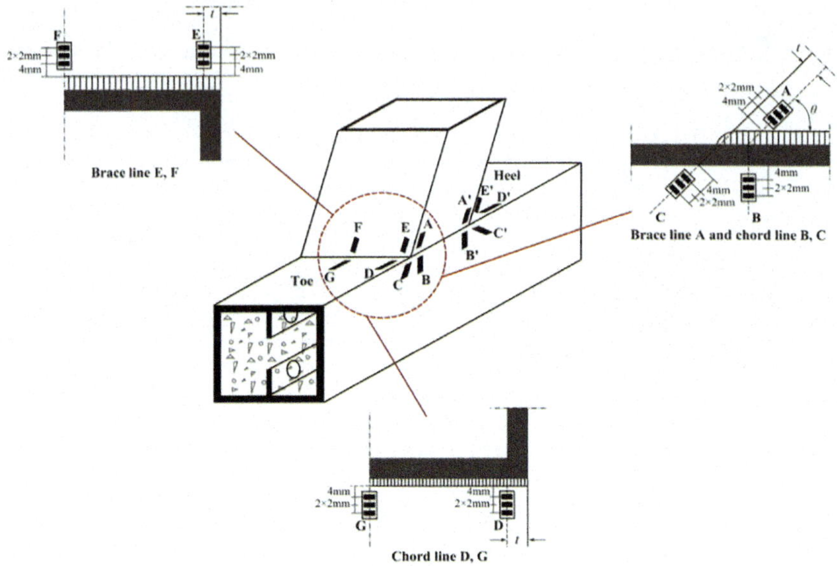

Figure 5.15 Hot-spot stress measurement points.

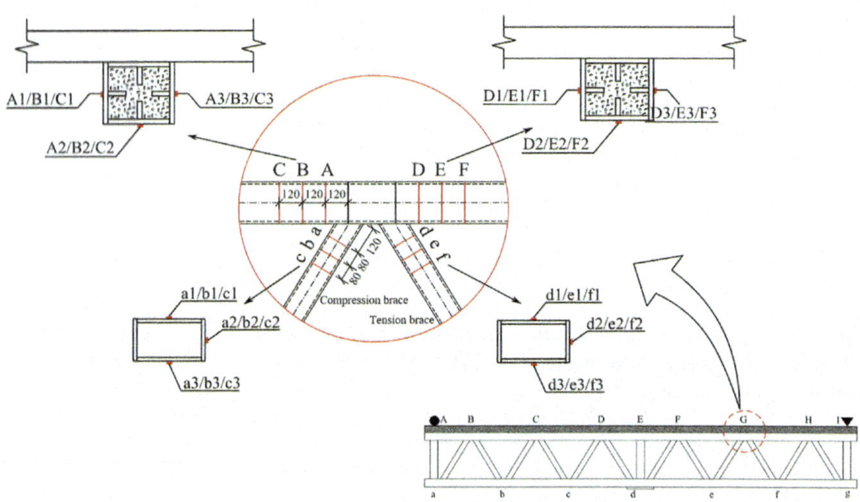

Figure 5.16 Nominal stress measurement point (unit: mm).

Fig. 5.16 shows the nominal stress measurement points for the truss joint. Three columns of nominal stress measurement points were arranged at both ends of the compression brace, tension brace, and the chord. The axial force in the member was determined from strain gauge data at the mid-length of the steel tube, while the bending moment was obtained from strain gauge data on the top and bottom plates. The bending stress at the weld toe of the brace and chord could be derived by linearly extrapolating the results from the three columns of nominal strain gauges. The distances of the three columns of strain gauges at the left and right ends of the chord from the brace centerline were 120 mm, 240 mm, and 360 mm, respectively. The distances of the three columns of strain gauges on the compression and tension braces from the weld toe were 120 mm, 200 mm, and 280 mm, respectively.

5.2.4 Test results of joint nominal stress

The distribution of hot-spot stress at the weld toe in a truss joint is closely related to the applied load cases. Therefore, it is necessary to analyze the nominal stresses on the surfaces of the chord and brace member tubes to determine the internal force distribution within each member. Furthermore, as this research focuses on fatigue, the analysis of hot-spot stress results should be confined to the elastic stage of the structure. Thus, it is also essential to identify the elastic phase throughout the entire loading process of the two trusses.

Based on the nominal stress measurement points shown in Fig. 5.16, the strains at corresponding locations on the surfaces of the chord and brace member tubes were measured during the test. The axial force at the weld toe section of the chord and brace members can be obtained using Eq. (5.7). The bending moment at the tested sections of the chord and brace members can be determined using Eqs. (5.8) and (5.9). Since the bending moment is linearly distributed along the member, the bending moment at the weld toe section of the chord and brace members can be obtained by linearly extrapolating the moments measured at the three test sections of each member.

$$P_i = EA_i\overline{\varepsilon_m} \tag{5.7}$$

$$M_c = 2EI_c\left(\frac{\varepsilon_b - \varepsilon_m}{h_0}\right) \tag{5.8}$$

$$M_b = EI_b\left(\frac{\varepsilon_b - \varepsilon_t}{h_1}\right) \tag{5.9}$$

Where P_i represents the axial force in the chord or brace member, EA_i is the axial stiffness of the brace member or chord, $\overline{\varepsilon_m}$ is the average strain of the measurement points at the mid-section of the steel tube (e.g., the average strain of points a2, b2, and c2 on the compression brace member in Fig. 5.16). M_c is the bending moment of the chord, EI_c is the flexural stiffness of the chord, ε_b and ε_m are the strain values at the midpoints of the bottom plate and side plate of the steel tube, respectively, EI_b is the flexural stiffness of the brace member, and ε_t is the strain value at the midpoint of the top plate of the steel tube.

The main truss is connected to the concrete deck slab via shear studs on the top chord, forming a composite truss girder. The top chord and the concrete deck slab can act together to resist axial force and bending moment, with the degree of composite action depending on the shear stiffness of the studs. For specimen CTB1, shear studs with locally reduced action were used, which have very low shear stiffness. Thus, the composite action between the top chord and the concrete deck slab can be considered minimal. For specimen CTB2, conventional shear studs were used, and its composite action should theoretically be greater than that of CTB1.

Although strains in the deck slab and ordinary reinforcement were measured during the test, it was impossible to accurately determine the axial force and bending moment resisted by the concrete deck slab due to shear lag in the slab, the significant discreteness of concrete strain measurements, and the difficulty in measuring strain in the prestressed tendons within CTB1.

Therefore, by analyzing the strain distribution in the chord and brace members using Eqs. (5.7-5.9), the internal force distribution at the sections of the chord and brace members near the truss joint intersection line can be obtained. Then, based on the equilibrium of axial force and bending moment at the composite joint location, as given by Eqs. (5.10-5.12), the axial force and bending moment resisted by the deck slab can be calculated. A typical distribution pattern is shown in Fig. 5.17.

$$\sum P_{i,x} = P_{d,L} + P_{c,L} - P_{d,R} - P_{c,R} + P_{b,C}\cos\theta - P_{b,T}\cos\theta = 0 \tag{5.10}$$

$$\sum P_{i,y} = P_{b,C}\sin\theta + P_{b,T}\sin\theta = 0 \tag{5.11}$$

$$\sum M_i = M_{d,L} + M_{d,R} + M_{c,L} + M_{c,R} + M_{b,C} + M_{b,T} = 0 \tag{5.12}$$

In this case, it is assumed that the tension is positive, the pressure is negative, the clockwise moment is positive, and the counterclockwise moment is not negative.

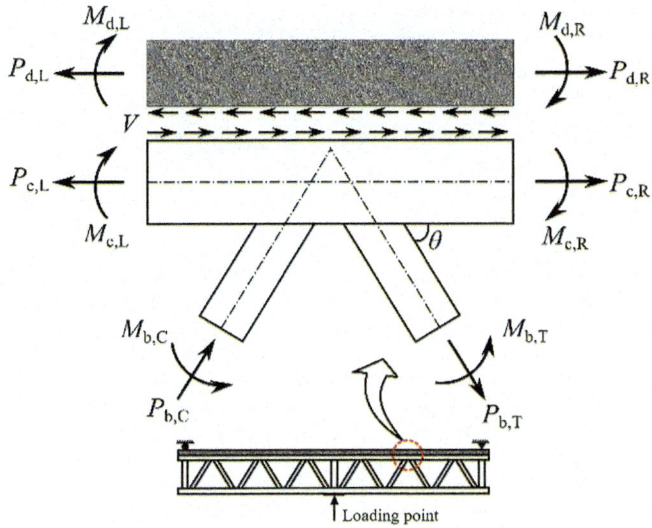

Figure 5.17 Typical internal force distribution in a truss joint.

Using Eqs. (5.7-5.9), the measured nominal strains of the chord and brace members were analyzed to determine the variations in axial force and bending moment in the chord and brace members under a loading point load ranging from 0 to 600 kN, as shown in Fig. 5.18. The Figure indicates that the directions of the axial forces and bending moments in the chord and brace members of both CTB1 and CTB2 are consistent with the typical internal force distribution illustrated in Fig. 5.17. Under the same loading point load, the axial forces and bending moments in the tension and compression brace members of CTB1 and CTB2 are essentially similar. However, the bending moments and axial forces carried by the chords differ between the two specimens.

Based on the measured internal forces in the chord and brace members and using the equilibrium relationships in Eqs. (5.10-5.12), the axial force and bending moment shared by the concrete deck slab can be obtained, as shown in Fig. 5.19. The Figure shows that the axial force shared by the deck slab in specimen CTB1 is greater than that in CTB2, due to the effect of the prestressed tendons in CTB1. Conversely, the bending moment shared by the deck slab in CTB1 is less than that in CTB2, because CTB1 uses shear studs with locally reduced action, resulting in a lower degree of composite action compared to CTB2.

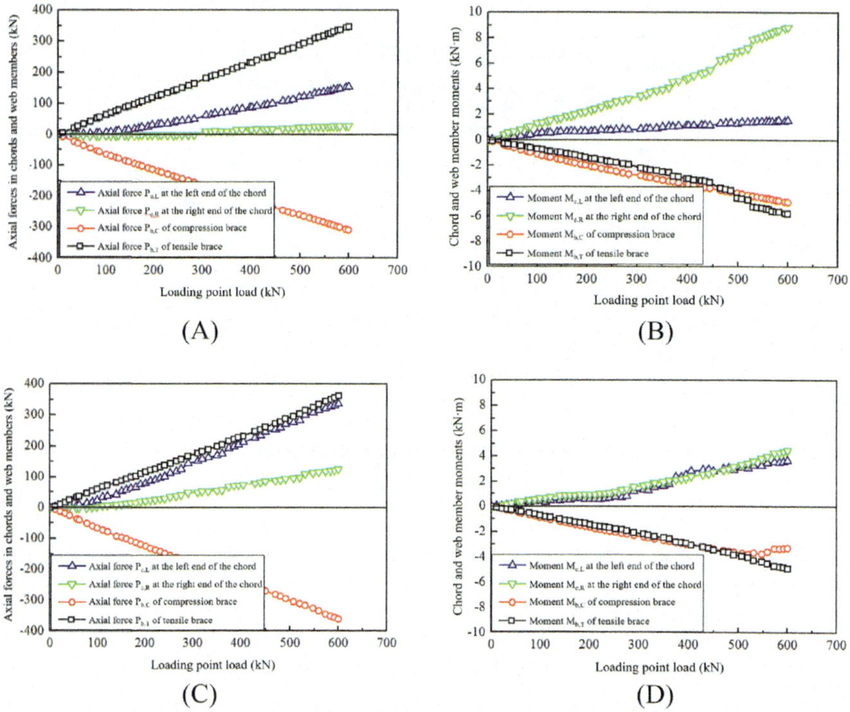

Figure 5.18 Axial forces and bending moments in the chord and brace members. (A) Axial force of CTB1 string rod and belly rod. (B) Bending moments of CTB1 string rods and belly rods. (C) Axial force of CTB2 chord and brace rod. (D) Bending moments of the chord and brace members of CTB2.

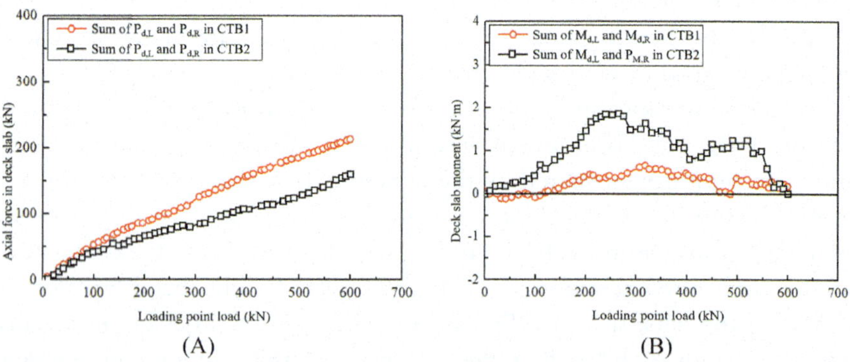

Figure 5.19 Axial force and bending moment in the deck slab. (A) Bridge panel axial force. (B) Bridge panel bending moment.

It is also noteworthy that the variation trend of the bending moment in the deck slab (Fig. 5.19(B)) is irregular. This further illustrates the significant discreteness in bending strain data, particularly for CFST composite truss girders. In such structures, composite action exists both between the top chord steel tube and the infill concrete, and between the top chord and the concrete deck slab, making it difficult to accurately measure the bending moment resisted by each individual component.

Based on the above analysis, when the loading point load reaches 200 kN, the truss structure remains within its elastic stage. Concurrently, analysis of the measured data indicates stable and accurate strain test results at this stage. Therefore, the hot-spot strain data measured at this moment were analyzed. Fig. 5.20 shows the internal force distribution for specimens CTB1 and CTB2 under a loading point load of 200 kN. Consistent with the conclusion from Section 3.4.2, the axial forces and bending moments in the brace members of both specimens are essentially similar. However, due to differences in the deck slab and shear stud configurations, the concrete deck slab in CTB1 resists more axial force but less bending moment compared to CTB2. Consequently, the chord in CTB1 carries less axial force but more bending moment. The hot-spot stress results at various points of the truss joints for CTB1 and CTB2 at this stage are listed in Table 5.7.

The geometric dimensions of the joints in specimens CTB1 and CTB2 are identical. Therefore, under the same load, the hot-spot stress distribution along the brace-chord intersection line should also be consistent. A comparison of the measured hot-spot stress test results from Table 5.7 is shown in Fig. 5.21.

Fig. 5.21(A) shows that the hot-spot stress distribution on the brace surfaces of the CTB1 and CTB2 truss joints is essentially the same. The magnitude of the hot-spot stress on the brace surface depends only on the level of nominal stress on the brace surface. According to the internal force analysis of the truss joint under the 200 kN loading point load shown in Fig. 5.20, the nominal stresses on the brace surfaces of both joint types are roughly equivalent. Consequently, the hot-spot stress levels on the brace surfaces are also comparable. The maximum tensile hot-spot stress in the braces of both CTB1 and CTB2 occurs at point A on the weld toe side of the tension brace, with values of 128.1 MPa and 128.5 MPa, respectively. The maximum compressive hot-spot stress in the braces occurs at point E on the weld toe side of the compression brace, with values of -251.4 MPa

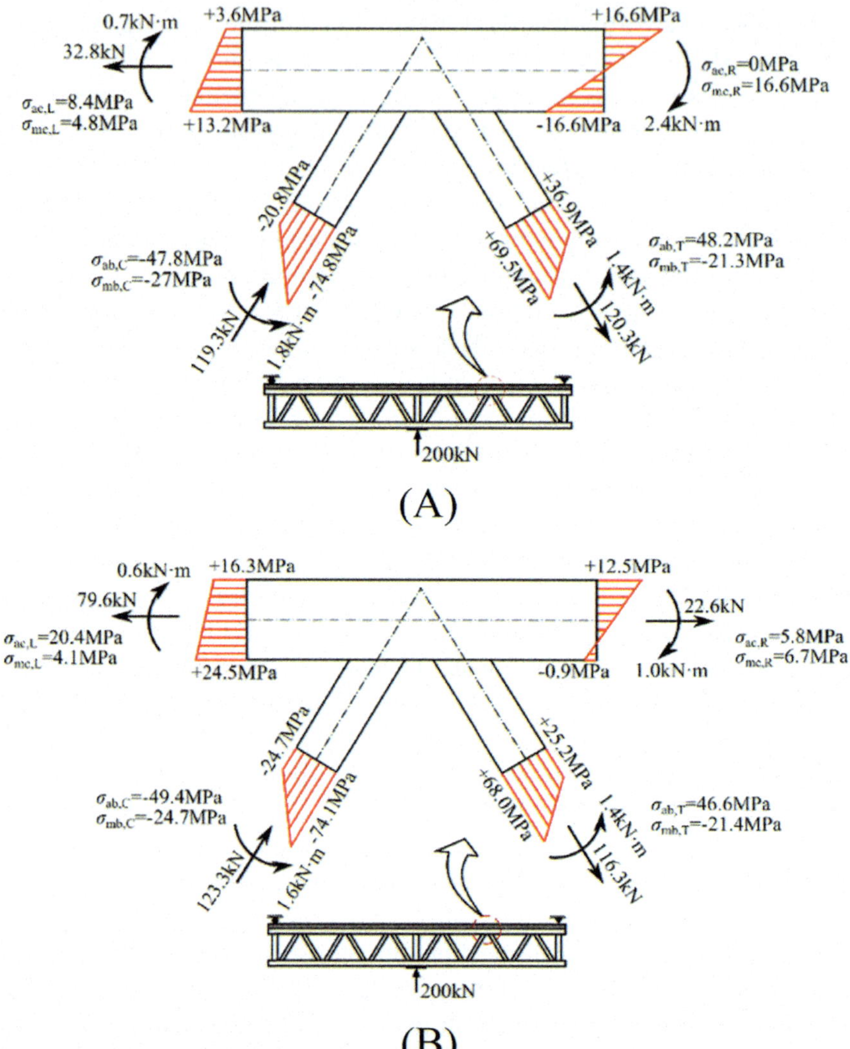

Figure 5.20 Internal force distribution in the truss joint under a loading point load of 200 kN. (A) CTB1 specimen nodes. (B) CTB2 specimen nodes.

and -259.9 MPa, respectively. Since fatigue failure at the weld toe of the brace is typically only caused by tensile stress, point A on the weld toe side of the tension brace is more susceptible to fatigue failure in both joints.

Table 5.7 Hot-spot stress results of the truss joint.

Specimen No	Location	Brace Hot-Spot Stress (MPa)			Chord Hot-Spot Stress (MPa)			
		A	E	F	B	C	D	G
CTB1	Tensile brace weld root side	71.3	54.2	45.3	39.7	20.2	-21.7	-2.1
	Tensile brace weld toe side	128.1	93.6	102.0	55.6	27.8	49.2	35.9
	Compressive brace weld root side	-21.1	-35.6	-19.5	-9.7	7.5	25.1	17.9
	Compressive brace weld toe side	-108.8	-251.4	-90.1	-47.6	-35.1	-23.2	-56.5
CTB2	Tensile brace weld root side	61.9	46.3	41.3	27.5	7.5	18.5	10.5
	Tensile brace weld toe side	128.5	94.2	100.5	50.7	45.7	72.1	70.0
	Compressive brace weld root side	-51.1	-33.0	-31.6	-7.4	-36.3	35.3	29.6
	Compressive brace weld toe side	-162.9	-259.9	-88.6	-83.8	-41.4	-62.3	-15.2

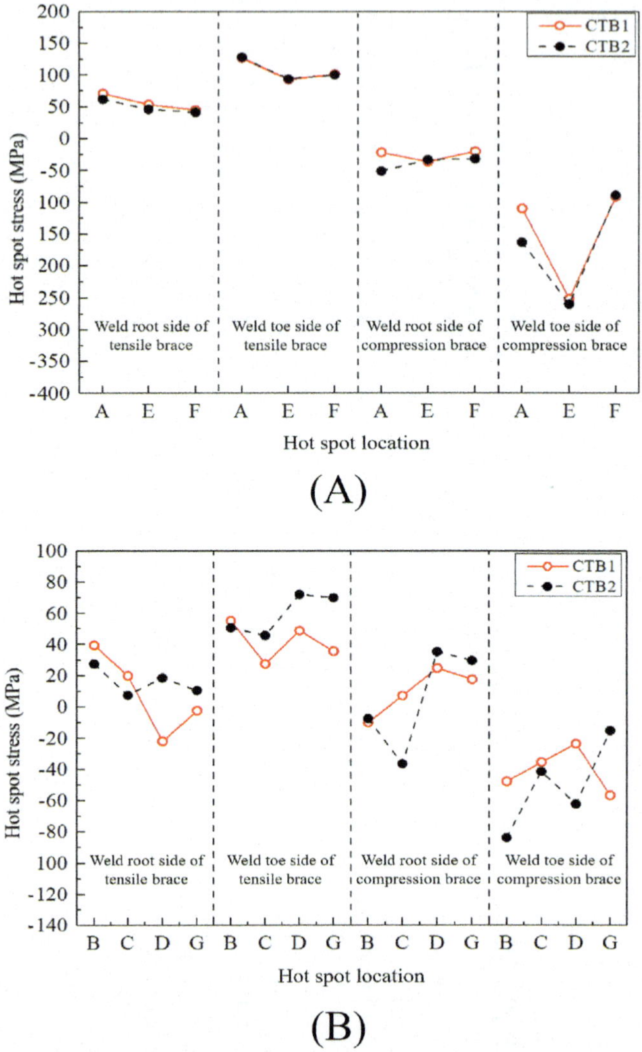

Figure 5.21 Comparison of hot-spot stress test results between CTB1 and CTB2. (A) Hot stress distribution of branch tube. (B) Supervise the distribution of stress on hot-spots.

Fig. 5.21(B) shows that the hot-spot stress distribution on the chord surfaces differs between the CTB1 and CTB2 joints. The hot-spot stress on the chord surface is related to the nominal stresses in both the brace and the chord. The internal force analysis in Fig. 5.20 indicates that due to differences in the deck slab and shear stud configurations, the internal forces in the chords of the two joints are different. The maximum tensile hot-spot stress on the chord of CTB1 occurs at point B on the weld toe side of the tension brace (55.6 MPa), while for CTB2 it occurs at point D (72.1 MPa). The maximum compressive hot-spot stress on the chord of CTB1 occurs at point G on the weld toe side of the compression brace (-56.5 MPa), while for CTB2 it occurs at point B (-83.8 MPa). Similarly, considering that fatigue failure at the weld toe of the chord is primarily caused by tensile stress, point B (for CTB1) or point D (for CTB2) on the weld toe side of the tension brace is more susceptible to fatigue failure.

5.3 Fatigue test study on joints of CFRST trusses

This section intends to conduct fatigue tests on 11 equal-width joints ($\beta = 1$). Different tensile stress ranges will be applied to the braces of each specimen to analyze the fatigue life and crack propagation mechanism of the joints. The aim is to establish an *S-N* curve for the fatigue calculation of CFRST joints based on the hot-spot stress method.

5.3.1 Overview of the test

5.3.1.1 Specimen design

Considering that equal-width joints ($\beta = 1$) are commonly used in bridges, this test focuses on Y-type joints with $\beta = 1$. A total of 11 specimens were designed, including 9 PBL-stiffened CFRST joints and 2 RST joints. The typical joint configuration is shown in Fig. 5.22, and the geometric parameters are summarized in

Table 5.8. The designation "YH" represents RST joints, while "YC" represents PBL-stiffened CFRST joints.

All specimens have a brace-chord angle of 60°. The 9 PBL-stiffened CFRST joints have identical geometric dimensions but are subjected to different stress ranges during the test. Specimens YH1 and YH2 serve as control specimens. They were originally designed to have the same dimensionless parameters as YC1~YC9. However, due to a fabrication error, the chord of specimen YH2 was made from a 4 mm thick steel plate instead of the originally designed 6 mm plate. The test setup is shown in Fig. 5.23.

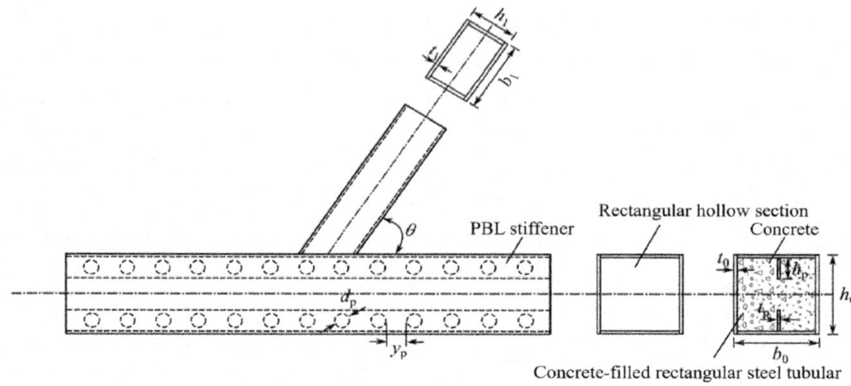

Figure 5.22 Typical joint configuration.

5.3.1.2 Static test procedure

Fatigue cracks in joints typically initiate at the location of maximum stress concentration at the weld toe. Therefore, before conducting fatigue tests, it is necessary to perform static tests to measure the hot-spot stress values at the weld toe and subsequently obtain the SCF to evaluate the joint's fatigue performance. The arrangement of hot-spot stress measurement points is shown in Fig. 5.24, and the hot-spot stress is determined using the quadratic extrapolation method. Calculating the SCF also requires knowing the nominal stress on the surfaces of the brace and chord. The arrangement of nominal stress measurement points is shown in Fig. 5.25.

The static test involved applying both brace tension and brace compression to the joint to determine the most critical load case. A load-controlled loading protocol was used for the test. Pre-loading was conducted before the formal test to eliminate joint eccentricity. Each loading step induced a nominal stress of 5 MPa in the brace, and the entire testing process ensured that the steel tube remained within its elastic range.

5.3.1.3 Fatigue test procedure

After the static test was completed, fatigue loading was applied to the joint to obtain its fatigue life and observe the crack propagation process. A sinusoidal stress range was applied during the test. The fatigue loading parameters are listed in Table 5.9. Among them, specimen YC6 was subjected to variable-amplitude loading with stress ranges of 20 MPa and 35 MPa, respectively.

Table 5.8 Geometry of specimens.

Number	Chord (mm)			Brace (mm)			PBL Stiffener (mm)				Dimensionless Parameters		
	b_0	h_0	t_0	b_1	h_1	t_1	h_p	d_p	y_p	t_p	β	2γ	τ
YH1	150	150	6	150	90	6	-	-	-	-	1	25	1
YH2	150	150	4	150	90	6	-	-	-	-	1	37.5	1.5
YC1~YC9	100	100	4	100	60	4	30	4	20	40	1	25	1

Figure 5.23 Testing arrangement.

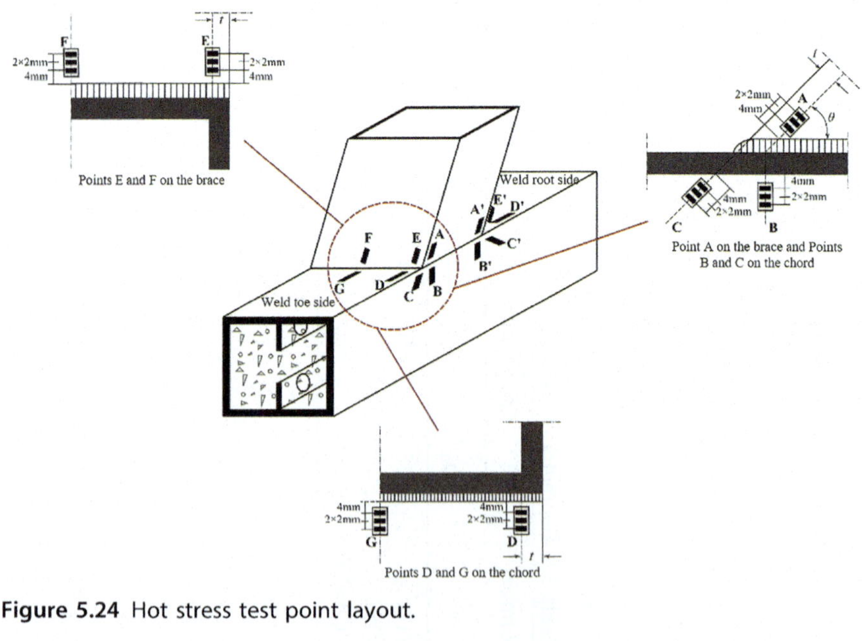

Figure 5.24 Hot stress test point layout.

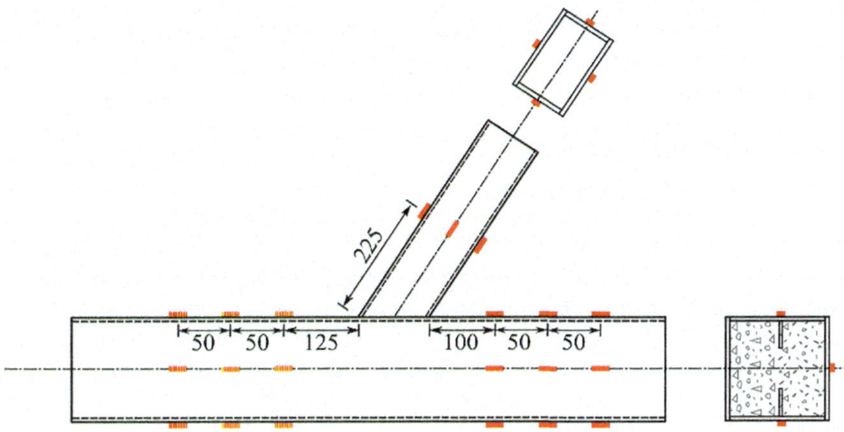

Figure 5.25 Nominal stress point layout (size unit: mm).

Table 5.9 Fatigue loading parameters.

Number	σ_{min} (MPa)	σ_{max} (MPa)	$\triangle\sigma$ (MPa)	Stress Ratio R	Frequency (Hz)
YH1	2.4	24.4	22.0	0.1	8
YH2	2.4	24.4	22.0	0.1	8
YC1	6.6	64.1	57.6	0.1	8
YC2	5.8	37.0	31.3	0.1	8
YC3	3.9	38.7	34.8	0.1	8
YC4	4.4	44.4	40.0	0.1	8
YC5	2.8	27.8	25.0	0.1	8
YC6	2.2/3.9	22.2/38.9	20.0/35.0	0.1	8
YC7	3.5	35.5	32.0	0.1	8
YC8	3.4	33.3	29.9	0.1	8
YC9	3.9	38.9	35.0	0.1	8

Under fatigue loading, the initiation and propagation of cracks at the joint weld toe is a process. The fatigue failure of the joint is typically determined by the number of cycles (N) at the following four characteristic moments: ①The number of cycles (N_1) when the strain value at a measurement point changes by 15%, ②The number of cycles (N_2) corresponding to crack initiation, ③The number of cycles (N_3) when the crack penetrates the steel tube thickness, ④The number of cycles (N_4) when the crack extends to half the perimeter of the brace. In this test, N_2 and N_4 were recorded.

5.3.2 Static test results

5.3.2.1 Nominal stress results

Axial tension and axial compression were applied at the end of the brace, respectively. The typical internal force distribution of the joint under brace tension is shown in Fig. 5.26, where N, M, and V represent the axial force

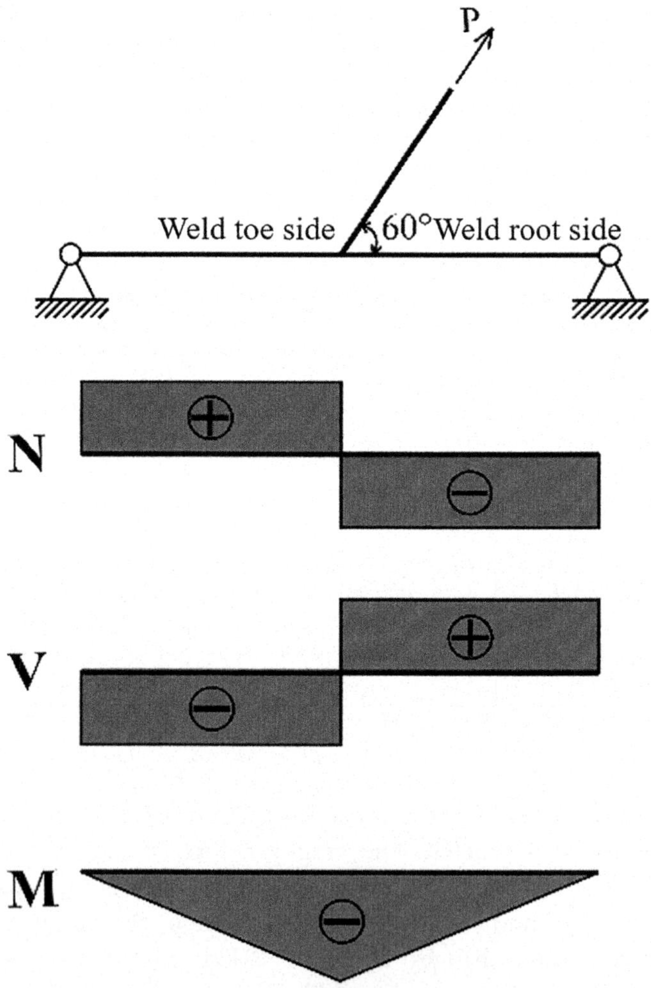

Figure 5.26 Internal force distribution in the joint under tension.

diagram, bending moment diagram, and shear force diagram, respectively. The corresponding stress distributions in the brace and chord are shown in Fig. 5.27. Under brace compression, the internal force and stress distributions are opposite in sign to those in Figs. 5.26 and 27. Consistent with the analysis in Section 5.3, since the chord ends are hinged in this test, the application of brace axial force induces additional bending moment and axial force in the chord.

The nominal stresses on the surfaces of the brace and chord of a RST joint can be calculated using simple beam theory. Under brace tension P, the tensile nominal stress σ_{AB} in the brace is equal to P/A_b, where A_b is the cross-sectional area of the brace. Under chord axial force N, the chord nominal stress σ_{AC} is equal to N/A_c. This results in tensile stress on the weld toe side of the chord and compressive stress on the weld heel side, where A_c is the cross-sectional area of the chord. Under chord bending moment M, the stresses in the top and bottom plates of the chord, $\sigma_{IC,\ t}$ and $\sigma_{IC,\ b}$, are equal to M/Z, where Z is the elastic section modulus of the chord. The nominal stresses for a PBL-stiffened CFRST joint can also be calculated using simple beam theory, by incorporating the contributions of the infill concrete and the PBL stiffeners into the elastic section modulus of the chord section.

Table 5.10 shows a comparison between the calculated and measured nominal stresses in the joint when the brace tensile stress σ_{AB} is 25 MPa. The Table indicates that the average ratios of the theoretical calculation results to the measured results for the brace nominal stress, the chord top plate nominal stress, and the chord bottom plate nominal stress are 1.00, 1.04, and 1.02, respectively. This demonstrates that simple beam theory can

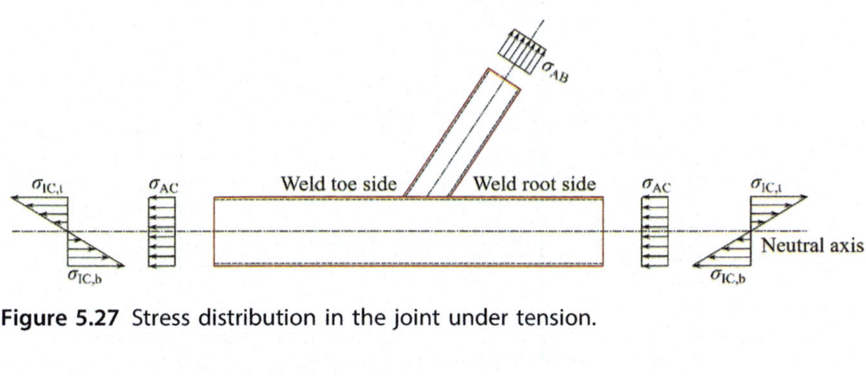

Figure 5.27 Stress distribution in the joint under tension.

Table 5.10 Calculated and measured nominal stresses in the joint under brace tension σ_{AB} = 25 Mpa.

Number	Theoretical Results (MPa)				Measured Results (MPa)			Comparison		
	$\sigma_{AB,1}$	σ_{AC}	$\sigma_{IC,t1}$	$\sigma_{IC,b1}$	$\sigma_{AB,2}$	$\sigma_{IC,t2}$	$\sigma_{IC,b2}$	$\frac{\sigma_{AB,1}}{\sigma_{AB,2}}$	$\frac{\sigma_{IC,t1}}{\sigma_{IC,t2}}$	$\frac{\sigma_{IC,b1}}{\sigma_{IC,b2}}$
YH1	25	4.9	65.9	-65.9	25.8	65.6	-66.0	1.03	1.00	1.00
YH2	25	7.3	94.9	-94.9	24.4	96.7	-96.3	0.98	1.02	1.01
YC1	25	4.9	111.4	-91.1	24.8	109.9	-83.6	0.99	0.99	0.92
YC2	25	4.9	111.4	-91.1	25.7	114.8	-89.9	1.03	1.03	0.99
YC3	25	4.9	111.4	-91.1	25.2	110.7	-98.9	1.01	0.99	1.09
YC4	25	4.9	111.4	-91.1	26.1	126.4	-92.6	1.04	1.13	1.02
YC5	25	4.9	111.4	-91.1	24.4	119.7	-93.4	0.98	1.07	1.02
YC6	25	4.9	111.4	-91.1	24.8	113.2	-93.6	0.99	1.02	1.03
YC7	25	4.9	111.4	-91.1	26.2	117.2	-94.8	1.05	1.05	1.04
YC8	25	4.9	111.4	-91.1	24.1	117.0	-90.6	0.96	1.05	1.00
YC9	25	4.9	111.4	-91.1	24.3	115.3	-99.8	0.97	1.03	1.10
Average value								1.00	1.04	1.02

effectively calculate the nominal stresses in the brace and chord. Based on the measured nominal stresses, the bending stress in the chord top plate ($\sigma_{IC,\ t}$) is greater than that in the chord bottom plate ($\sigma_{IC,\ b}$). This indicates that although the infill concrete may crack in tension during loading, the concrete in the lower compression zone can still contribute, sharing part of the compressive stress and thereby reducing the stress level in the steel bottom plate. Furthermore, although specimens YC1 to YC9 have identical geometric dimensions and material properties, there are still some variations in the measured bending stresses on the chord surface. These differences are likely due to variations in the concrete infill. Therefore, the measured nominal stresses are still selected for SCF calculations.

5.3.2.2 Hot-spot stress results

Based on the measured hot-spot strains and referring to the calculation method described in Section 5.3, the SCF values can be obtained and are listed in Table 5.11. Analysis of the data in Table 5.11 leads to the following conclusions:

1. Similar to the nominal stress results in Table 5.10, although specimens Y1 to Y9 have identical geometric dimensions and material properties, there are still minor differences in the measured SCF values. This indicates that, even though pre-loading was conducted before the test, the measured SCF values are highly sensitive to factors such as load concentration, and the properties of the concrete and steel.
2. Analyzing the SCF values on the chord surface reveals that for RST joints, the SCF values under brace tension and brace compression are essentially similar. This indicates that both load cases are equally critical for the fatigue performance of RST joints. After the chord is filled with concrete, the SCF values under brace tension are significantly higher than those under brace compression. This demonstrates that the concrete infill provides internal support, restricting the compressive deformation of the chord and thereby reducing its SCF.
3. Experimental research on RST joints has shown that for non-equal width joints ($\beta < 1$), the bending stress in the width direction of the chord is relatively large, and points B and C may exhibit the highest SCF values. However, the results of this study on equal-width RST joints ($\beta = 1$) indicate that the load from the brace is transferred directly to the side plates of the chord through its own side plates.

Table 5.11 Static test results.

Number	Weld toe side							Weld root side						
	Brace			Chord				Brace			Chord			
	A	E	F	B	C	D	G	A′	E′	F′	B′	C′	D′	G′
Load Case: Compressive Force in Brace (AC)														
YH1	4.21	2.12	1.05	0.70	0.92	2.65	1.99	3.78	1.51	1.03	0.61	0.90	1.79	1.43
YH2	4.12	2.46	1.04	0.56	0.61	2.65	1.75	3.30	1.82	1.04	0.55	0.56	1.52	1.50
YC1	2.03	1.10	1.28	/	/	1.70	1.34	2.07	1.32	1.13	/	/	1.66	1.20
YC2	1.17	2.37	2.37	/	/	1.82	1.02	1.19	2.21	2.11	/	/	1.25	1.16
YC3	2.03	1.36	1.06	/	/	1.35	1.56	1.24	1.11	1.44	/	/	1.11	1.04
YC4	1.79	2.01	2.08	/	0.27	1.12	1.37	1.21	1.03	1.34	/	/	1.16	1.05
YC5	2.01	1.72	2.26	0.18	0.32	2.06	1.73	1.27	1.39	1.74	/	/	1.64	1.00
YC6	1.93	1.41	1.26	0.27	0.43	1.57	1.52	1.62	1.02	1.09	0.21	0.27	1.41	1.06
YC7	2.24	2.01	2.50	/	/	1.44	1.51	1.16	1.08	1.30	0.12	/	1.02	1.32
YC8	2.25	1.84	2.01	/	/	1.45	1.08	1.90	1.21	1.14	/	/	1.21	1.16
YC9	1.14	1.24	2.05	/	/	1.82	1.73	1.07	1.24	1.26	/	0.19	1.28	1.13
Average value YC1~YC9	1.84	1.67	1.87	/	/	1.59	1.43	1.41	1.29	1.39	/	/	1.30	1.12
Load Case: Tensile Force in Brace (AT)														
YH1	4.16	2.26	1.04	0.74	1.29	2.30	1.95	3.68	1.53	1.17	0.53	1.26	1.58	1.36
YH2	3.97	2.45	1.03	1.01	0.95	2.99	1.75	3.09	1.68	1.05	0.87	0.63	1.60	1.49
YC1	2.29	1.45	1.53	/	/	2.03	1.39	1.48	1.23	1.57	/	/	1.37	1.17
YC2	1.93	2.83	3.17	/	/	2.16	1.04	1.24	1.81	2.24	/	/	2.19	1.72
YC3	2.40	1.61	1.79	/	/	2.08	1.38	1.36	1.31	1.21	/	/	1.18	1.08

YC4	3.06	2.03	2.81	0.21	0.31	2.25	1.10	1.43	1.73	1.69	/	/	1.16	1.12
YC5	2.63	1.85	3.15	/	0.47	2.35	1.97	1.31	1.02	2.09	/	/	1.54	1.02
YC6	2.14	1.71	1.63	0.26	0.43	2.12	1.59	1.67	1.03	1.05	/	/	1.18	1.09
YC7	2.76	2.76	3.05	0.53	/	1.75	1.64	1.37	1.05	1.17	/	/	1.00	1.06
YC8	3.35	2.24	2.80	/	/	1.30	1.07	1.29	1.21	1.08	/	/	1.72	1.12
YC9	1.50	1.42	2.60	/	/	1.99	1.88	1.47	1.22	1.48	/	/	1.33	1.14
Average value YC1~YC9	2.45	1.99	2.50	/	/	2.00	1.45	1.40	1.29	1.51	/	/	1.41	1.17

Consequently, the SCF values at points B and C are very small and can be neglected.

To further compare the difference in stress concentration levels between PBL-stiffened CFRST joints and RST joints, the average SCF values from specimens YC1 to YC9 (PBL-stiffened CFRST) were compared with the SCF values of the RST joint YH1, as shown in Fig. 5.28 It is important to note that YH2, having different dimensionless dimensions, cannot be used for this comparison. Under brace axial compression loading, the SCF values at most points on the surfaces of both the brace and chord of the PBL-stiffened CFRST joints are significantly reduced compared to those of the RST joint, except for points F on the weld toe side and F' on the weld heel side. This indicates

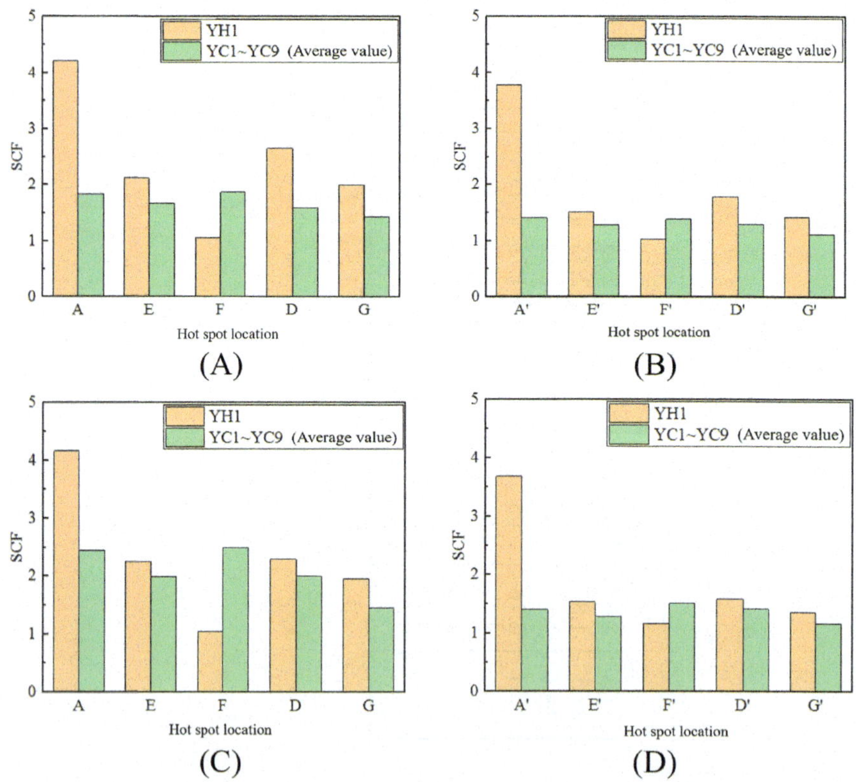

Figure 5.28 Comparison of SCF between RST joints and PBL-stiffened CFRST joints. (A) Weld toe side under brace compression. (B) Weld heel side under brace compression. (C) Weld toe side under brace tension. (D) Weld heel side under brace tension.

that the infill concrete effectively supports the steel top plate, leading to a more uniform stress distribution along the intersection line and consequently reducing stress concentration. The slight increase in SCF at points F and F' on the brace surface might be related to local effects near the stiffener. Under brace axial tension loading, the differences in SCF values at various points are generally similar to those under compression loading. However, the primary reason for the reduction is attributed to the PBL stiffeners inside the chord. These stiffeners reduce the unsupported length of the chord's top plate, thereby decreasing its outward bulging bending deformation and mitigating stress concentration.

5.3.3 Fatigue test results

5.3.3.1 Overview

The measured SCF values in Table 5.11 indicate that after the chord is filled with concrete, the brace tension load case is more critical than the brace compression load case. Therefore, a tensile stress range was applied during the fatigue tests. The number of cycles corresponding to crack initiation (N_2), the number of cycles when the crack extended to half the perimeter of the brace (N_4), and the initial crack location were recorded and are listed in Table 5.12.

The initial cracks all occurred at the weld toe on the chord surface. Combined with the levels of the various nominal stresses in Table 5.11, it can be concluded that the chord bending moment controls the fatigue cracking of the joint. Furthermore, it is observed that the crack initiation location coincides with the point of maximum SCF. Table 5.12 also lists the brace nominal stress range ($\Delta\sigma_T$), the chord nominal bending stress range (S_{nom}), and the corresponding chord hot-spot stress range (S_{hs}).

5.3.3.2 Crack initiation and propagation

Fig. 5.29 shows the fatigue failure modes of the various joints. As can be seen from the Figure, the crack initiation locations for the 11 joints fall into two categories: point D on the weld toe side and point D on the weld heel side. This paper selects these two failure modes and uses specimens YC4 and YC8 as examples to describe the entire crack propagation process.

5.3.3.2.1 Fatigue cracking at point D on the weld toe side

Taking specimen YC4 as an example, Fig. 5.30 shows the entire crack propagation process corresponding to the failure mode of fatigue cracking

Table 5.12 Fatigue test results.

Number	$\triangle\sigma$(MPa)	S_{nom}(MPa)	SCF	S_{hs}(MPa)	N_2	N_4	N_4/N_2	Crack initiation		S_{hs} (t_0=16) (MPa)
								Location	t_0(mm)	
YH1	22.0	58.0	2.65	153.7	981990	1413896	1.44	Point D at weld toe side	6	107.0
YH2	22.0	89.6	2.99	267.9	465898	889280	1.91	Point D at weld toe side	4	163.3
YC1	57.6	237.3	2.03	481.6	-	53653	-	Point D at weld toe side	4	325.0
YC2	31.3	128.8	2.19	282.1	1183850	1234229	1.04	Point D at weld root side	4	170.0
YC3	34.8	143.4	2.08	298.2	-	345568	-	Point D at weld toe side	4	188.1
YC4	40.0	164.7	2.25	370.6	412709	451330	1.09	Point D at weld toe side	4	231.6
YC5	25.0	103.0	2.35	242.1	1024297	1123252	1.10	Point D at weld toe side	4	146.4
YC6	20.0/35.0	82.3/144.4	2.12	183.0	-	3108085	-	Point D at weld toe side	4	106.6
YC7	32.0	131.9	1.75	230.7	462435	481187	1.04	Point D at weld toe side	4	143.8
YC8	29.9	123.4	1.72	212.2	2450971	2536411	1.03	Point D at weld root side	4	124.6
YC9	35.0	144.4	1.99	287.3	419248	428916	1.02	Point D at weld toe side	4	179.9

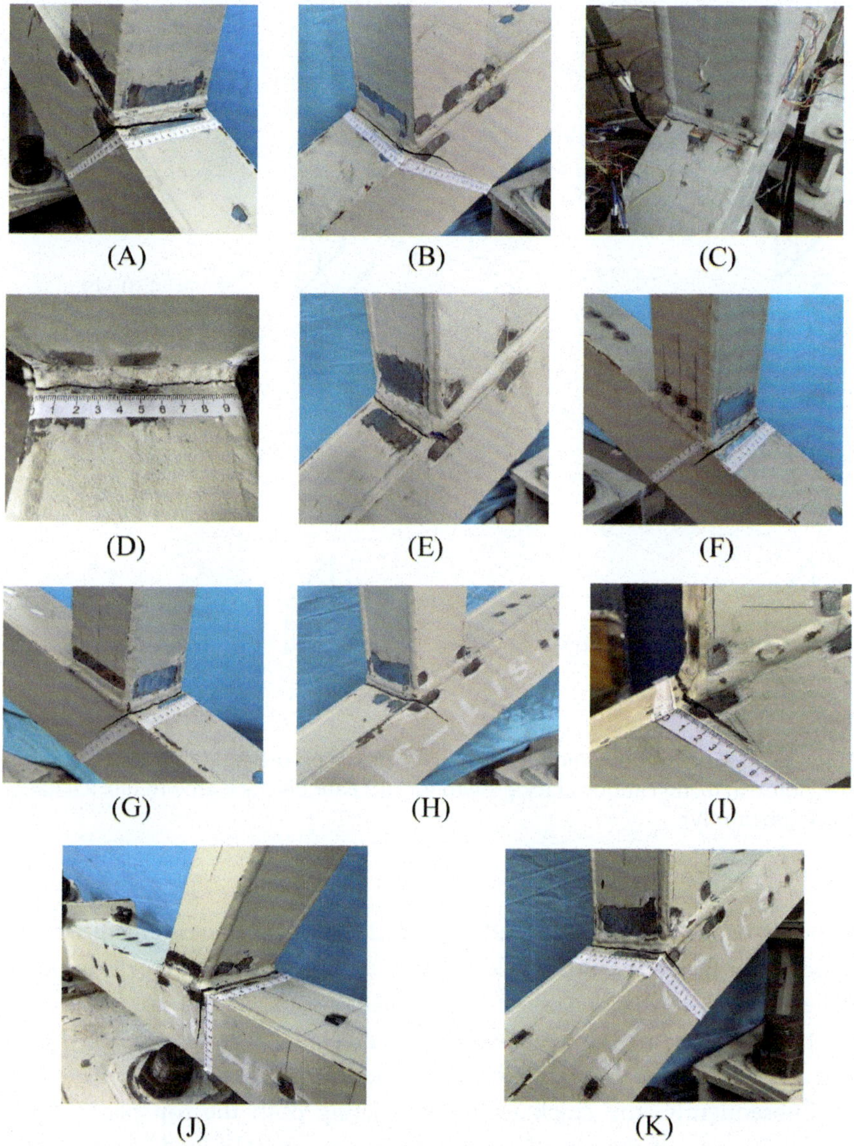

Figure 5.29 Fatigue failure modes of the joints. (A) YH1 (point D at weld toe side). (B) YH2 (point D at weld toe side). (C) YC1 (point D at weld toe side). (D) YC2 (point D at weld toe side). (E) YC3 (point D at weld toe side). (F) YC4 (point D at weld toe side). (G) YC5 (point D at weld toe side). (H) YC6 (point D at weld toe side). (I) YC7 (point D at weld toe side). (J) YC8 (point D at weld root side). (K) YC9 (point D at weld toe side).

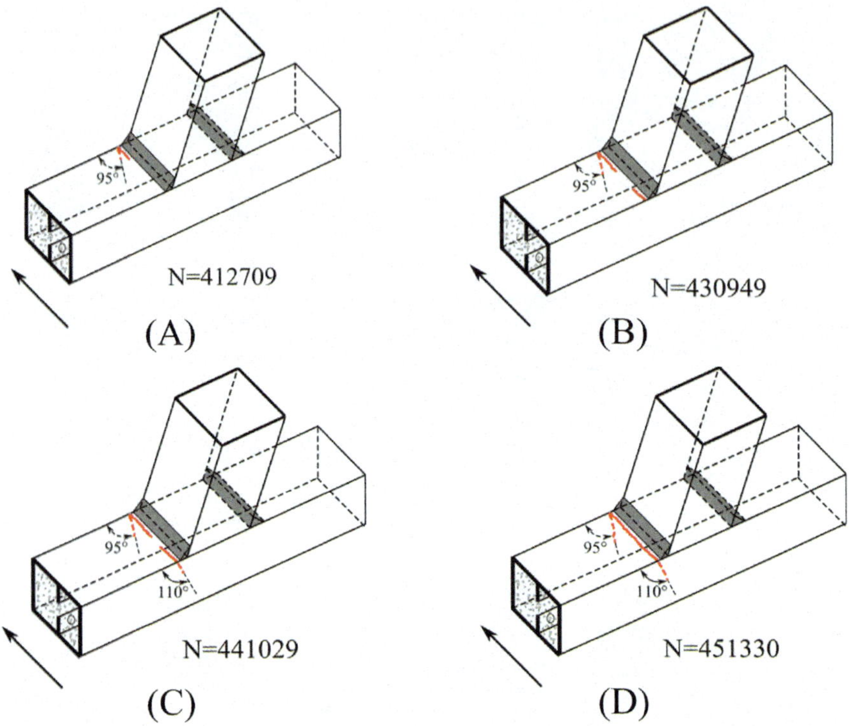

Figure 5.30 Crack propagation (cracking at point D on the weld toe side). (A) Fatigue crack growth (I). (B) Fatigue crack growth ((II). (C) Fatigue crack growth (III). (D) Fatigue crack growth (IV).

at point D on the weld toe side. At 412,709 load cycles, an initial crack appeared at point D on the weld toe side. At this stage, the crack on the top plate of the chord was 1.5 cm long, and it propagated at an angle of 95° relative to the chord. Simultaneously, a crack 1 cm long appeared on the side plate at the back of the chord, as shown in Fig. 5.30(A).

Subsequently, the crack propagated along both the top plate and the back side plate of the chord. Furthermore, at 430,949 load cycles, a crack appeared on the top plate at the front of the chord, with a length of 1.3 cm, as shown in Fig. 5.30(B).

As the number of cycles increased, the cracks continued to propagate. By 441,029 load cycles, the cracks on the back top plate and side plate of the chord had reached lengths of 3.7 cm and 1.7 cm, respectively. The crack on the front of the chord propagated from the top plate towards the side plate at an angle of 110° relative to the chord, with lengths of 1.8 cm on the top plate and 0.8 cm on the side plate at this point.

Finally, at 451,330 load cycles, the crack penetrated through the entire top plate of the chord. The cracks on the back and front side plates were 3.3 cm and 2.5 cm long, respectively. At this stage, the joint was considered to have failed by fatigue.

5.3.3.2.2 Fatigue cracking at point D on the weld heel side

Taking specimen YC8 as an example, Fig. 5.31 shows the entire crack propagation process corresponding to the failure mode of fatigue cracking at point D on the weld heel side. At 2,450,971 load cycles, an initial crack appeared at point D on the weld heel side. It propagated along both the top

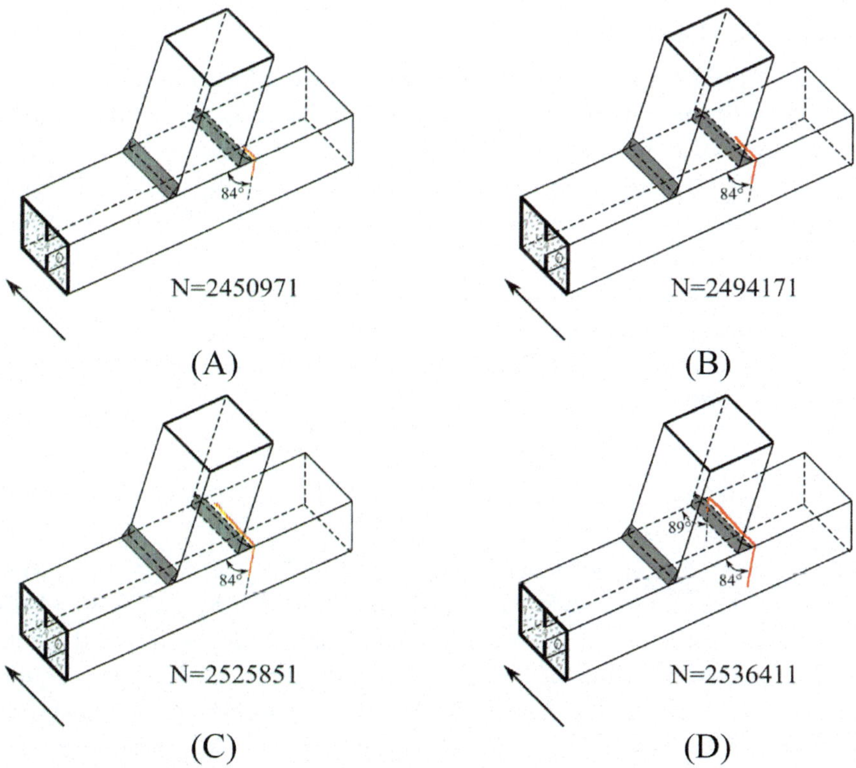

Figure 5.31 Crack propagation (cracking at point D on the weld heel side). (A) Fatigue crack growth (I). (B) Fatigue crack growth (II). (C) Fatigue crack growth (III). (D) Fatigue crack growth (IV).

plate of the chord and the front side plate of the chord, extending at an angle of 84° relative to the chord on the front side plate. The lengths of the crack on the chord's top plate and side plate were 2.2 cm and 1.3 cm, respectively, as shown in (A).

Subsequently, the crack continued to propagate along the chord's top plate and the front side plate, as shown in Fig. 5.31(B) and (C).

Finally, at 2,536,411 load cycles, the crack penetrated through the chord's top plate and extended to the back side plate of the chord. The joint was considered to have failed by fatigue at this point. The lengths of the cracks on the front and back side plates of the chord were 7.4 cm and 2.1 cm, respectively.

5.3.4 Fatigue design S_{hs}-N curve

5.3.4.1 S_{hs}-N *data analysis*

The data points from Table 5.12 are plotted on a logarithmic coordinate system, with the hot-spot stress range (S_{hs}) as the ordinate and the number of cycles (N) as the abscissa. The fatigue strength S_{hs}-N curve can be obtained by fitting the data. The hot-spot stress range S_{hs} can be calculated as the product of the nominal stress range (S_{nom}) and the corresponding maximum SCF.

For specimen YC6, which was subjected to variable-amplitude fatigue loading during the test, the equivalent nominal stress range ($S_{nom,\ equ}$) can be calculated according to Miner's linear cumulative damage rule using the following formula:

$$S_{nom,equ} = \left(\frac{\Sigma n_i \cdot S_i^m}{\Sigma n_i} \right)^{1/m} \tag{5.13}$$

Where $m = 3$. For specimen YC6, $S_1 = 82.3$ MPa, $S_2 = 144.4$ MPa, $n_1 = 3000000$, and $n_2 = 108085$. Calculations show that $S_{nom,\ equ} = 86.3$ MPa. S_1 and S_2 are the nominal stress amplitudes applied for the first and second cycles, respectively, while n_1 and n_2 represent the corresponding cycle counts.

The S_{hs}-N curve is related to the wall thickness of the joint's steel tube. In the CIDECT code, the S_{hs}-N curve is based on a reference wall thickness of 16 mm. Multiple S_{hs}-N curves related to different wall thicknesses can be derived by applying a thickness correction to this baseline. This paper also uses 16 mm as the reference wall thickness. The hot-spot stress ranges for joints with different wall thicknesses are equivalently converted to those corresponding to the reference wall thickness. The thickness correction formula from the CIDECT code is selected for this equivalent conversion of the hot-spot stress range.

$$S_{t=16} = S_{t=T} \cdot \left(\frac{T}{16} \right)^{0.06 \cdot \log N} \tag{5.14}$$

In the equation, $S_t = 16$ is the hot-spot stress range corresponding to a thickness $t = 16$ mm, $S_t = T$ is the hot-spot stress range corresponding to a joint with another wall thickness, and T is the wall thickness at the crack initiation location. For this test, all cracks occurred on the chord surface, therefore $T = t_0$.

Table 5.12 lists the number of cycles corresponding to crack initiation (N_2) and the number of cycles when the crack extended to half the perimeter of the brace (N_4). For the PBL-stiffened CFRST joints YC1 to YC9, N_4 and N_2 are very close, with a ratio of less than 1.1. In contrast, for the RST joints YH1 and YH2, the ratios of N_4 to N_2 are 1.44 and 1.91, respectively, which are greater than those of the PBL-stiffened CFRST joints. This indicates that after initial crack formation, the RST joints exhibit better ductility.

Overall, for all 11 specimens, N_4 and N_2 are relatively close. The difference between the S-N curves derived from regression analysis based on N_4 and those based on N_2 is not significant. Therefore, the subsequent analysis is based on N_4.

5.3.4.2 Proposal of the fatigue design S_{hs}-N curve

Based on the statistically analyzed test data, regression analysis using the least squares method is employed to obtain the S_{hs}-N curve. The regression can be performed using either a free regression method or a forced regression method. Due to the limited number of fatigue test data points and their significant scatter, the free regression method often fails to yield the correct curve slope. Therefore, the forced regression method is adopted. Referring to the S_{hs}-N curve in the CIDECT code, a slope of -3 is specified. The regression yields the lower bound line equation with 95% confidence level as the S_{hs}-N curve. The final expression is:

$$\lg N = A - 3 \lg S_{hs} - 2\sigma_{\lg N} \tag{5.15}$$

In the equation, A is a constant obtained through regression analysis, and $\sigma_{\lg N}$ is the standard deviation of the statistical data with respect to lg$N$.

Fig. 5.32 shows the regression results of the S_{hs}-N curve. As can be seen from the Figure, the fatigue strength of the RST joints and the PBL-stiffened CFRST joints are essentially the same. Therefore, regression analysis was performed on the results of all 11 specimens to obtain the S_{hs}-N curve for the reference wall thickness:

When $10^4 < N < 10^7$:

$$\log N = 12.119 - 3 \log S_h \tag{5.16}$$

Considering the wall thickness correction, it can be written as:

When $10^4 < N < 10^7$:

$$\log S_h = \frac{1}{3}\cdot(12.119 - \log N) + 0.06\cdot \log N\cdot \log\left(\frac{16}{T}\right)$$
$$or \log N = \frac{12.119 - 3\cdot \log S_h}{1 - 0.18\cdot \log(16/T)} \tag{5.17}$$

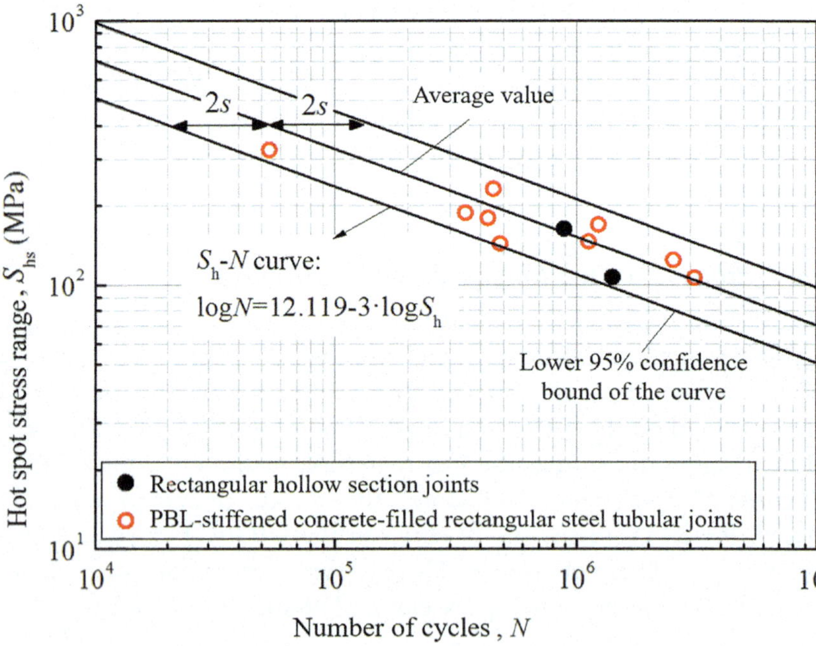

Figure 5.32 Shs-N curve analysis.

5.4 Parametric analysis of SCF for CFRST truss joints

The finite element simulation method is intended to be used to conduct a parametric analysis of the SCF for CFRST T-type, X-type, and K-type joints. For T-type and X-type joints, the influence of the dimensionless geometric parameters β, 2γ, and τ on the SCF will be analyzed. For K-type joints, the influence of the dimensionless geometric parameters β, 2γ, τ, g', and θ on the SCF will be analyzed. The aim is to propose calculation formulas for the SCF parameters specific to T-type, X-type, and K-type joints.

5.4.1 Finite element modeling method and verification

The general-purpose software ABAQUS was used for simulation. Model validation was performed by comparing the results with the measured SCF data from CFRST Y-type joints. In actual engineering, joints are subjected to complex loading, including both in-plane and out-of-plane loads. However, this study considers only the in-plane loads on CFRST joints, which is consistent with the load cases for RST joints specified in the CIDECT code.

For T-type and X-type joints, four basic load cases were analyzed: brace axial force, brace in-plane bending moment, chord axial force, and chord in-plane bending moment. For K-type joints, three basic load cases were analyzed: balanced brace tension-compression loading, chord axial force, and chord bending moment.

5.4.1.1 Element type and mesh size

Based on research into finite element simulation methods for RST joints, it was determined that 20-node reduced integration of solid elements (C3D20R) would be used for steel tubes, infill concrete, and welds. The mesh was generated using the "Structure" meshing technique in ABAQUS. Refined meshing was applied in the extrapolation region near the welds to ensure computational accuracy, while in areas far from the extrapolation region, larger mesh sizes were used to improve computational efficiency.

The seeding strategy for the refined regions was defined as follows: For thicker steel plates, i.e., when the chord satisfies $b_0/t_0 \leq 20$ or the brace satisfies $b_1/t_1 \leq 20$, four layers of elements were arranged through the plate thickness. For thinner steel plates, i.e., when the chord satisfies $b_0/t_0 > 20$

or the brace satisfies $b_1/t_1 > 20$, two layers of elements were arranged through the plate thickness. The finite element model of the CFRST Y-type joint used in the test is shown in Fig. 5.33.

5.4.1.2 Material properties

SCF analysis belongs to elastic analysis, therefore, only the elastic parameters of the materials are defined in the simulation. The material properties of the steel and concrete are consistent with the test measurements. The elastic modulus of steel (E_s) is taken as 2.05×10^5 MPa, with a Poisson's ratio (v_s) of 0.283. The elastic modulus of concrete (E_c) is taken as 3.45×10^4 MPa, with a Poisson's ratio (v_c) of 0.167. The steel tube and the welds are assigned the same steel properties.

5.4.1.3 Weld configuration

The weld structure in the finite element model is referred to the "*Code for Welding of Steel Structures*" (GB 50661-2011), and the specific structure is shown in Fig. 5.34.

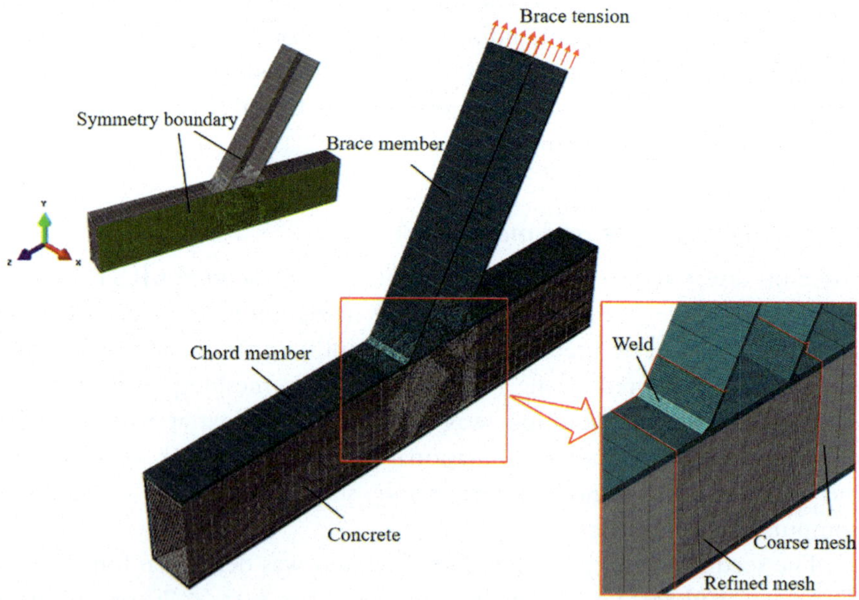

Figure 5.33 FEM of the tested CFRST Y-type joint.

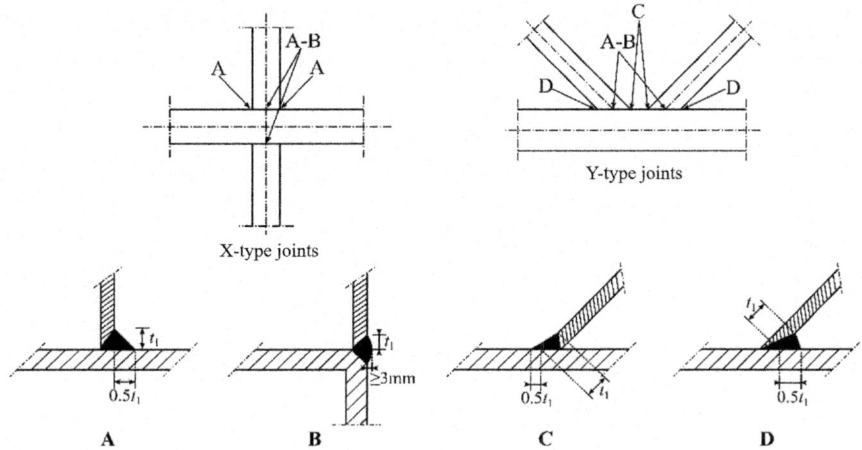

Figure 5.34 Weld configuration.

5.4.1.4 Loads and boundary conditions

The CFRST joints simulated in this study are symmetric structures. To reduce computational costs, a half-model was established, as shown in Fig. 5.33.

Symmetry boundary conditions ($U_1=UR_2=UR_3=0$) were applied on the symmetry plane. Hinged supports ($U_1=U_2=U_3=0$) were applied at both ends of the chord, consistent with the test boundary conditions. An axial tensile load was applied at the end of the brace.

5.4.1.5 Steel-concrete interface relationship

The interaction between the steel tube and concrete is complex, requiring consideration of both normal and tangential contact behavior in the finite element model. Since the bond effect at the steel-concrete interface is very small, it can be neglected in the model. Normal behavior is simulated using "Hard" contact, meaning the steel tube and concrete cannot penetrate each other. The steel tube surface is defined as the master surface, and the concrete surface as the slave surface. The tangential behavior is modeled using a Coulomb friction relationship with a slip curve, and the friction coefficient is set to 0.3.

5.4.1.6 Finite element model verification

Using the finite element simulation method described above, the SCF values were calculated and compared with the measured SCF values from the CFRST joint YC tests in Section 5.3, as shown in Fig. 5.35. The Figure compares the SCF on both the weld toe side and the weld heel side for both the brace and the chord. The results show good agreement between the finite element results (SCF_{FE}) and the experimental results (SCF_{EXP}). Furthermore, the mean value (μ) of the ratio SCF_{FE}/SCF_{EXP} is 1.053, with a standard deviation (σ) of 0.135 and a coefficient variation (σ/μ) of 0.128. This indicates that the finite element simulation method is reliable.

5.4.2 Parametric analysis of CFRST T-type and X-type joints

5.4.2.1 Parameter design

The CIDECT code considers that T-type and X-type RST joints have the same SCF values. Therefore, CFRST T-type and X-type joints should be analyzed uniformly, and a unified calculation formula for SCF parameters should be provided. Using the X-type joint as an example, its typical configuration is shown in Fig. 5.36.

In the CIDECT code, the applicable range for the SCF parameter calculation formulas of T-type and X-type RST joints is: $0.35 \le \beta \le 1.0$, $12.5 \le 2\gamma \le 25.0$ and $0.25 \le \tau \le 1.0$. The same range of dimensionless geometric parameters was selected for the parametric design, as detailed in Table 5.13. The geometric dimensions are based on the Huangyan Bridge joint to ensure the calculation results are applicable to real engineering scenarios.

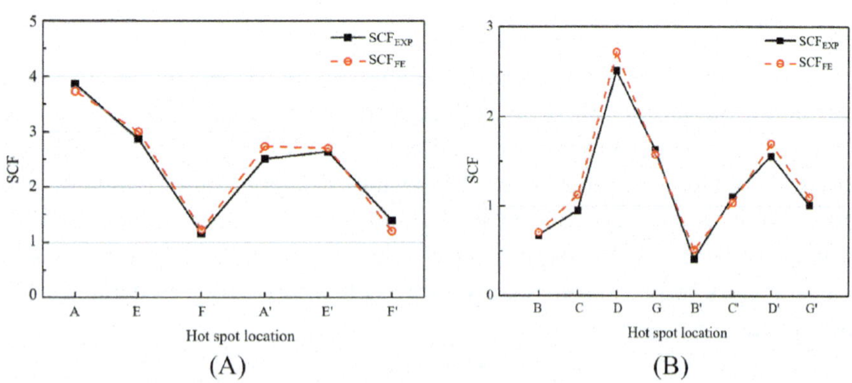

Figure 5.35 Comparison of experimental results and finite element calculation results. (A) Brace. (B) Chord.

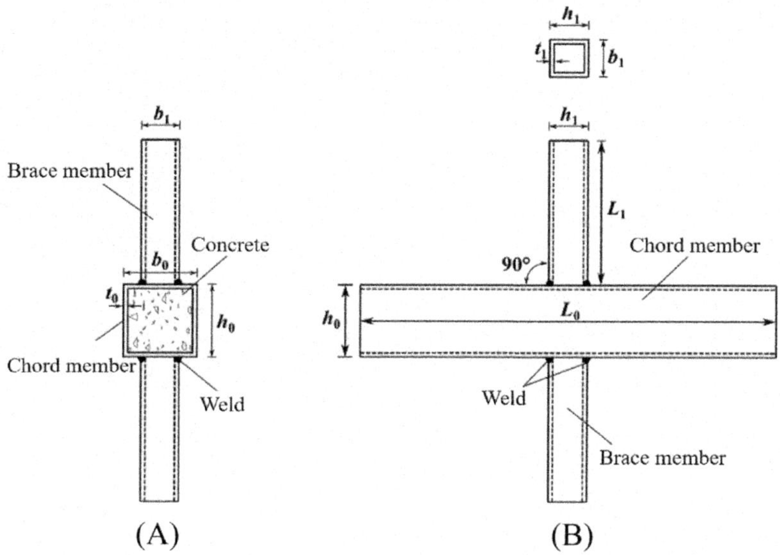

Figure 5.36 Rectangular steel tube concrete X-type node structure. (A) Side view. (B) Elevation view.

Table 5.13 Parametric design for T-type and X-type joints.

β	2γ	τ
0.40	12.5	0.25
0.55	16.0	0.50
0.70	20.0	0.75
0.82	25.0	1.00
1.00		

The chord width and height are kept constant at $b_0 = h_0 = 400$ mm. The chord thickness ranges from $t_0 = 16 \sim 32$ mm, and the chord length is $L_0 = 3000$ mm. The brace width and height are equal, ranging from $b_1 = h_1 = 160 \sim 400$ mm. The brace thickness ranges from $t_1 = 6.25 \sim 32$ mm, and the brace length is $L_1 = 1200$ mm.

Table 5.14 provides a summary of the models used in parametric analysis. Taking the first specimen as an example, the specimen designation has the following meaning: "C" represents the chord, "400 × 32" indicates the chord cross-sectional dimensions (b_0=h_0=400 mm, t_0=32 mm),

Table 5.14 FE calculation results of SCF for CFRST T-type and X-type joints.

Number	β	2γ	τ	SCF_{FE}				
				A	B	C	D	E
C400×32-B160×8	0.40	12.5	0.25	4.04	2.42	2.18	1.36	3.77
C400×32-B160×16	0.40	12.5	0.50	4.95	4.20	3.77	2.30	3.90
C400×32-B160×24	0.40	12.5	0.75	4.22	5.71	5.10	2.78	3.01
C400×32-B160×32	0.40	12.5	1.00	3.44	6.37	5.83	3.41	2.27
C400×25-B160×6.25	0.40	16.0	0.25	5.14	3.18	2.61	1.83	5.09
C400×25-B160×12.5	0.40	16.0	0.50	7.23	6.23	5.66	3.58	5.81
C400×25-B160×18.8	0.40	16.0	0.75	5.77	7.35	6.79	3.95	4.56
C400×25-B160×25	0.40	16.0	1.00	5.10	9.32	11.19	6.45	4.32
C400×20-B160×5	0.40	20.0	0.25	6.10	4.82	3.95	2.74	6.37
C400×20-B160×10	0.40	20.0	0.50	10.14	9.11	8.49	5.40	6.29
C400×20-B160×15	0.40	20.0	0.75	8.64	12.30	11.77	6.86	6.48
C400×20-B160×20	0.40	20.0	1.00	7.14	16.51	13.85	8.94	4.94
C400×16-B160×4	0.40	25.0	0.25	6.91	6.99	5.90	4.11	7.65
C400×16-B160×8	0.40	25.0	0.50	12.63	13.17	13.23	8.11	10.85
C400×16-B160×12	0.40	25.0	0.75	11.97	17.95	17.57	10.44	9.18
C400×16-B160×16	0.40	25.0	1.00	10.01	24.12	20.93	13.58	7.10
C400×32-B220×8	0.55	12.5	0.25	4.53	3.74	2.23	1.12	4.02
C400×32-B220×16	0.55	12.5	0.50	5.39	3.78	3.29	1.89	4.14
C400×32-B220×24	0.55	12.5	0.75	4.83	5.03	3.26	2.22	3.34
C400×32-B220×32	0.55	12.5	1.00	4.17	6.81	5.89	2.74	3.62
C400×25-B220×6.25	0.55	16.0	0.25	5.23	2.91	2.17	1.48	5.12

C400×25-B220×12.5	0.55	16.0	0.50	7.55	5.41	4.81	2.97	5.93
C400×25-B220×18.8	0.55	16.0	0.75	6.69	7.05	6.86	3.67	4.76
C400×25-B220×25	0.55	16.0	1.00	5.80	9.63	7.94	4.71	4.73
C400×20-B220×5	0.55	20.0	0.25	6.05	4.15	3.25	2.29	6.29
C400×20-B220×10	0.55	20.0	0.50	10.31	8.15	7.32	4.59	8.23
C400×20-B220×15	0.55	20.0	0.75	9.07	10.91	10.00	5.76	6.59
C400×20-B220×20	0.55	20.0	1.00	7.74	14.60	11.48	7.42	5.11
C400×16-B220×4	0.55	25.0	0.25	9.87	7.27	6.52	4.95	10.19
C400×16-B220×8	0.55	25.0	0.50	12.62	12.08	11.16	7.11	10.64
C400×16-B220×12	0.55	25.0	0.75	12.26	16.35	15.44	9.01	9.11
C400×16-B220×16	0.55	25.0	1.00	10.46	21.73	17.98	11.53	7.08
C400×32-B280×8	0.70	12.5	0.25	3.68	2.57	2.30	0.61	3.35
C400×32-B280×16	0.70	12.5	0.50	4.88	2.77	2.44	1.20	3.70
C400×32-B280×24	0.70	12.5	0.75	4.62	2.63	2.99	1.37	3.16
C400×32-B280×32	0.70	12.5	1.00	4.19	3.10	3.50	1.81	3.63
C400×25-B280×6.25	0.70	16.0	0.25	4.43	2.13	1.80	0.96	4.29
C400×25-B280×12.5	0.70	16.0	0.50	6.58	3.25	3.72	1.83	5.05
C400×25-B280×18.8	0.70	16.0	0.75	6.14	6.74	5.92	2.31	4.26
C400×25-B280×25	0.70	16.0	1.00	5.53	6.31	5.49	3.03	3.48
C400×20-B280×5	0.70	20.0	0.25	4.99	2.97	2.55	1.45	5.14
C400×20-B280×10	0.70	20.0	0.50	8.74	5.51	5.22	2.97	6.79
C400×20-B280×15	0.70	20.0	0.75	8.08	7.30	7.66	3.93	5.68
C400×20-B280×20	0.70	20.0	1.00	7.13	11.49	10.21	4.66	4.54
C400×16-B280×4	0.70	25.0	0.25	5.43	4.38	4.29	2.32	6.02
C400×16-B280×8	0.70	25.0	0.50	10.51	8.46	7.42	4.74	8.63
C400×16-B280×12	0.70	25.0	0.75	10.67	11.40	11.20	5.96	7.64

(*Continued*)

Table 5.14 (Continued)

Number	β	2γ	τ	SCF_{FE}				
				A	B	C	D	E
C400×16-B280×16	0.70	25.0	1.00	9.42	16.06	12.59	7.64	6.10
C400×32-B340×8	0.85	12.5	0.25	2.56	0.62	0.45	0.52	2.48
C400×32-B340×16	0.85	12.5	0.50	2.72	1.06	0.62	0.93	3.30
C400×32-B340×24	0.85	12.5	0.75	2.64	1.08	0.79	1.36	3.15
C400×32-B340×32	0.85	12.5	1.00	2.55	1.32	1.21	1.34	2.88
C400×25-B340×6.25	0.85	16.0	0.25	3.38	0.71	0.54	0.75	3.20
C400×25-B340×12.5	0.85	16.0	0.50	4.81	1.26	0.93	1.04	3.75
C400×25-B340×18.8	0.85	16.0	0.75	4.76	1.96	1.43	1.76	3.41
C400×25-B340×25	0.85	16.0	1.00	4.56	2.12	1.65	1.98	3.03
C400×20-B340×5	0.85	20.0	0.25	3.61	1.10	1.06	1.01	3.57
C400×20-B340×10	0.85	20.0	0.50	6.09	2.36	1.32	1.72	4.67
C400×20-B340×15	0.85	20.0	0.75	6.06	3.30	2.54	2.87	4.24
C400×20-B340×20	0.85	20.0	1.00	5.76	2.93	2.65	3.04	3.71
C400×16-B340×4	0.85	25.0	0.25	3.74	1.59	1.41	1.59	3.99
C400×16-B340×8	0.85	25.0	0.50	7.04	3.79	2.63	2.39	5.59
C400×16-B340×12	0.85	25.0	0.75	7.71	5.23	4.45	4.28	5.33
C400×16-B340×16	0.85	25.0	1.00	7.34	5.23	4.45	4.18	4.64
C400×32-B400×8	1.00	12.5	0.25	1.35	1.24	1.03	1.29	1.93
C400×32-B400×16	1.00	12.5	0.50	1.68	1.43	1.13	1.53	2.58
C400×32-B400×24	1.00	12.5	0.75	1.74	1.54	1.25	1.41	2.56
C400×32-B400×32	1.00	12.5	1.00	1.62	1.33	1.12	1.40	2.41
C400×25-B400×6.25	1.00	16.0	0.25	1.55	1.24	0.96	0.69	2.18

C400×25-B400×12.5	1.00	16.0	0.50	1.92	1.86	1.54	1.81	2.93
C400×25-B400×18.8	1.00	16.0	0.75	2.00	1.93	1.64	1.69	2.97
C400×25-B400×25	1.00	16.0	1.00	2.05	1.85	1.42	1.63	2.79
C400×20-B400×5	1.00	20.0	0.25	1.74	1.52	1.39	0.80	2.28
C400×20-B400×10	1.00	20.0	0.50	1.76	1.49	1.24	2.09	3.30
C400×20-B400×15	1.00	20.0	0.75	2.20	2.04	1.87	1.99	3.34
C400×20-B400×20	1.00	20.0	1.00	2.31	2.21	2.03	1.94	3.15
C400×16-B400×4	1.00	25.0	0.25	1.96	1.87	1.71	0.94	2.36
C400×16-B400×8	1.00	25.0	0.50	2.44	2.31	2.20	2.41	3.41
C400×16-B400×12	1.00	25.0	0.75	2.51	2.42	2.29	2.35	3.69
C400×16-B400×16	1.00	25.0	1.00	2.57	2.43	2.26	2.46	3.59

"B" represents the brace, "160 × 8" indicates the brace cross-sectional dimensions (b_1=h_1=160 mm, t_1 = 8 mm).

5.4.2.2 Loads and boundary condition

Fig. 5.37 shows the four basic load cases: brace axial force (AB), brace in-plane bending moment (IB), chord axial force (AC), and chord in-plane bending moment (IC). Among these, for the brace axial force load case, a T-type joint would induce additional bending moment and axial force in the chord. Therefore, using an X-type joint for simulation is a common practice among researchers. For the brace in-plane bending moment load case, to prevent inducing additional loads in the chord, the entire bottom of the chord should be fixed.

In the model, an axial force or bending moment that induces a nominal stress of 1 MPa is applied at the end of the brace or chord. The resulting hot-spot stress value at the weld toe is then the SCF. A total of 320 finite element models were established. The established finite element models for the CFRST T-type and X-type joints are shown in Fig. 5.38.

5.4.2.3 Calculation result analysis

The hot-spot locations in CFRST T-type and X-type joints are consistent with those in RST T-type joints, as shown in Fig. 5.35(A). Table 5.14 lists the finite element calculation results for the SCF of CFRST T-type and X-type joints. The Table shows that the location of the maximum SCF varies depending on the geometric parameters. Therefore, it is necessary to propose calculation formulas for the SCF parameters specific to points A through E. This will facilitate designers in calculating the maximum SCF value and identifying its location.

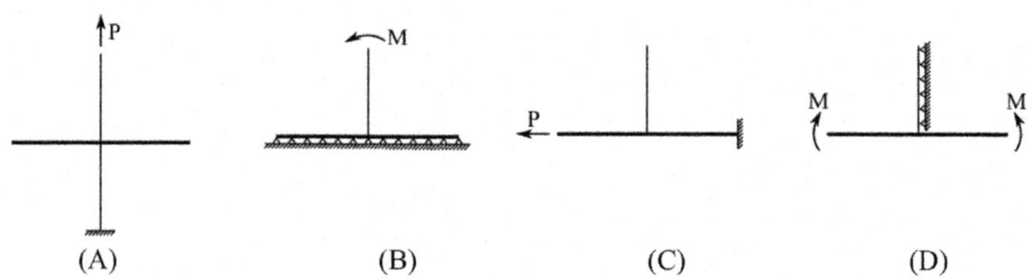

Figure 5.37 Load cases and boundary conditions for T-type and X-type joints. (A) Axial force in brace. (B) In-plane bending moment in brace. (C) Axial force in brace. (D) In-plane bending moment in brace.

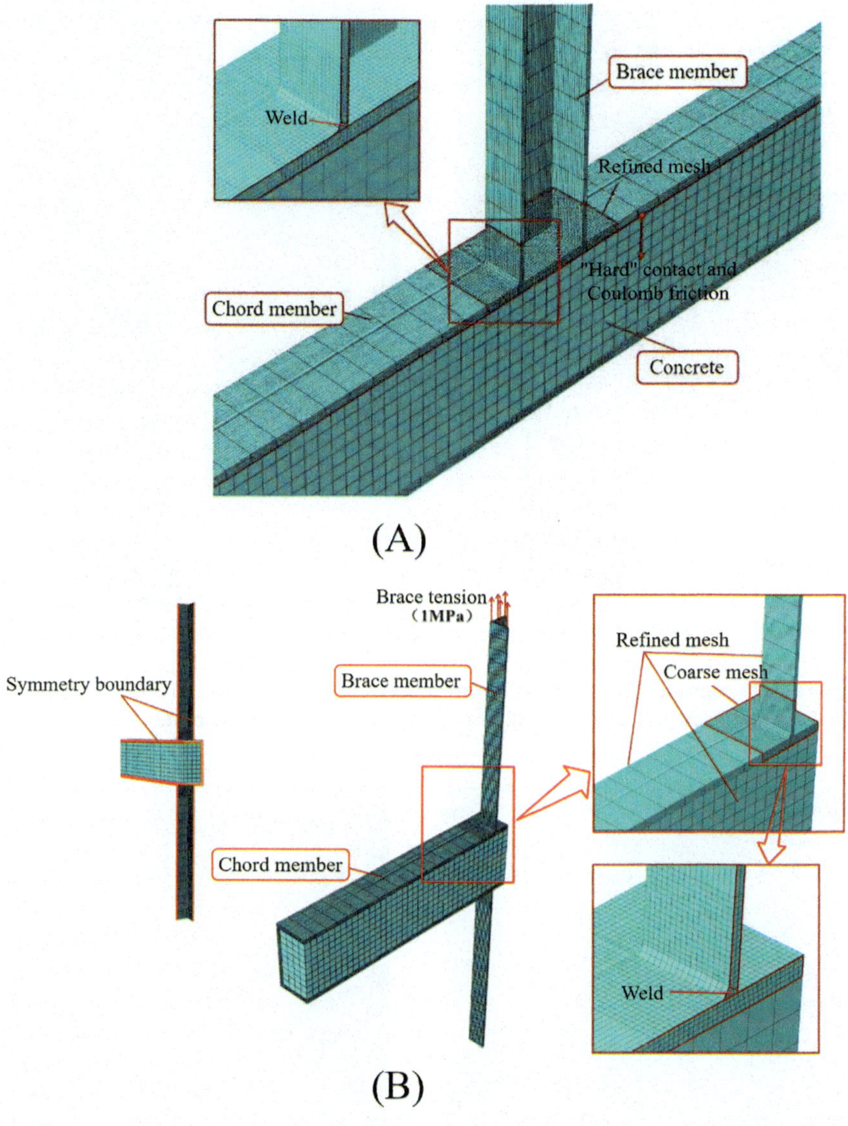

Figure 5.38 FEM of CFRST T-type and X-type joints. (A) T-type joint. (B) X-type joint.

To analyze the effect of concrete filling in the chord, a comparison of deformation and stress results between a CFRST joint and a RST joint with identical parameters (β=0.7, 2γ=20, τ=1) is presented, as shown in Figs. 5.39 and 5.40.

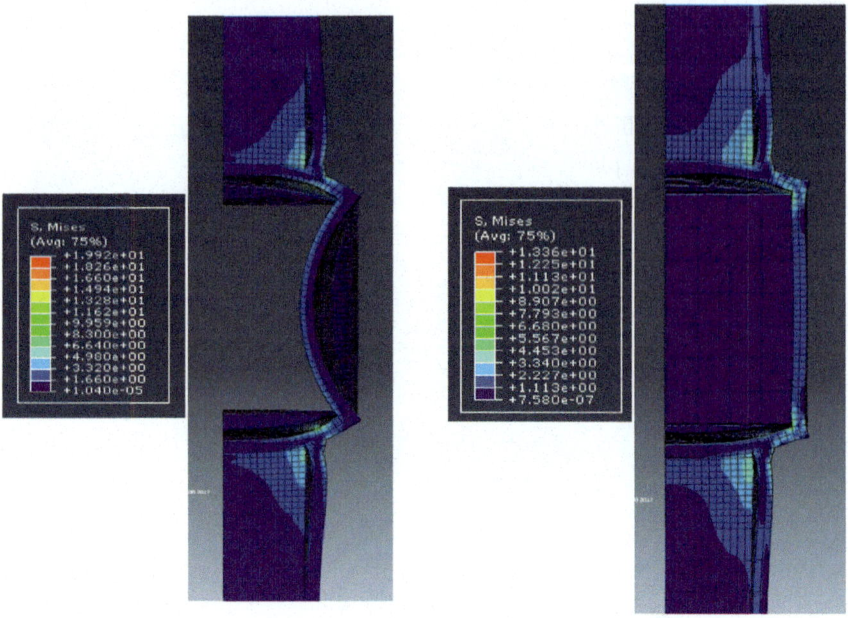

(A) (B)

Figure 5.39 Comparison of joint deformation (β=0. 7, 2γ=20, τ=1). (A) Rectangular steel tubular joint. (B) Concrete-filled rectangular steel tubular joint.

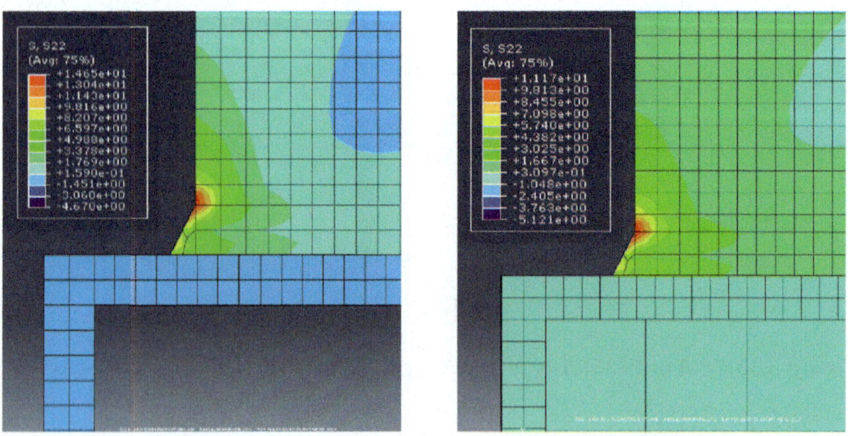

(A) (B)

Figure 5.40 Comparison of stress distribution in the brace direction (β=0.7, γ=20, τ=1). (A) Rectangular steel tubular joint (maximum stress: 14.7 MPa). (B) Concrete-filled rectangular steel tubular joint (maximum stress: 11.2 MPa).

Fig. 5.39 shows that under brace tension, the top and bottom plates of the hollow chord experience tensile forces leading to outward bulging deformation, while the side walls exhibit inward buckling deformation. After the chord is filled with concrete, the outward bulging deformation of the top and bottom plates is reduced. Furthermore, due to the supporting effect of the concrete, the side walls of the chord do not undergo inward buckling deformation.

Fig. 5.40 shows that the maximum hot-spot stress in both the RST joint and the CFRST joint occurs at the weld toe on the brace surface, with values of 14.7 MPa and 11.2 MPa, respectively. The presence of the infill concrete effectively restricts the deformation of the joint, thereby reducing the degree of stress concentration at the intersection line (Fig. 5.41).

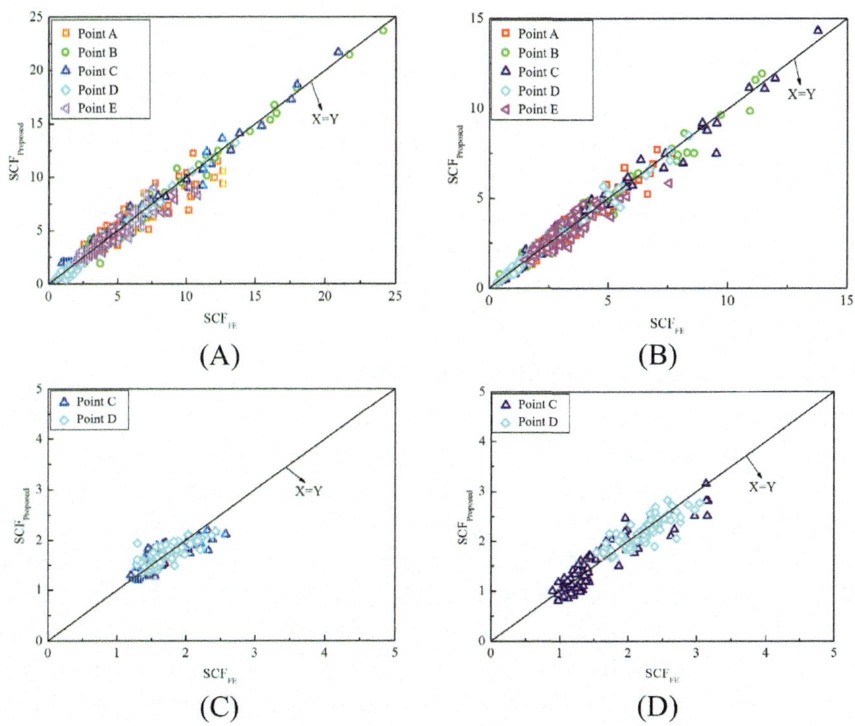

Figure 5.41 Comparison between FE calculation results and fitting formula calculation results. (A) Axial force in brace. (B) In-plane bending moment in brace. (C) Axial force in chord. (D) n-plane bending moment in chord.

5.4.3 Calculation formulas for SCF parameters of CFRST T-type and X-type joints

5.4.3.1 Fitting of SCF parameter calculation formulas

The SCF values at points A to E were calculated for the 320 joint models. Multiple regression analysis was employed to derive the calculation formulas for the SCF parameters of CFRST T-type and X-type joints under different load cases. van Wingerde et al. and Tong et al. proposed forms for the SCF expressions of hollow section joints. Based on the FE calculation results and considering these existing expression forms, a suitable expression was determined as follows:

For brace loading:

$$SCF = (c_1 + c_2\beta + c_3\beta^2 + c_4 \cdot 2\gamma)(2\gamma)^{(c_5 + c_6\beta + c_7\beta^2)}\tau^{(c_8 + c_9\beta)} \quad (5.18)$$

For chord loading:

$$SCF = c_1(2\gamma)^{c_5\beta}\tau^{c_8} \quad (5.19)$$

Where $c_1 \sim c_9$ are constant value, obtained through regression analysis. For points A, B, C, and E, c_4=0, for point D, c_4 may not be 0.

Through multiple regression analysis, the SCF parameter formula of rectangular steel tube concrete T-type and X-type nodes is fitted and obtained, which is summarized in Table 5.15.

5.4.3.2 Validation of the fitted formulas

The formulas in Table 5.15 were used to calculate the SCF at points A to E on the surfaces of the 80 specimens listed in Table 5.14. These calculated results (SCFproposed) were compared with the finite element results (SCFFE) from Table 5.14, as shown in. The comparison reveals the following: For brace axial force, the mean ratio (μ) of SCFproposed to SCFFE is 1.016, with a standard deviation (σ) of 0.177 and a coefficient of variation (σ/μ) of 0.174. For brace in-plane bending moment, the mean ratio (μ) is 1.006, with a standard deviation (σ) of 0.129 and a coefficient of variation (σ/μ) of 0.128. For chord axial force, the mean ratio (μ) is 1.010, with a standard deviation (σ) of 0.106 and a coefficient of variation (σ/μ) of 0.105. For chord in-plane bending moment, the mean ratio (μ) is 1.000, with a standard deviation (σ) of 0.120 and a coefficient of variation (σ/μ) of 0.120. These results indicate that the calculation formulas obtained through regression analysis can accurately calculate the SCF at points A to E.

Table 5.15 SCF parameter formulas for CFRST T-type and X-type joints.

Location	SCF formular
Load Case (I): Branch Axial Force (AB)	
Chord(B&C&D)	$SCF_B = (0.131 - 0.095\beta - 0.052\beta^2)\cdot(2\gamma)^{(1.512+0.734\beta-0.343\beta^2)}\cdot\tau^{0.927-0.128\beta}$
	$SCF_C = (-0.069 + 0.537\beta - 0.526\beta^2 + 0.0005\cdot 2\gamma)\cdot(2\gamma)^{(2.205-1.566\beta+1.161\beta^2)}\cdot\tau^{0.774+0.047\beta}$
	$SCF_D = (0.108 - 0.241\beta + 0.150\beta^2)\cdot(2\gamma)^{(0.934+3.324\beta-2.651\beta^2)}\cdot\tau^{0.918-0.314\beta}$
Brace (A&E)	$SCF_A = (-0.870 + 3.533\beta - 2.585\beta^2)\cdot(2\gamma)^{(2.372-3.380\beta+2.143\beta^2)}\cdot\tau^{-0.002+0.374\beta}$
	$SCF_E = (-0.143 + 0.429\beta + 0.224\beta^2)\cdot(2\gamma)^{(2.276-2.205\beta+0.547\beta^2)}\cdot\tau^{-0.297+0.425\beta}$
Load case (II): In-plane bending moment in brac(IB)	
Chord(B&C&D)	$SCF_B = (-0.109 + 0.437\beta - 0.357\beta^2)\cdot(2\gamma)^{(3.750-5.933\beta+4.799\beta^2)}\cdot\tau^{0.740-0.139\beta}$
	$SCF_C = (-0.436 - 1.121\beta + 1.617\beta^2 + 0.097\cdot 2\gamma)\cdot(2\gamma)^{(-0.431+3.976\beta-3.708\beta^2)}\cdot\tau^{0.803-0.177\beta}$
	$SCF_D = (0.133 - 0.355\beta + 0.281\beta^2)\cdot(2\gamma)^{(0.008+6.305\beta-5.407\beta^2)}\cdot\tau^{0.825-0.394\beta}$
Brace(A&E)	$SCF_A = (0.635 - 1.849\beta + 1.495\beta^2)\cdot(2\gamma)^{(-1.349+8.650\beta-6.739\beta^2)}\cdot\tau^{-0.304+0.291\beta}$
	$SCF_E = (0.196 - 0.998\beta + 1.689\beta^2)\cdot(2\gamma)^{(1.629-0.815\beta-0.498\beta^2)}\cdot\tau^{-0.274+0.113\beta}$

(Continued)

Table 5.15 (Continued)

Location	SCF formular
Load case (III): Axial force in chord(AC)	
Chord(B&C&D)	$SCF_B = 0$
	$SCF_C = 1.006\cdot(2\gamma)^{0.243\beta}\cdot\tau^{0.047}$
	$SCF_D = 1.416\cdot(2\gamma)^{0.135\beta}\cdot\tau^{0.133}$
Brace(A&E)	$SCF_A = SCF_E = 0$
Load case (IV): In-plane bending moment in chord(IC)	
Chord(B&C&D)	$SCF_B = 0$
	$SCF_C = 0.491\cdot(2\gamma)^{0.508\beta}\cdot\tau^{-0.164}$
	$SCF_D = 1.343\cdot(2\gamma)^{0.219\beta}\cdot\tau^{-0.030}$
Brace(A&E)	$SCF_A = SCF_E = 0$
Scope of application: $0.35 \le \beta \le 1.0$; $12.5 \le 2\gamma \le 25.0$; $0.25 \le \tau \le 1.0$	

5.4.3.3 SCF graphology

Based on the SCF calculation formulas provided in Table 5.15, the relationship curves between the SCF at points A to E for CFRST T-type and X-type joints and the parameters β, 2γ, and τ are plotted in Figs. 5.42–5.45, respectively. These charts effectively illustrate the variation

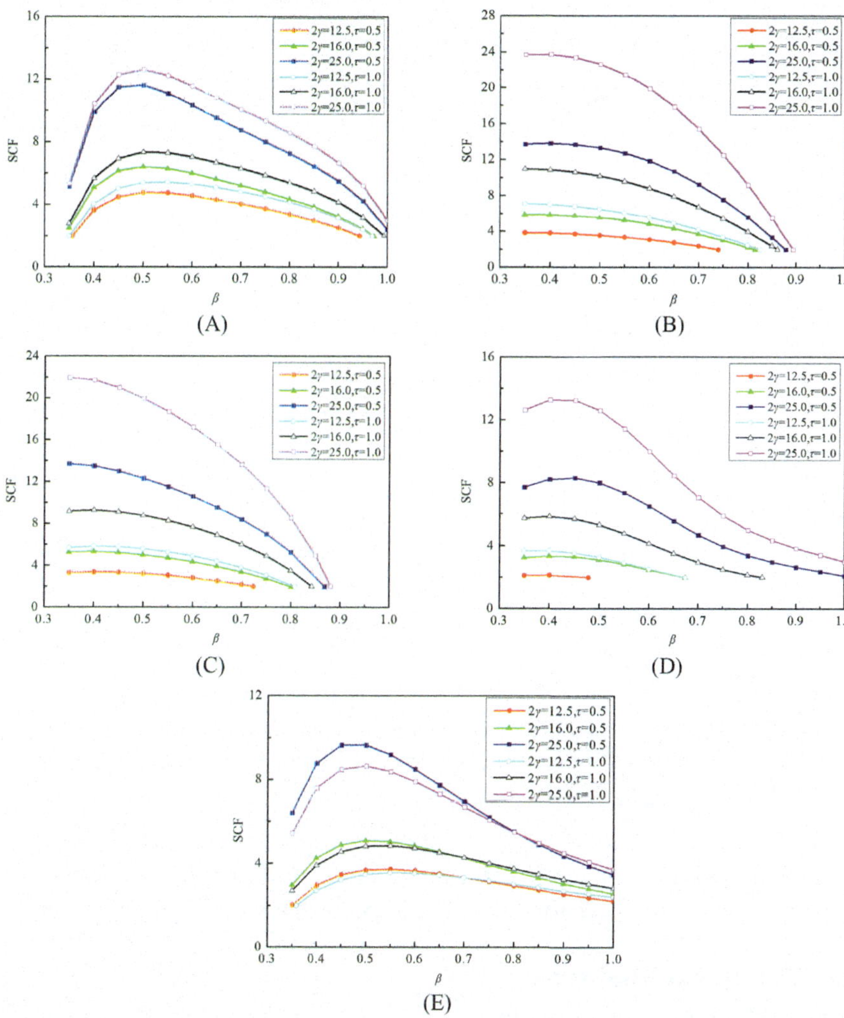

Figure 5.42 SCF chart method under brace axial force load case. (A) Point A. (B) Point B. (C) Point C. (D) Point D. (E) Point E.

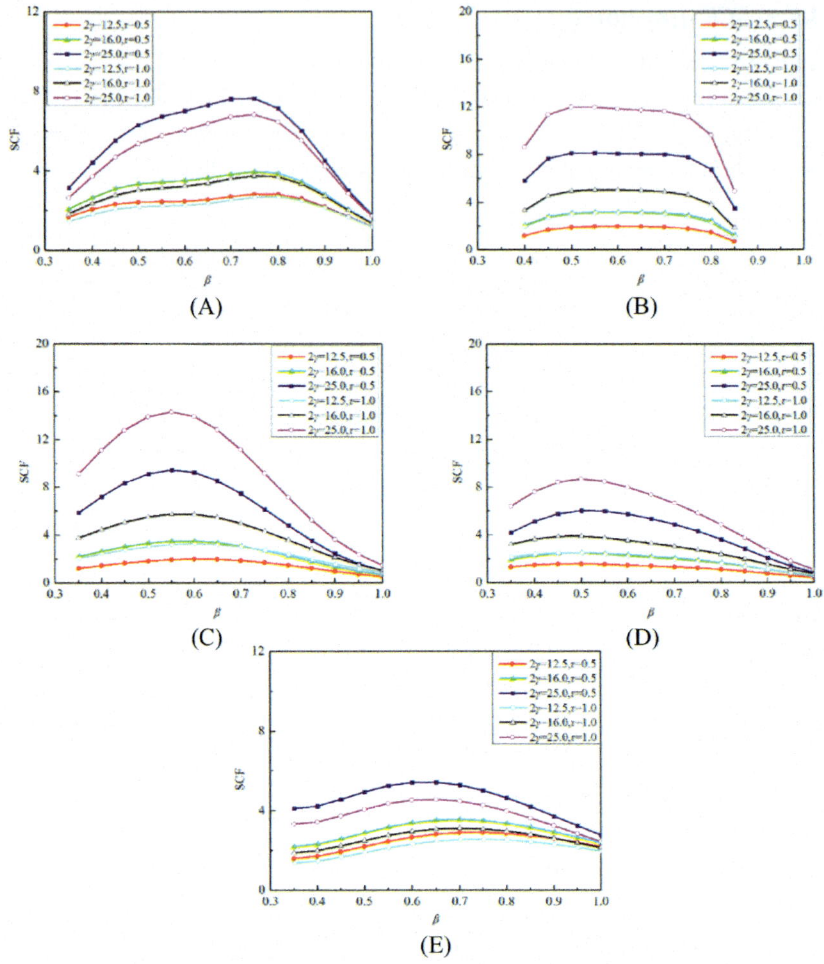

Figure 5.43 SCF chart method under brace in-plane bending moment load case. (A) Point A. (B) Point B. (C) Point C. (D) Point D. (E) Point E.

patterns of SCF with geometric parameters. Additionally, designers can directly refer to these charts for value lookup during SCF calculations.

5.4.3.3.1 Brace axial force

Fig. 5.42 shows the variation of SCF with dimensionless geometric parameters under the brace axial force load case. The following conclusions can be drawn from the Figure:

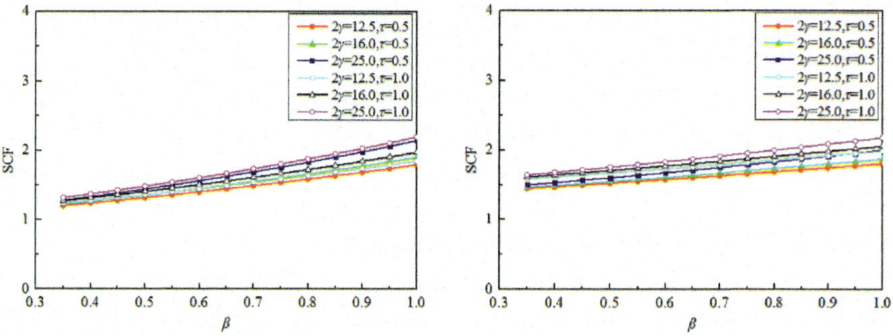

(A) (B)

Figure 5.44 SCF chart method under chord axial force load case. (A) Point C. (B) Point D.

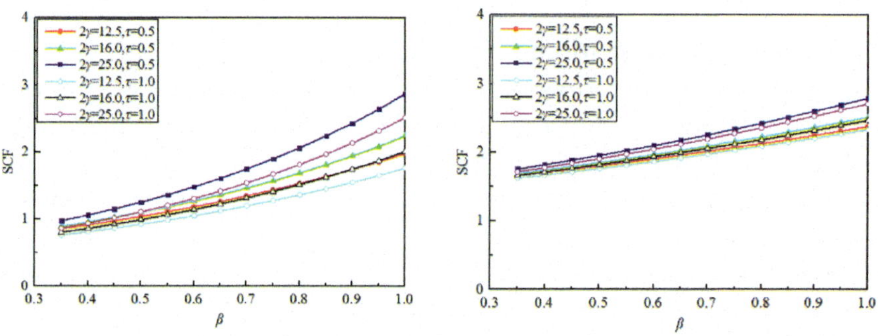

(A) (B)

Figure 5.45 SCF chart method under chord in-plane bending moment load case. (A) Point C. (B) Point D.

a. Influence of β: For point A on the brace, point E, and point D on the chord, the SCF changes parabolically with β, first increasing and then decreasing, reaching its maximum value between β=0.4 and 0.5. For points B and C on the chord, the SCF decreases as β increases, which is related to the magnitude of the transverse bending stress in the chord's top plate.

b. Influence of 2γ: A larger value of 2γ indicates a thinner steel tube wall, leading to greater bending deformation and a correspondingly larger SCF. When 2γ is large (e.g., 2γ=25), the maximum SCF occurs at point B or C. When 2γ is small (e.g., $2\gamma < 16$), the SCF values at various points on the brace and chord are generally similar and should be determined through calculation.

c. The influence of τ: The SCF is positively correlated with τ, a larger τ value corresponds to a larger SCF value. Point E on the brace, however, shows the opposite trend.

5.4.3.3.2 Brace in-plane bending moment

Fig. 5.43 shows the variation of SCF with dimensionless geometric parameters under the brace in-plane bending moment load case. As can be seen from the Figure, the trends of SCF variation with β, 2γ, and τ are basically consistent with those under the brace axial force load case. The differences are that for points B and C on the chord, the SCF also changes parabolically with β, and for point A on the brace, the SCF shows a negative correlation with τ.

5.4.3.3.3 Chord axial force

Fig. 5.44 shows the variation of SCF with dimensionless geometric parameters under the chord axial force load case. The following conclusions can be drawn from the Figure.

a. Influence of β: SCF increases with β values, maintaining an approximate linear relationship.

b. Influence of 2γ and τ: Both SCF and $2\gamma/\tau$ demonstrate positive correlations.

c. Minimal impact of the three dimensionless geometric parameters on SCF values.

5.4.3.3.4 Chord in-plane bending moment

Fig. 5.45 shows the variation of SCF with dimensionless geometric parameters under the chord in-plane bending moment load case. As can be seen from the Figure, the trends of SCF variation with β and 2γ are basically consistent with those under the chord axial force load case. However, the trend with τ is opposite to that under chord axial force, showing a negative correlation. Furthermore, the degree of influence of the three dimensionless geometric parameters on the SCF is greater than that under brace axial force.

5.4.3.4 Comparison of SCF_{max} between RST joint and RST concrete joint

The CIDECT code provides formulas for calculating the SCF parameters of T-type and X-type RST joints. To quantitatively analyze the change in the SCF at the joint weld toe after the chord is filled with concrete, calculations were performed for the 80 joints listed in Table 5.14 using both the SCF formulas from the CIDECT code and the formulas in Table 5.15. A comparison of the calculated SCFs for the RST joints and the CFRST joints is shown in Fig. 5.46. In the Figure, for each load condition, the SCF_{max} values of both the brace and the chord for the two types of joints are compared.

Statistical analysis of the results in Fig. 5.46 reveals the following: Under the brace axial load condition, the average ratios of $SCF_{proposed}$ to SCF_{CIDECT} for the brace and chord surfaces of the CFRST joints are 0.875 and 0.734, respectively. Under the brace in-plane bending moment condition, the average ratios are 0.879 and 0.692 for the brace and chord surfaces, respectively. Under the chord axial load condition, the average

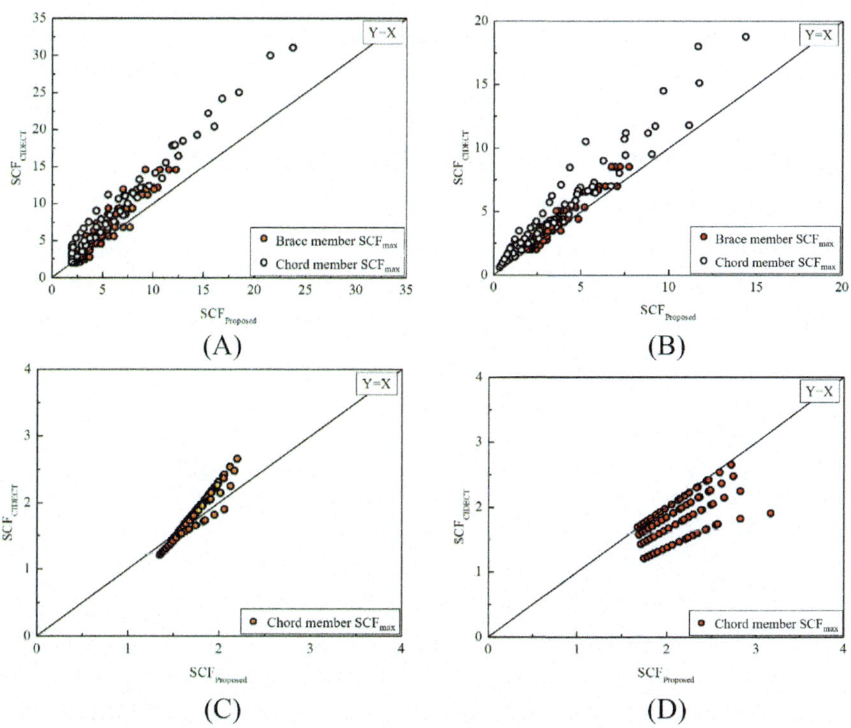

Figure 5.46 Comparison of SCFmax between RST and CFRST T-type and X-type Joints. (A) Brace axial force. (B) In-plane bending moment in brace. (C) Chord axial force. (D) In-plane bending moment in chord.

ratio for the chord surface is 0.966. Under the chord in-plane bending moment condition, the average ratio for the chord surface is 1.196.

In summary, after filling the chord of the RST joint with concrete, the SCF changes as follows: under brace axial load, the SCF_{max} of the brace and chord decreases by 12.5% and 26.6%, respectively, under brace in-plane bending moment, the SCF_{max} decreases by 12.1% and 30.8%, respectively. Under chord axial load, the SCF_{max} of the chord decreases by 3.4%. Under chord in-plane bending moment, the SCF_{max} of the chord increases by 19.6%.

It is noteworthy that the magnitude of SCF generated under brace loading (axial load and in-plane bending moment) is significantly greater than that under chord load (chord axial load and in-plane bending moment). Although the SCF_{max} of CFRST joints shows an increasing trend compared to RST joints under chord in-plane bending moment, the SCF_{max} on both the brace and chord surfaces decreases significantly under brace loading and chord axial loading. This indicates that filling the chord with concrete can effectively reduce the degree of stress concentration.

5.4.4 Parametric analysis of CFRST K-type joints

5.4.4.1 Parameter design

The typical configuration of a CFRST K-type joint is illustrated in Fig. 5.47. In the CIDECT code, the applicable range for the SCF parameter calculation formulas of RST K-type joints is: $0.35 \le \beta \le 1.0$,

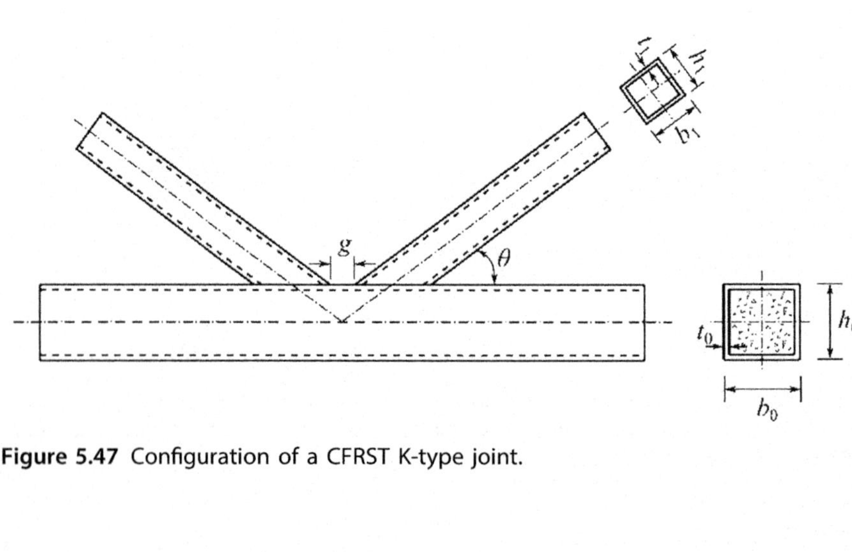

Figure 5.47 Configuration of a CFRST K-type joint.

$10 \le 2\gamma \le 35$, $0.25 \le \tau \le 1.0$, $30° \le \theta \le 60°$. The same range of dimensionless geometric parameters was selected for the parametric design, as detailed in Table 5.16.

Meanwhile, the following assumptions were made during the parametric analysis of the K-type joints:

1. Neglecting the geometric eccentricity of the joint, the centerlines of the two braces intersect at the centerline of the chord.
2. The two braces have the same length.
3. The angles between the two braces and the chord are equal ($\theta_1=\theta_2$).
4. Neglecting the influence of the height-to-width ratio (h/b) of the steel tube on the SCF, assuming both the chord and braces have square cross-sections.
5. In bridge engineering, gap joints ($g > 0$) are typically used, therefore, overlap joints ($g < 0$) were not considered in this study.

The ratio of the gap g between the two braces of a K-type joint to the chord thickness t_0 is defined as the dimensionless parameter g′. This value can be calculated once the joint's geometric dimensions are specified. The geometric dimensions of the joint are based on the benchmark dimensions of the Huangyan Bridge joint: the chord side length is set as $b_0 = h_0 = 400$ mm, the chord thickness ranges from t_0=11 to 40 mm, the brace side length ranges from $b_1 = h_1 = 160$ to 400 mm, and the brace thickness ranges from $t_1 = 3$ to 40 mm.

According to the parametric design in Table 5.16, 360 models should theoretically be established for each load case. However, for certain parameter combinations resulting in overlapping joints ($g < 0$), these should be disregarded. For example, when $\beta = 0.55$, $2\gamma = 30$, $\tau = 1$, and $\theta = 60°$, $g = -23$ mm. Consequently, 190 valid models are established for each load case. With three load cases analyzed, a total of 570 models are ultimately created.

Table 5.16 Parametric design of K-type joints.

β	2γ	τ	θ
0.40	10	0.25	30°
0.55	15	0.50	45°
0.70	20	0.75	60°
0.82	25	1.00	/
1.00	30	/	/
/	35	/	/

5.4.4.2 Load cases and boundary conditions

The parametric analysis of K-type joints should consider three load cases: balanced axial loading (tension and compression) in the braces, axial load in the chord, and bending moment in the chord, as illustrated in Fig. 5.48. The consideration of these load cases is consistent with the provisions for RST K-type joints in the CIDECT code. The established finite element model is shown in Fig. 5.49.

5.4.5 Calculation formulas for SCF parameters of CFRST K-type joints

5.4.5.1 Fitting and validation of SCF parameter calculation formulas

After calculating the SCF values of the joints, multiple regression analysis was employed to derive the calculation formulas for the SCF parameters of

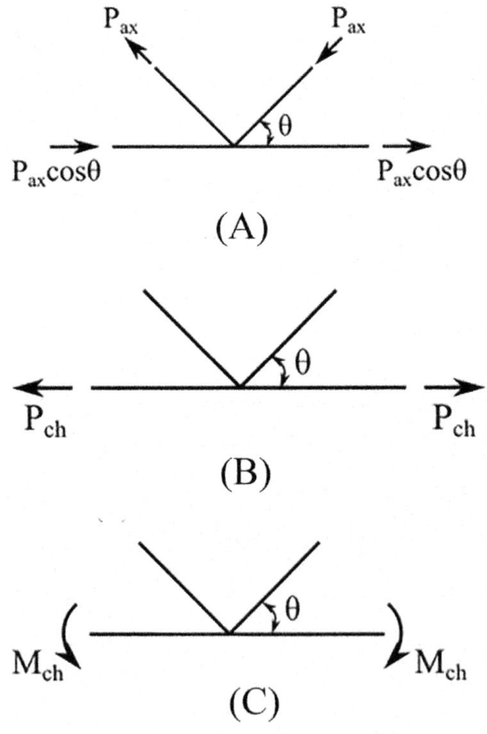

Figure 5.48 Load cases for K-type joints. (A) Self-equilibrating brace load. (B) Chord axial force. (C) Bending moment in chord.

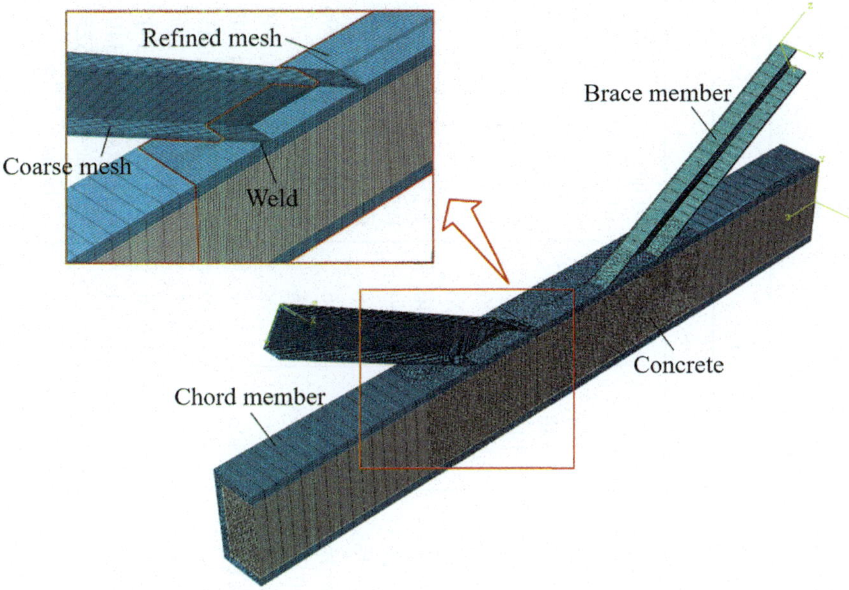

Figure 5.49 FEM of CFRST K-type Joints.

CFRST K-type joints under different load cases. With reference to the SCF parameter calculation formulas for RST K-type joints proposed by van Wingerde et al., the following formula was adopted:

$$\mathrm{SCF} = (c_1 + c_2\beta + c_3\beta^2)\cdot(2\gamma)^{c_4}\cdot\tau^{c_5}\cdot(g')^{c_6}\cdot(\sin\theta)^{c_7} \tag{5.20}$$

Through multiple regression analysis, the formulas for the SCF parameters of CFRST K-type joints were fitted and are summarized in Table 5.17.

To validate the reliability of the fitted formulas, the SCF for all 570 models were calculated using the formulas provided in Table 5.17 and compared with the finite element results, as shown in Fig. 5.50. It is noteworthy that under different basic load cases, compressive stresses occur at the following hotspot locations: at the weld heel side (D and G points) of the tension brace and at both the weld heel (A, B, E, F points) and weld toe (A, B, D, E, F, G points) sides of the compression brace under balanced brace tension-compression loading, at the weld heel side (point B) of the chord under chord axial loading. Consequently, the calculated SCF at these

Table 5.17 Formulas for SCF parameters of CFRST K-type Joints.

Location	Coefficient						
	c_1	c_2	c_3	c_4	c_5	c_6	c_7
Load case (I): self-equilibrating brace load							
Point A at tensile brace weld root side	-0.113	0.795	-0.640	1.175	-0.405	-0.046	1.315
Point B at tensile brace weld root side	0.013	0.169	-0.215	1.857	0.629	-0.031	1.804
Point C at tensile brace weld root side	0.027	-0.014	-0.007	2.144	0.375	0.063	2.607
Point D at tensile brace weld root side	-0.626	1.908	-1.575	1.338	0.600	-0.110	1.869
Point E at tensile brace weld root side	-0.033	0.875	-0.605	1.145	-0.123	0.001	0.701
Point A at tensile brace weld toe side	0.064	1.242	-1.186	0.665	-0.064	0.100	1.264
Point B at tensile brace weld toe side	0.016	0.682	-0.647	1.050	0.796	0.685	2.638
Point C at tensile brace weld toe side	-0.028	1.054	-0.676	0.729	0.440	1.194	4.437
Point D at tensile brace weld toe side	0.217	0.621	-0.703	0.915	0.590	0.500	2.708
Point E at tensile brace weld toe side	0.083	1.741	-1.785	0.802	0.158	0.117	1.701
Point A at compressive brace weld root side	-0.806	-0.097	-0.166	0.088	0.056	-0.00004	-0.096
Point B at compressive brace weld root side	-7.810	0.023	-0.254	-0.268	0.675	-0.0001	1.950
Point C at compressive brace weld root side	0.075	0.044	0.018	0.596	0.358	-0.00005	-0.349
Point D at compressive brace weld root side	0.140	-0.039	0.147	0.505	0.567	-0.00005	-1.347
Point E at compressive brace weld root side	-1.377	-0.268	-0.459	0.193	0.240	-0.00005	0.085
Point A at compressive brace weld toe side	-1.756	-0.003	-0.059	-0.050	0.105	-0.00006	1.109
Point B at compressive brace weld toe side	-0.331	-0.195	0.013	0.354	0.874	-0.00005	1.241
Point C at compressive brace weld toe side	0.051	0.014	-0.070	1.136	0.134	-0.00004	0.340
Point D at compressive brace weld toe side	-0.371	0.304	0.004	1.228	0.628	-0.00004	2.525
Point E at compressive brace weld toe side	-1.340	0.188	-0.081	0.494	0.253	-0.00004	2.642

Load case (II): axial force in chord							
Brace (points A,E)	SCF=0						
Point B at chord weld root side	-0.695	-0.914	0.682	0.003	-0.276	0.006	0.018
Point C at chord weld root side	0.511	-0.118	-0.046	0.362	-0.473	0.007	0.130
Point D at chord weld root side	1.376	0.527	0.249	0.145	-0.238	0.071	0.329
chord weld toe side(points B, C & D)	SCF=0						
Load case (III): bending moment in chord							
Brace (points A,E)	SCF=0						
Point B at chord weld root side	SCF=0						
Point C at chord weld root side	0.633	0.012	-0.354	0.179	-0.503	-0.046	-0.518
Point D at chord weld root side	1.524	-0.605	-0.102	0.219	-0.127	-0.049	-0.472
chord weld toe side (points B, C & D)	SCF=0						
Scope of application							
$0.35 \le \beta \le 1.0$; $10 \le 2\gamma \le 35$; $0.25 \le \tau \le 1.0$; $30° \le \theta \le 60°$; $2\tau \le g'$; $e/h_0=0$							

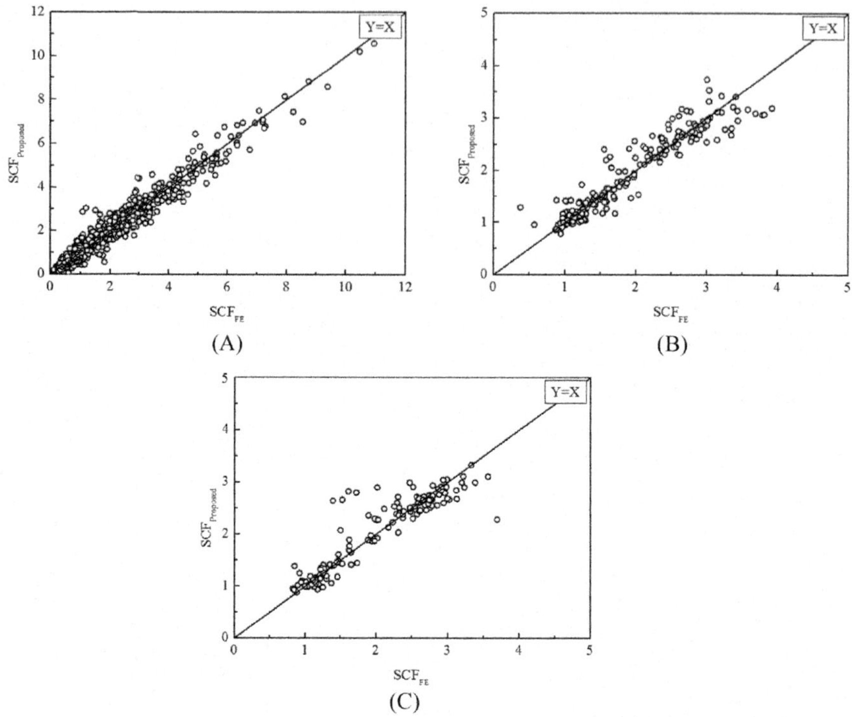

Figure 5.50 Comparison between FE and calculated results from fitted formulas. (A) Self-equilibrating brace load. (B) Brace axial force. (C) Bending moment in chord.

locations are negative values, and their absolute values were taken for comparison in Fig. 5.50.

As can be seen from the comparison in Fig. 5.50, for balanced brace tension-compression loading, the mean value (μ) of the ratio between the calculated $SCF_{proposed}$ from the fitted formulas and the finite element result SCF_{FE} is 1.014, with a standard deviation (σ) of 0.148 and a coefficient of variation (σ/μ) of 0.146. For chord axial loading, the mean ratio (μ) is 1.034, with a standard deviation (σ) of 0.206 and a coefficient of variation (σ/μ) of 0.200. For chord bending moment, the mean ratio (μ) is 1.022, with a standard deviation (σ) of 0.115 and a coefficient of variation (σ/μ) of 0.113. These results indicate that the calculated values from the fitted SCF parameter formulas are reliable.

5.4.5.2 Fitting and validation of SCF_{max} parameter calculation formula

To cover all hotspot locations in the K-type joint, a total of 26 SCF parameter calculation formulas are provided in Table 5.17. Using these formulas for calculation and comparing the results can determine the location of SCF_{max}. However, the computational burden is substantial for designers. Therefore, the number of SCF parameter formulas was simplified, and the fitted SCFmax parameter formulas are listed in Table 5.18.

Similarly, to validate the reliability of the fitted SCF_{max} formulas, the calculated results $SCF_{proposed}$ from the fitted formulas were compared with the finite element results SCF_{FE}, as shown in Fig. 5.51. The comparison reveals that under balanced brace tension-compression loading, the mean value (μ) of the ratio $SCF_{proposed}/SCF_{FE}$ is 1.010 for the chord surface, with a standard deviation (σ) of 0.243 and a coefficient of variation (σ/μ) of 0.240, for the brace surface, the mean ratio (μ) is 1.017, with a standard deviation (σ) of 0.179 and a coefficient of variation (σ/μ) of 0.175. Under chord axial loading, the mean ratio (μ) is 1.007, with a standard deviation (σ) of 0.256 and a coefficient of variation (σ/μ) of 0.254. Under chord in-plane bending moment, the mean ratio (μ) is 1.014, with a standard deviation (σ) of 0.236 and a coefficient of variation (σ/μ) of 0.233. These results demonstrate that the fitted SCF_{max} calculation formulas are accurate and reliable.

5.4.5.3 Comparison of SCF_{max} between RST joints and CFRST joints

The CIDECT code provides formulas for calculating the SCFmax parameters of RST K-type joints under three basic load cases. Using the proposed formulas for SCFmax parameters of CFRST K-type joints listed in Table 5.18, calculations were performed and compared with the results obtained from the CIDECT code formulas. This comparison aims to investigate the influence of concrete filling in the chord on the SCF of K-type joints, as illustrated in Fig. 5.52.

As can be seen from Fig. 5.52, under balanced brace tension-compression loading, the mean ratios of $SCF_{proposed}$ (calculated for CFRST joints) to SCF_{CIDECT} (calculated for RST joints using CIDECT formulas) are 0.798 and 0.816 for the brace and chord surfaces, respectively. This indicates that because of the infill concrete, the SCFmax on the brace and chord surfaces decreases by 20.2% and 18.4%, respectively. Under chord axial load and chord bending moment, the mean ratios of $SCF_{proposed}$ to

Table 5.18 Formulas for SCF_{max} parameters of CFRST K-type joint.

Location	**Maximum SCF formula**
Load case (I): self-equilibrating brace load	
Chord	$SCF_{max} = (0.130 - 0.211\beta + 0.099\beta^2)\cdot(2\gamma)^{1.730}\cdot\tau^{0.577}\cdot(g')^{-0.075}\cdot(\sin\theta)^{0.810}$
Brace	$SCF_{max} = (-0.003 + 0.590\beta - 0.407\beta^2)\cdot(2\gamma)^{1.160}\cdot\tau^{-0.130}\cdot(g')^{-0.007}\cdot(\sin\theta)^{0.703}$
Load case (II): axial force in chord	
Chord	$SCF_{max} = (1.239 + 0.007\beta - 0.673\beta^2)\cdot(2\gamma)^{0.210}\cdot\tau^{-0.452}\cdot(g')^{-0.002}\cdot(\sin\theta)^{0.162}$
Brace	$SCF_{max} = 0$
Load case (III): bending moment in chord	
Chord	$SCF_{max} = (0.645 + 0.007\beta - 0.184\beta^2)\cdot(2\gamma)^{0.154}\cdot\tau^{-0.516}\cdot(g')^{-0.002}\cdot(\sin\theta)^{-0.299}$
Brace	$SCF_{max} = 0$
Scope of application: $0.35 \le \beta \le 1.0$; $10 \le 2\gamma \le 35$; $0.25 \le \tau \le 1.0$; $30° \le \theta \le 60°$; $2\tau \le g'$; $e/h_0=0$	

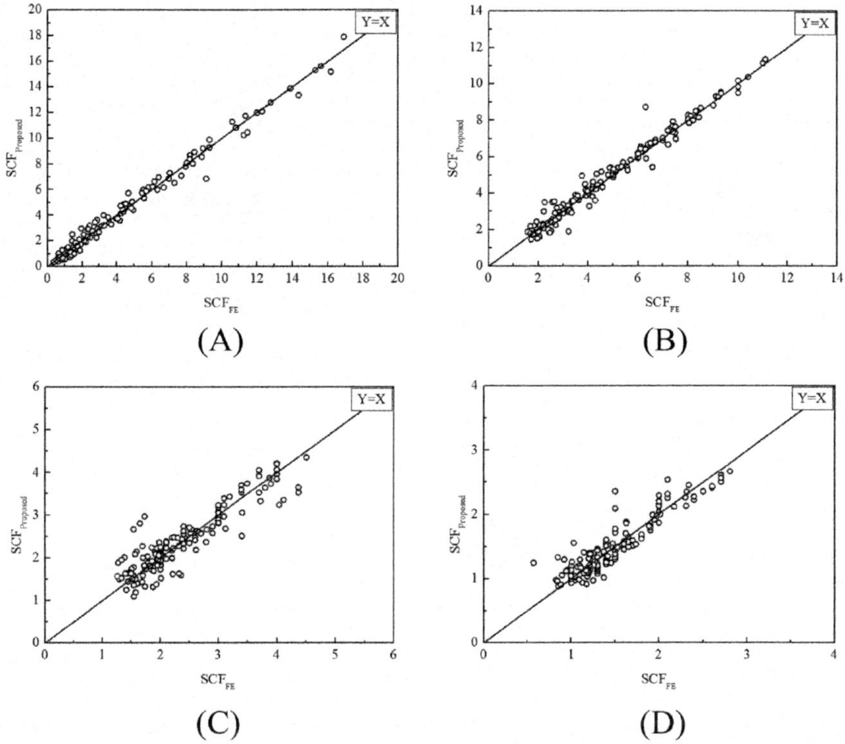

Figure 5.51 Comparison between FE and calculated results from fitted formulas for SCF_{max}. (A) Hot-spot in chord under self-equilibrating brace load. (B) Hot-spot in brace under self-equilibrating brace load. (C) Axial force in chord. (D) Bending moment in chord.

SCF_{CIDECT} are 1.204 and 0.766, respectively. The concrete filling increases the SCF_{max} by 20.4% under chord axial load but decreases it by 23.4% under chord bending moment.

However, it is noteworthy that under balanced brace tension-compression loading, the SCF on both the brace and chord surfaces is significantly higher than that under chord loading. Therefore, balanced brace tension-compression loading is more critical for joint fatigue, and filling the chord with concrete significantly alleviates stress concentration in the joint.

Simultaneously, these results also demonstrate that the current CIDECT code formulas cannot accurately predict the SCF_{max} of CFRST K-type joints. The calculations are overly conservative for SCF_{max} under balanced brace tension-compression loading and chord bending moment, but non-conservative (potentially unsafe) for SCF_{max} under chord axial load.

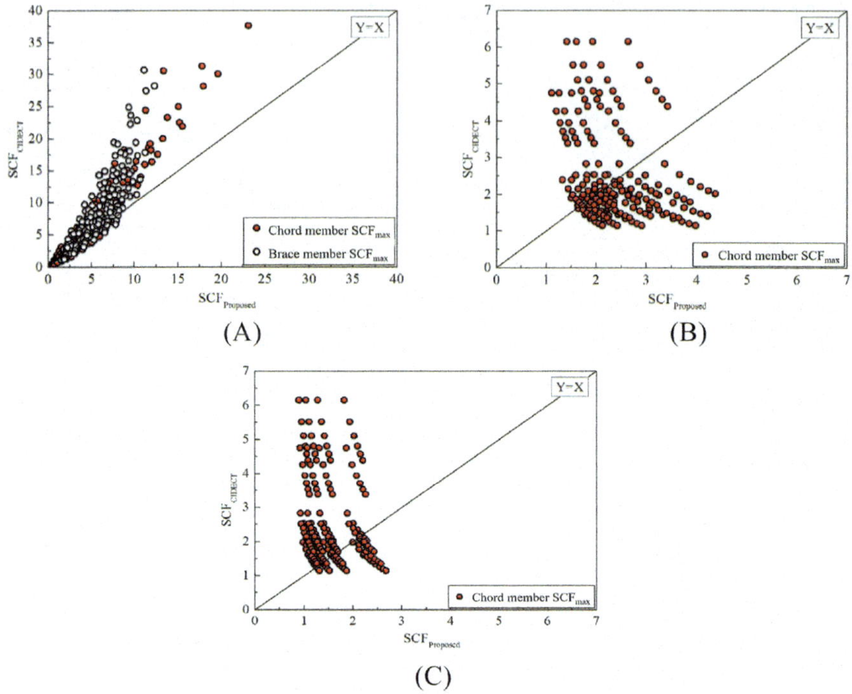

Figure 5.52 Comparison of SCF_{max} between RST and CFRST K-type joints. (A) Self-equilibrating brace load. (B) Axial force in chord. (C) Bending moment in chord.

5.5 Conclusion

This Chapter through fatigue tests on CFST joints, the crack initiation locations were identified. Using the propagation of cracks through the tube wall thickness as the fatigue failure criterion, a hot-spot stress range *S–N* curve was proposed. The characteristics of stress distribution in the hot-spot regions of the joints and the influence of the infill concrete were thoroughly understood. Based on this, calculation formulas for SCF parameters were developed, and a computational model for the fatigue effects in bridge joints using CFST was established. This led to the formation of a hot-spot stress method for joint fatigue assessment. This method simplifies the structural details used in the traditional nominal stress method by employing a single hot-spot stress range *S–N* curve to represent the fatigue strength of different joint types.

Extended Reading

Jiang L, Liu YJ, et al. Experimental and Numerical Analysis on the SCF of Concrete-filled Square Hollow Section Y-type joints Stiffened with Perfobond Ribs[J]. Advances in Structural Engineering 2019;23(5):869–83.

Jiang L, Liu YJ, Fam A. Stress Concentration Factors in Concrete-filled Square Hollow Section Joints with Perfobond Ribs[J]. Engineering Structures 2019;181:165–80.

Jiang L, Liu YJ, Fam A. Stress Concentration Factors in Joints of Square Hollow Section (SHS) Brace and Concrete-filled SHS Chord Under Axial Tension in Brace[J]. Thin-Walled Structures 2018;132:79–92.

Jiang L, Liu YJ, Fam A, et al. Fatigue Behavior of Integral Built-up Box Y-type joints Between Concrete-filled Chords with Perfobond Ribs and Hollow Braces[J]. Journal of Structural Engineering 2020;146(3).

Jiang L, Liu YJ, Fam A, et al. Fatigue Behaviour of Non-integral Y-type joint of CFRST Continuous Chord Stiffened with Perfobond Ribs[J]. Engineering Structures 2019;191:611–24.

Jiang L, Liu YJ, Fam A, et al. Stress Concentration Factor Parametric Formulae for CFRST K-type joints with Perfobond Ribs[J]. Journal of Constructional Steel Research 2019;160:579–97.

Jiang L, Liu YJ, Long X, et al. Experimental Study on Stress Concentration Characteristics of Joints in Concrete-filled Rectangular Steel Tubular Truss[J]. Journal of Building Structures 2020;1-18. Available from https://doi.org/10.14006/j.jzjgxb.2019.0879. -04-20.

Jiang L, Liu YJ, Wang KN. Development of Welded Tubular Joints and Comparison of Fatigue Behaviour[J]. Journal of Building Structures 2019;3:180–91.

Liu YJ, Jiang L, Wang KN. Review of Fatigue Behavior in Welded Tubular Joints[J]. Journal of Architecture and Civil Engineering 2017;34(5):1–20.

Liu YJ, Xiong ZH, Feng YC, et al. CFRST X-type joint with Perfobond Leister Rib Structural Performance Study: ultimate and fatigue experimental investigation[J]. Steel and Composite Structures 2017;24(4):455–65.

CHAPTER 6

Flexural performance of concrete-filled rectangular steel tube composite truss girders

6.1 Overview

When concrete-filled rectangular steel tube trusses are connected to concrete bridge decks, they form a composite truss girder system. The resulting composite action between the deck and the truss significantly enhances the flexural stiffness and load-bearing capacity of the girder. This chapter presents a model test study on the flexural performance of concrete-filled rectangular steel tube composite truss girders subjected to negative bending moments. Furthermore, full-scale load tests were conducted on a composite truss girder bridge under various loading conditions, including those producing maximum positive moment, maximum negative moment, and maximum shear force. The study investigates the overall mechanical behavior of the bridge, as well as the stress distribution patterns in the deck, truss members, and joints. Finally, a calculation method for the flexural capacity of the concrete-filled steel tube (CFST) composite truss girders is proposed, which accounts for the composite action between the deck and the truss.

6.2 Experiment on composite truss girders with concrete-filled rectangular steel tube in negative moment region

Deck-type composite truss girders with concrete-filled rectangular steel tube (CFRST) exhibit relatively weak mechanical performance in the negative moment region, where the concrete slab in tension is prone to cracking. To address this issue, a static loading test was conducted on two simply supported composite truss girders subjected to negative bending moments. A reverse concentrated load was applied at mid-span to simulate the support reaction of a continuous beam. The mechanical performance of the girders was subsequently analyzed.

Composite Truss Bridges With Concrete-filled Rectangular Steel Tube Members
DOI: https://doi.org/10.1016/B978-0-443-44011-3.00007-X

6.2.1 Test overview

6.2.1.1 Specimen design

Specimen CTB1 was a rectangular steel tube truss with a prestressed concrete slab and concrete infill in both the top and bottom chords. Perfobond Leister ribs (PBL) shear connectors were continuously welded along the internal surfaces of the chord tubes. The top and bottom chords had a width of 140 mm and a height of 120 mm, while the brace members measured 140 mm in width and 80 mm in height. The chord plates were 8 mm thick, and the brace member plates and PBL connector plates were all 6 mm thick. The shear connection between the concrete slab and the truss utilized shear studs with a localized stiffness-reduction feature, achieved by wrapping a 5-mm-thick layer of low-elastic-modulus foam adhesive around them. Each shear stud consisted of a shank (13 mm diameter, 48 mm length) and a cap (40 mm diameter, 12 mm length), connected via threads and arranged in clustered configurations. Double rows of these studs were installed on the top surface of the upper chord with longitudinal and transverse spacings of 70 mm. The prestressed concrete slab was constructed using the posttensioning method. It contained 6 Φs 15.24 prestressed steel strands and 18 10-mm-diameter ordinary threaded reinforcing bars, with 6-mm-diameter stirrups. The prestressed reinforcement ratio was 0.84%, and the ordinary reinforcement ratio was 1.41%. Each strand was tensed to 120 kN, imparting a precompressive stress of 5.3 MPa to the concrete slab, with a prestress introduction efficiency of 82%.

Specimen CTB2 shared identical design parameters with specimen CTB1, except for the type of shear studs and the form of the concrete slab (Fig. 6.1). CTB2 employed conventional shear studs with dimensions of Φ13 × 60, arranged in the same pattern as in CTB1. This specimen utilized an ordinary reinforced concrete slab, which contained thirty 10-mm-diameter ordinary threaded reinforcing bars, resulting in a reinforcement ratio of 2.36%.

6.2.1.2 Material properties

The measured material properties of concrete are listed in Table 6.1, while the material properties of steel and reinforcement bars are provided in Table 6.2.

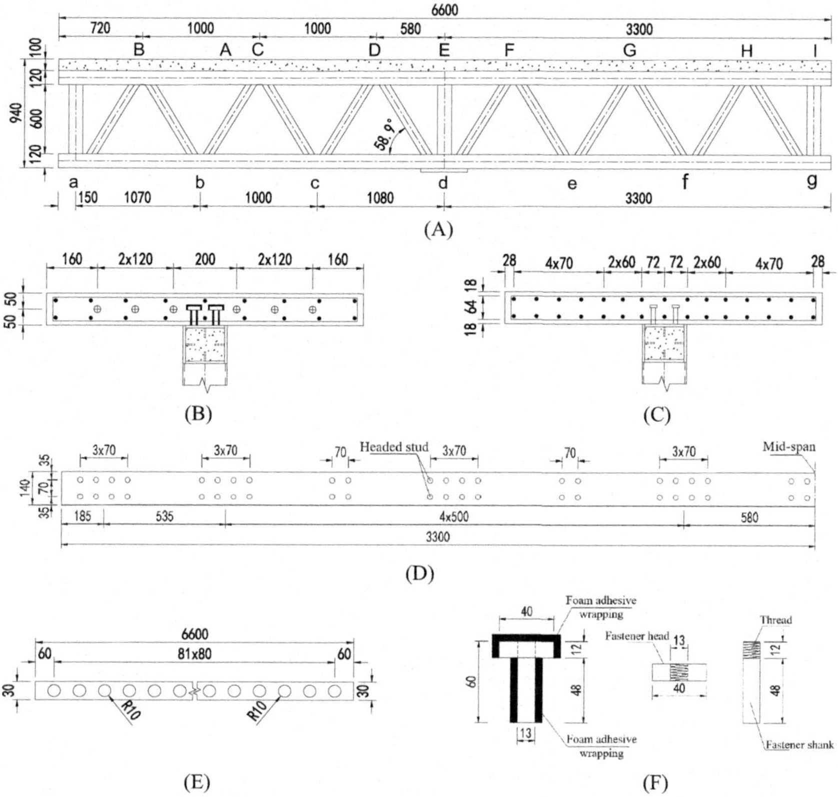

Figure 6.1 Specimen configuration and geometric dimensions. (A) General arrangement. (B) CTB1 prestressed concrete slab. (C) CTB2 conventional reinforced concrete slab. (D) Shear stud arrangement. (E) PBL shear key. (F) Partially debonded shear stud.

Table 6.1 Concrete material properties.

Name	Average cubic compressive strength f_{cu}/MPa	Average splitting tensile strength f_{ts}/MPa	Average elastic modulus E_c/MPa
Concrete slab	58.6	3.8	3.78×10^4
Concrete-filled chord	56.3	3.6	3.57×10^4

Table 6.2 Steel material properties.

Type	Plate thickness t/mm	Average yield strength f_y/MPa	Average tensile strength f_u/MPa	Average elastic modulus E_s/MPa	Elongation δ/%
Chord	8	355	525	2.06×10^5	28
Brace, PBL shear key	6	325	436	2.02×10^5	29
Longitudinal reinforcement	10	452	590	1.92×10^5	28
Stirrup	6	352	462	1.95×10^5	27

6.2.1.3 Loading scheme

The test setup is illustrated in Fig. 6.2. A reaction beam was used to restrain the ends of the concrete slab. Between the reaction beam and the concrete slab, one end employed a steel round bar support to simulate a movable hinge bearing, while the other end used a steel round bar support welded with reinforcing bars to simulate a fixed hinge bearing. To prevent lateral instability of the specimen before reaching its ultimate load-carrying capacity during testing, lateral restraint devices were installed near the mid-span region.

The test employed a single-point loading method using a jack at the mid-span applying force from bottom to top. Preloading was initially conducted to ensure proper contact between all components of the specimen. A load-controlled loading scheme was adopted during the test. Below 65% of the estimated ultimate load (before the specimen entered the elastoplastic stage), the load increment was 40 kN per step. Above 65% of the estimated ultimate load (after the specimen entered the elastoplastic stage), the load increment was reduced to 20 kN per step. Loading continued until no further load increase was possible, indicating that the specimen had reached its ultimate load.

6.2.2 Test results

6.2.2.1 Failure phenomena

For specimen CTB1 (Fig. 6.3), when the load reached 440 kN, cracks appeared in the concrete slab at the mid-span. At 600 kN, two penetrating cracks formed on the top surface of the mid-span slab. Upon loading to 840 kN, the cracks in the slab had extended over one panel length of the

Figure 6.2 Test setup.

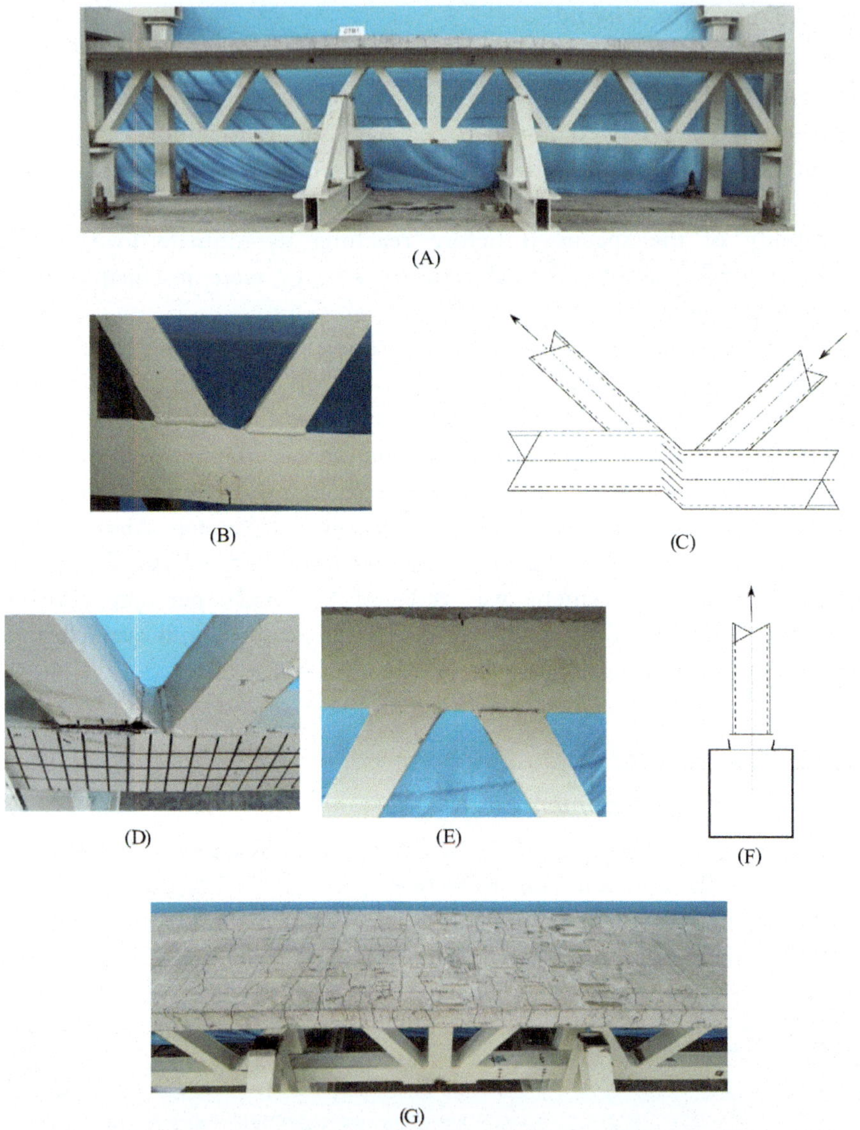

Figure 6.3 Failure mode of specimen CTB1. (A) Global bending of truss girder. (B) Shear failure in chord at joint E. (C) Shear failure mechanism in chord of tubular joint. (D) Weld tearing failure at joint C. (E) Weld tearing failure at joint G. (F) Weld tearing failure mechanism in tubular joint. (G) Cracking failure in concrete slab.

top chord, with a maximum crack width of 0.16 mm. At 933 kN, the maximum crack width reached 0.2 mm, indicating that the structure had reached its serviceability limit state. When the load increased to 1282 kN, the ordinary reinforcement in the mid-span slab began to yield, with the corresponding axial force in the slab being 1102 kN. At the ultimate load of 1302 kN, the steel tube of the bottom chord was near yield ($0.98f_y$, where f_y is the yield strength of the chord), and both the top chord and brace members had entered the elastoplastic stage. Cracking of the concrete slab and yielding of the members led to significant internal force redistribution and increased plastic deformation. Final failure modes included a chord shear failure at joint e, and weld tearing failures at the tension braces of joints c and G. By this stage, the cracked area of the concrete slab had extended over three panel lengths of the top chord, with a maximum crack width of 0.56 mm.

For specimen CTB2 (Fig. 6.4), the first crack appeared at the junction of the top surface and side face of the mid-span concrete slab when the load reached 120 kN, accompanied by a sharp increase in strain of the reinforcement within the slab. At 200 kN, cracks had propagated within the slab over one panel length of the top chord. Subsequently, the strain in the top chord and the crack widths in the slab continued to increase. Upon loading to 520 kN, cracks had extended over three panel lengths of the top chord, with a maximum crack width of 0.08 mm. At 792 kN, the maximum crack width reached 0.2 mm, indicating the structure had reached its serviceability limit state; the corresponding axial force in the concrete slab was 786 kN. At 840 kN, cracks had spread across the entire width of the slab, with a maximum width of 0.23 mm. When the load increased to 1020 kN, the ordinary reinforcement in the mid-span slab yielded, with a slab axial force of 1012 kN. At 1180 kN, the entire cross-section of the bottom chord at mid-span yielded, carrying an axial force of 2048 kN; due to the concrete infill, no significant local buckling was observed on the steel tube surface. Upon reaching the ultimate load of 1321 kN, both the top chord and brace members were near yield.

Compared to specimen CTB1, CTB2 exhibited more pronounced internal force redistribution and greater plastic deformation. The final failure modes included chord shear failure at joint e and weld tearing failures at joints C and G. Additionally, the concrete infill at joint c was crushed. At ultimate failure, the maximum crack width in the concrete slab reached 3 mm.

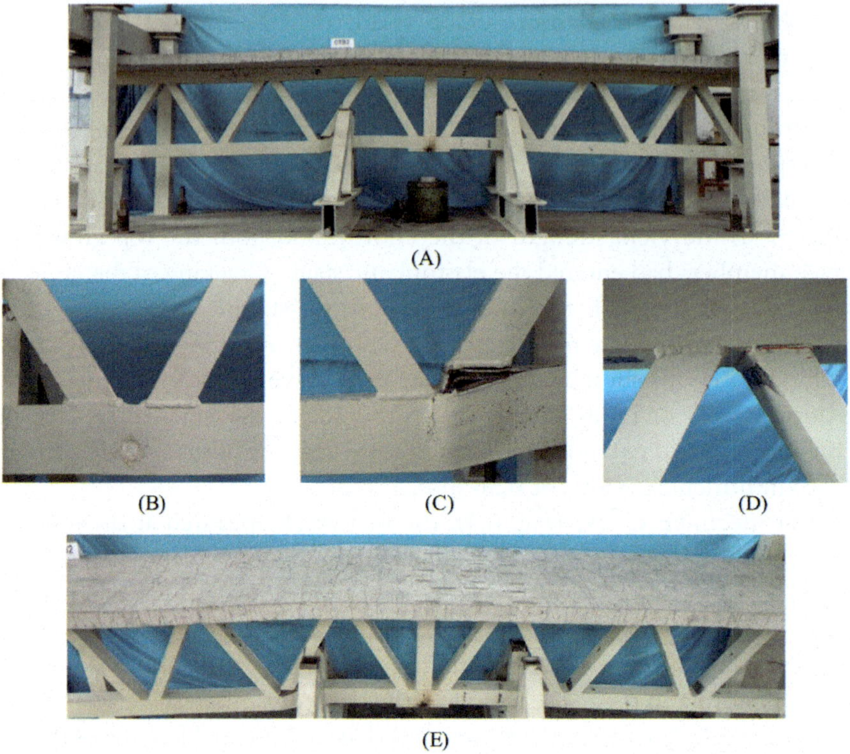

Figure 6.4 Failure mode of specimen CTB2. (A) Global bending of truss girder. (B) Shear failure in chord at joint E. (C) Weld tearing failure at joint C. (D) Weld tearing failure at joint G. (E) Cracking failure in concrete slab.

6.2.2.2 Load-displacement curves

The load-displacement curves for the mid-span section of specimens CTB1 and CTB2 are shown in Fig. 6.5. As seen in Fig. 6.5, during the elastic stage, the stiffness of CTB1 was slightly lower than that of CTB2. However, from the elastoplastic to the plastic stage, the stiffness of the two specimens was approximately the same. The primary reason for this is that in the initial loading phase, before cracks in the concrete slab had fully developed, the initial sectional stiffness of CTB1 was lower due to the use of shear studs with localized stiffness reduction. As the load increased and the specimens entered the elastoplastic stage, the stiffness of both specimens decreased owing to concrete cracking and the gradual yielding of members.

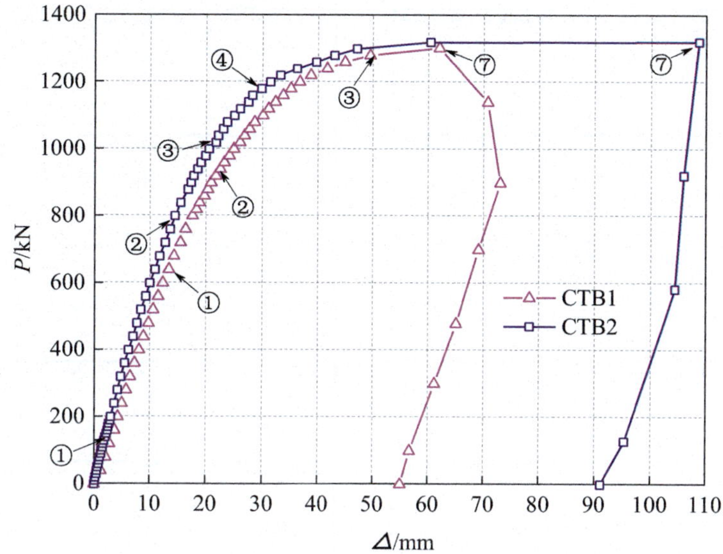

Figure 6.5 Load-displacement curves.

At the ultimate limit state, specimen CTB1 reached an ultimate load of 1302 kN with a maximum mid-span deflection of 62 mm, while specimen CTB2 reached an ultimate load of 1321 kN with a maximum mid-span deflection of 109 mm. The cracking load, the load corresponding to a 0.2 mm crack width, and the yield load of the ordinary reinforcement were all higher for CTB1 than for CTB2. In contrast, the ultimate loads of the two specimens were essentially identical. This indicates that using shear studs with localized stiffness reduction has minimal impact on the ultimate load-carrying capacity of the composite truss girder in the negative moment region but significantly enhances the crack resistance of the concrete slab in this area.

6.2.2.3 Concrete slab cracking

The crack patterns in the concrete slabs of specimens CTB1 and CTB2 under different load levels are shown in Fig. 6.6. The figure presents developed views of the slab's top surface and two side surfaces, covering a range of 1.7 m on either side of the mid-span. It can be observed that cracks in both CTB1 and CTB2 originated at the junction of the top surface and side faces. However, crack development was much more rapid in CTB2. When the maximum crack width reached 0.2 mm, the number of cracks in

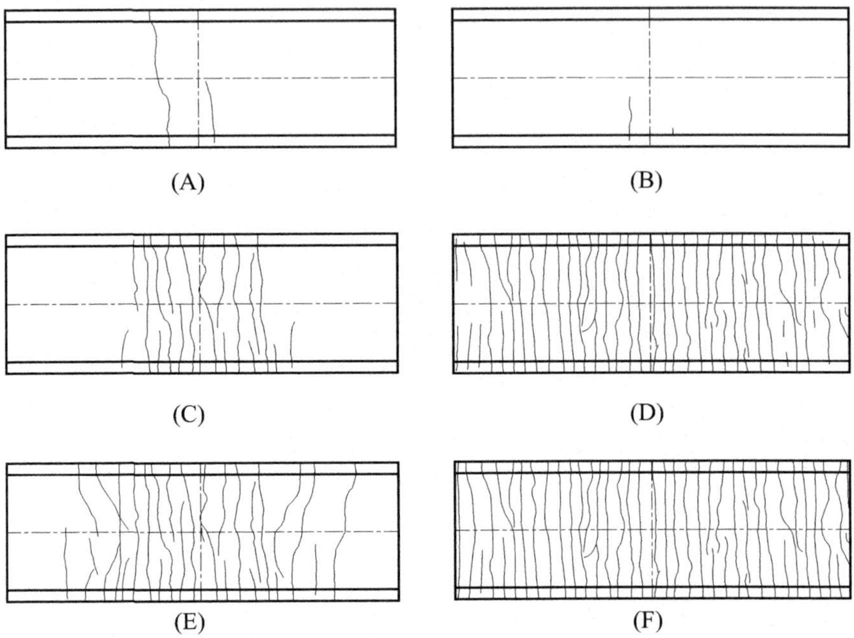

Figure 6.6 Concrete slab crack propagation. (A) CTB1440 kN (0.04 mm). (B) CTB2120 kN (0.04 mm). (C) CTB1933 kN (0.2 mm). (D) CTB2792 kN (0.2 mm). (E) CTB11,302 kN (0.56 mm). (F) CTB21,321 kN (3 mm).

CTB2 was significantly greater than in CTB1, and the crack spacing in CTB2 was denser. At the ultimate state, cracks in CTB1 were distributed within only a 3-m region around the mid-span, whereas cracks in CTB2 had spread across the entire width of the concrete slab.

The curve illustrating the relationship between the maximum crack width and the applied load is shown in Fig. 6.7. It can be observed from the figure that under identical load levels, the concrete crack width in specimen CTB1 was significantly smaller than that in specimen CTB2. This is primarily attributed to the precompressive stress applied to the concrete slab of CTB1.

For specimen CTB1, during the initial cracking stage, the maximum crack width increased in an approximately linear manner with the load. When the load increased from 920 kN until structural failure, the maximum crack width began to develop rapidly, reaching 0.56 mm at the ultimate limit state.

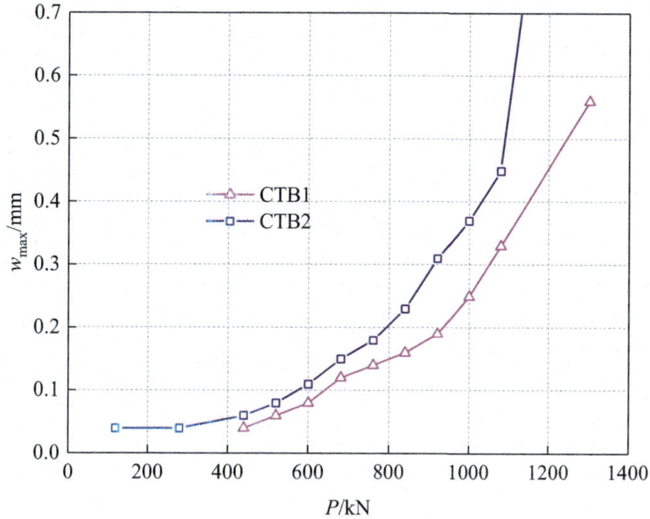

Figure 6.7 Variation curve of maximum crack width with load.

For specimen CTB2, the development of crack width was relatively slow in the initial cracking stage, but the area of crack propagation expanded rapidly. By the time the load reached 840 kN, cracks had already propagated across the entire width of the concrete slab, with a maximum width of 0.23 mm. Subsequently, the rate of crack width development increased slightly, and the maximum crack width reached 3 mm at the ultimate limit state.

The relationship curves among the applied load, the strain in the top reinforcement at the mid-span section, and the maximum crack width are shown in Fig. 6.8.

Fig. 6.8 indicates that for a given specimen, the development trends of the maximum crack width and the load with increasing reinforcement strain are similar. At the same load level or maximum crack width, the reinforcement strain in specimen CTB2 was greater than that in specimen CTB1. This is because no precompressive stress was applied to CTB2, leading to earlier cracking of the concrete slab and a consequent reduction in its load-carrying contribution. Throughout the loading process, the crack width in specimen CTB2 developed in an approximately linear relationship with the reinforcement strain. This linearity is attributed to the closely spaced reinforcement arrangement in CTB2. For specimen CTB1, the crack width increased slowly with reinforcement strain until the

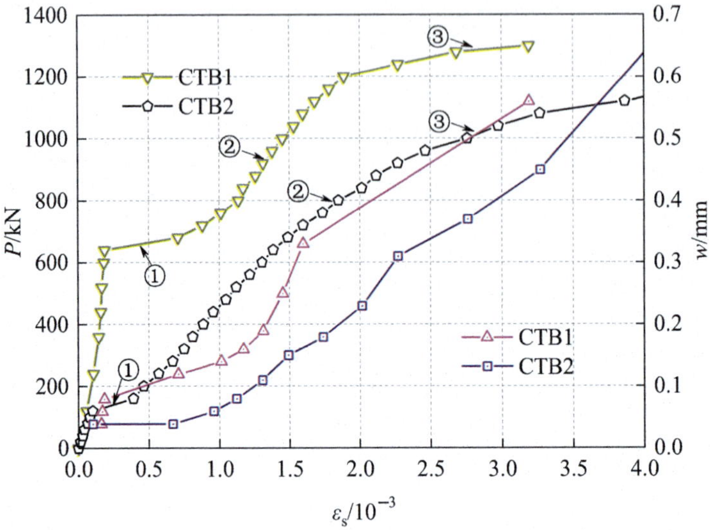

Figure 6.8 Variation curves of load and maximum crack width with reinforcement strain.

maximum crack width reached 0.2 mm (at 933 kN), after which the rate of crack width development accelerated.

For specimen CTB1, the reinforcement strain increased linearly with the load during the initial loading stage. When the load reached approximately 660 kN, concrete cracking occurred, leading to a significant increase in reinforcement strain. Subsequently, the development of reinforcement strain gradually transitioned from the elastic stage to the elastoplastic stage, reaching yield at 1282 kN. The overall development trend of reinforcement strain in specimen CTB2 was similar to that of CTB1, with the transition point occurring at 140 kN and the yield load at 1020 kN. This paper defines the transition point on the load-reinforcement strain curve as the cracking load of the concrete slab. Thus the cracking loads for specimens CTB1 and CTB2 are determined to be 660 and 140 kN, respectively.

6.2.2.4 Strain distribution in concrete slab

Figs. 6.9 and 6.10 show the longitudinal strain distribution on the top surface of the concrete slab at the panel section (A–A) and the joint section (J–J) for specimens CTB1 and CTB2, respectively. The horizontal

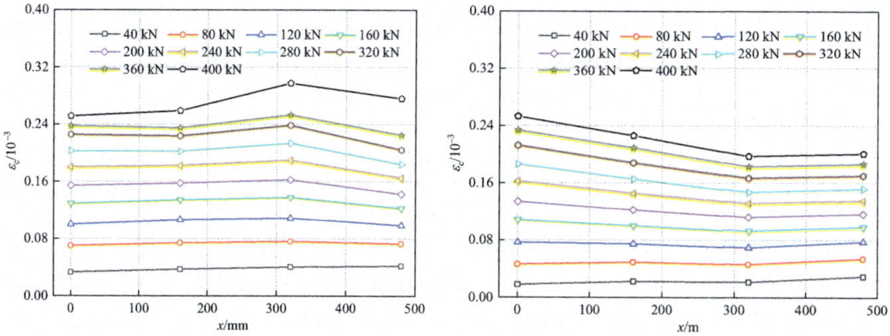

(A) (B)

Figure 6.9 Longitudinal strain distribution in concrete slab (specimen CTB1). (A) Cross-section at panel point (A–A). (B) Cross-section at panel point (J–J).

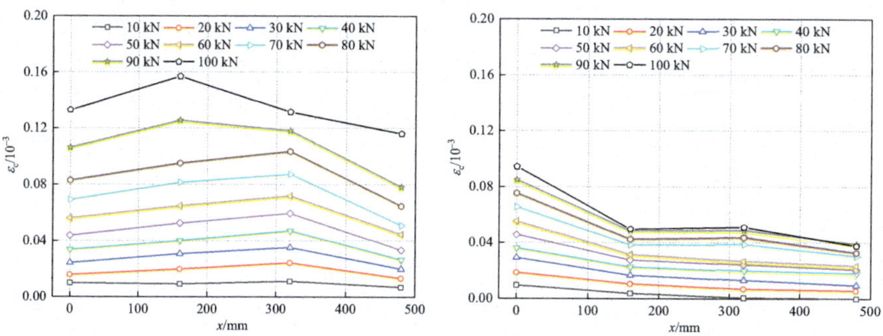

(A) (B)

Figure 6.10 Longitudinal strain distribution in concrete slab (specimen CTB2). (A) Cross-section at panel point (A–A). (B) Cross-section at panel point (J–J).

coordinate (x) represents the distance of the concrete measurement point from the axis of the top chord. The figure indicate that the concrete strain distribution follows similar trends for both specimens, whether at the panel section or the joint section.

For the A–A section (panel section), the strain distribution across the width is relatively uniform. In contrast, for the J–J section (joint section), the strain distribution is significantly nonuniform. The maximum concrete strain occurs near the axis of the top chord and decreases with increasing distance from it. This phenomenon is attributed to the location of the J–J section at the intersection of the diagonal brace member and the top chord, where shear forces from the joint induce a pronounced shear lag effect in the concrete slab.

6.2.2.5 Load-strain curves of steel truss girder

The load-strain curves for the top and bottom chords of the steel truss girder at the mid-span section (A–A) of specimens CTB1 and CTB2 are shown in Fig. 6.11. The horizontal coordinate represents the average value of the strain measurement points on the chords. The figure shows that under the same load, the strain in the bottom chord is slightly greater than that in the top chord. This is because the concrete slab and reinforcement share part of the tensile force, resulting in the bottom chord, despite being assisted by concrete infill in compression, exhibiting a higher strain level.

During the elastic stage, the strain in the top chord of specimen CTB2 was greater than that in CTB1, while the strains in the bottom chords were essentially the same. The reason for this is that the concrete slab of CTB2 cracked earlier, leading to a greater proportion of the axial tensile force being allocated to its top chord, with minimal impact on the bottom chord.

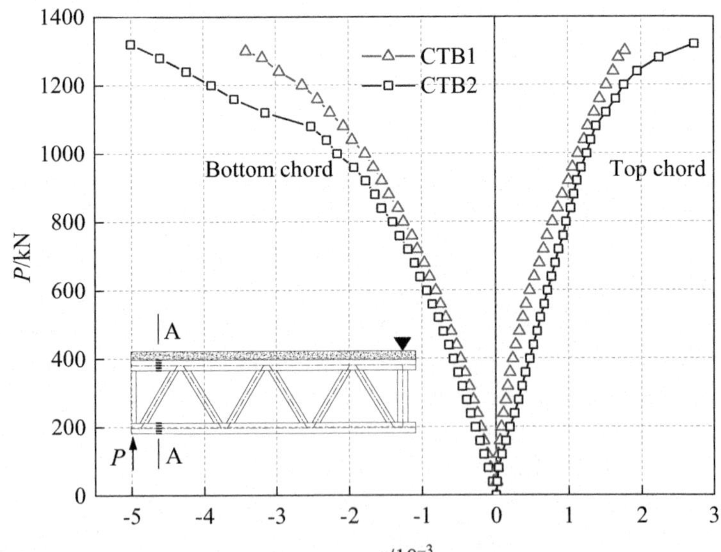

Figure 6.11 Load-strain curve of chord member at mid-span section.

From the elastoplastic stage until structural failure, cracks in the concrete slab of specimen CTB1 had not fully developed, meaning the slab still contributed significantly to the sectional stiffness. In contrast, specimen CTB2, with its earlier and more extensive concrete cracking, exhibited significant internal force redistribution, and the chords essentially reached full-section yield.

The load-strain curves for the brace members of specimens CTB1 and CTB2 are shown in Fig. 6.12. As concentrated loading was applied at the mid-span, the absolute shear force values in the brace members of the composite truss girder were approximately the same. Therefore the average measured axial strain of the tension and compression brace members, respectively, is used as the horizontal coordinate. The figure indicates that the strain levels and development trends of the brace members in both specimens are generally similar. The strain in the tension brace members increased almost linearly, while the strain growth in the compression brace members exhibited a more pronounced nonlinear segment. During the elastic stage, the strains in the tension and compression brace members were essentially identical. After entering the elastoplastic stage, the strain in the compression brace

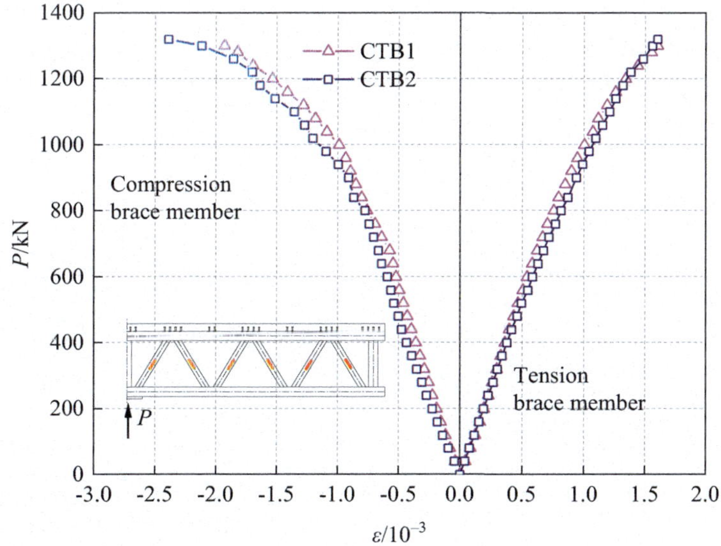

Figure 6.12 Load-strain curves of brace members.

became significantly greater than that in the tension brace members, indicating that in the later loading stages, the compression brace members developed sufficient ductility, and some members experienced local buckling.

Fig. 6.13 shows the load-slip relationship curves at the steel-concrete interface for specimens CTB1 and CTB2 throughout the loading process. The horizontal coordinate (y) represents the distance of the slip measurement point from the midspan. The figure indicates that the slip values at different measurement points for the same specimen were relatively similar. This is primarily because the shear force transmitted at each joint was essentially identical.

Under the same load level, the interface slips in specimen CTB1, which used shear studs with localized stiffness reduction, were significantly greater than that in specimen CTB2, which used conventional shear studs.

Fig. 6.14 presents the load versus maximum slip curves for the steel-concrete interface of both specimens. The maximum slip for CTB1 occurred at joint G, while for CTB2 it was at joint F.

The curves show that prior to the failure of the natural bond at the interface, there was negligible slip in both specimens. When the load reached approximately 160 kN, the slip in both specimens began to increase, indicating the failure of the natural bond. After this point, the rate of slip increase accelerated noticeably in both specimens. At the ultimate limit state, the maximum interface slips for CTB1 and CTB2 were 2.84 and 0.71 mm, respectively.

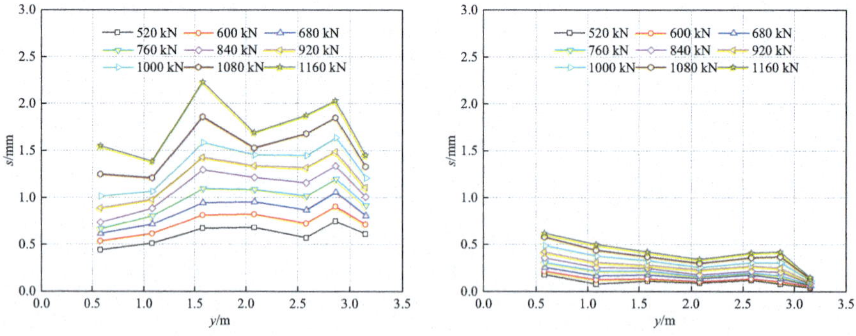

Figure 6.13 Load-slip curve. (A) Specimen CTB1. (B) CTB2 specimen CTB2.

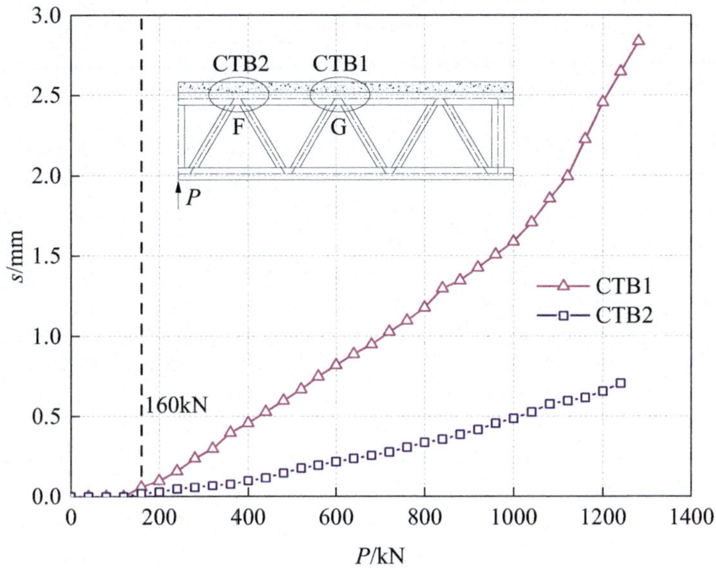

Figure 6.14 Load-maximum slip curves.

6.3 Full-scale test on a concrete-filled rectangular steel tube composite truss girder bridge

6.3.1 Test overview

The Huangyan Expressway Overpass Bridge is a continuous rigid-frame bridge constructed with CFRST. The span arrangement is 24 + 40 + 24 m, with a deck width of 5.5 m. The girder height is 2.5 m (excluding the 10 cm pavement layer). The longitudinal panel length is 4 m, and the transverse spacing between the main trusses is 2.78 m. The depth-to-span ratio of the main girder is 1/16, and the width-to-span ratio is 1/14.4. The layout of the Huangyan Expressway Overpass Bridge is shown in Fig. 6.15, with detailed information provided in Chapter 8 of this book.

6.3.1.1 Load cases and loading method

A full-scale load test was conducted on the Huangyan Expressway Overpass Bridge after its completion. The test utilized a total of three loading trucks, each with a gross weight of 400 kN. The axle weights and axle spacing are illustrated in Fig. 6.16. The trucks were weighed prior to the test, and the measured vehicle parameters are listed in Table 6.3.

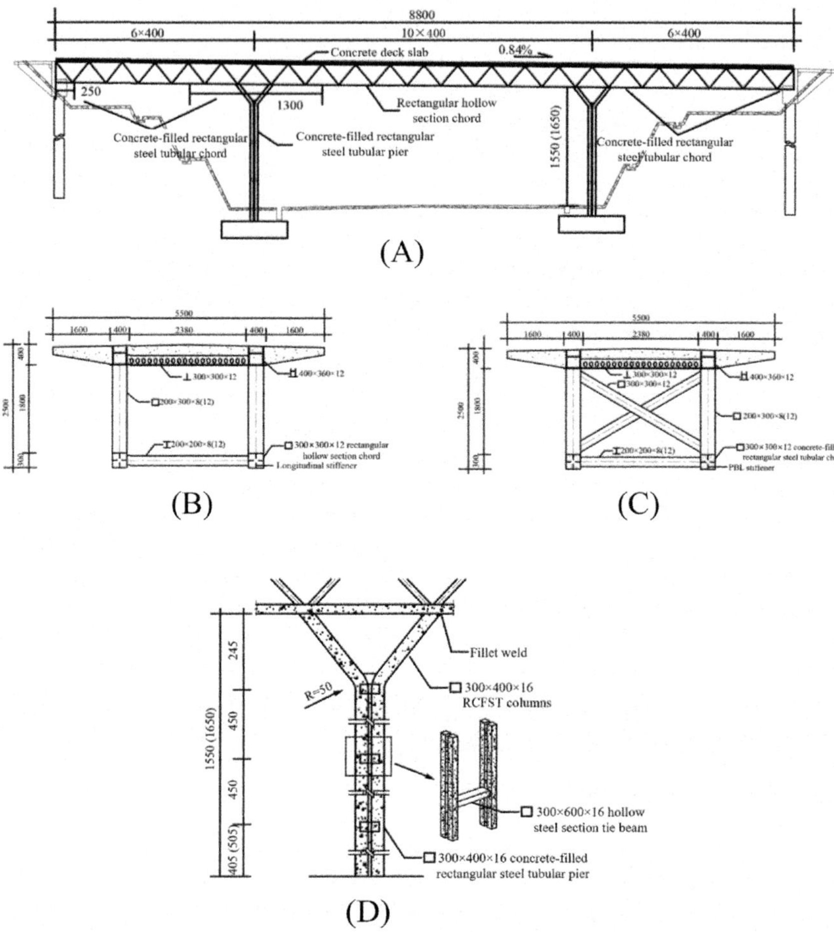

Figure 6.15 Schematic layout of the Huangyan Expressway Overpass Bridge. (A) Elevation layout (unit: cm). (B) Mid-span cross-section (unit: mm). (C) Cross-section layout at pier top (unit: mm). (D) Pier layout (unit: cm).

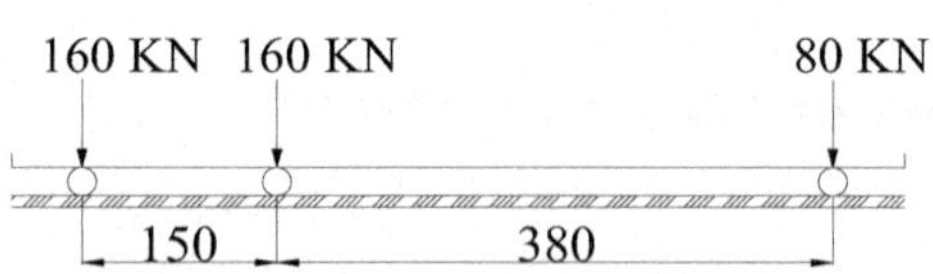

Figure 6.16 Schematic diagram of loading vehicle axle loads and spacing (unit: cm).

Table 6.3 Measured loading vehicle axle loads (unit:kN).

Number	Axle spacing/m	Axle load/kN		Total weight/kN
	Front and rear axles	Front axle	Rear axle	
1	1.50 + 3.80	80.1	320.0	401.1
2	1.50 + 3.80	81.6	322.1	403.7
3	1.50 + 3.80	79.8	326.8	406.6

The load test was planned for four load cases: maximum positive moment in the mid-span, maximum positive moment in the side span, maximum negative moment at the pier top, and maximum shear in the brace members. Since the loading vehicle positions for the maximum negative moment at the pier top and the maximum shear in the brace members were similar, these two cases were combined into a single test case (case 3). Each test case was loaded according to the influence lines of the controlling members. As the Huangyan Expressway Overpass Bridge was designed as a test bridge with a relatively conservative approach, the load efficiency for each static load case was appropriately increased during the full-scale test to enable a more thorough investigation of the mechanical behavior of this new bridge type.

A finite element model using beam elements was established in Midas Civil. CFST members were simulated using dual elements, nodes of the steel truss girder were connected sharing nodes, the bridge deck slab was transversely discretized and connected by virtual cross beams, and the deck slab was also connected to the top chord nodes of the main truss sharing nodes, with the pier bottom fixed. The finite element model is shown in Fig. 6.17. Based on the calculation results from the finite element model, the load efficiency for each case ranged from 1.90 to 3.05. Detailed test case contents are shown in Table 6.4.

Load case 3 of the test was divided into seven loading steps. Cases 1 and 2 were simplified into three and two loading steps respectively, due to the proximity of loading positions between different steps in these cases. Therefore the full-scale test involved a total of 3 load cases and 12 loading steps. For each step, after the loading vehicles reached the designated positions on the deck, the load was maintained for 2–5 minutes. Data acquisition was performed only after the measured readings stabilized, before proceeding to the next loading step. After completing each load

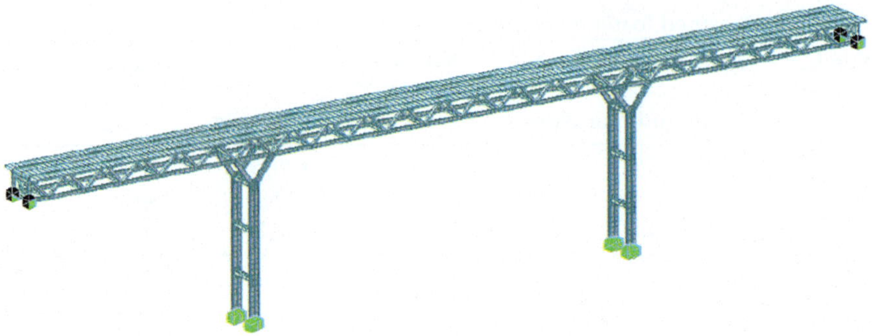

Figure 6.17 Finite element model.

case, all vehicles were unloaded before commencing the next case. Given the large number of loading steps across all cases, this paper only presents the test photographs for the seven loading steps of case 3 (Fig. 6.18).

6.3.1.2 Test instruments and sensor arrangement

External electrical resistance strain gauges were used for strain measurement in this full-scale bridge test, while dial gauges were employed for deflection measurement. A TDS-303 static data acquisition system was used to collect strain and displacement data throughout the test. A total of 2 deflection measurement points and 54 strain measurement points were installed.

The top chords of the test bridge are embedded within the concrete deck, forming an integral part with it, making sensor installation difficult. Therefore strain and deflection measurement points for the static load test were all arranged on the brace members and bottom chords of the main trusses. The test sections for all load cases were located at the mid-span sections of the members. (Fig. 6.19).

Longitudinal PBL stiffeners run continuously inside the bottom chords of the main truss. Consequently, strain gauges were placed at the quarter-points and mid-points of the top and side plates. The brace members are hollow rectangular steel tubes, and strain gauges for their test sections were installed at the center of the top and bottom plates. Due to on-site constraints, no strain gauges were installed on the bottom plates of the bottom chords.

Deflection measurement points were also set up on the bottom chord sections for case 1 and case 2 (Figs. 6.20A and 6.21). For case 3, strain measurements were taken on the bridge deck. Strain gauges were placed on

Table 6.4 Full-scale bridge test load cases.

Number	Load case	Critical member and corresponding axial force (kN)	Load efficiency	Test content	Loading steps
Load case 1	Maximum positive bending moment at mid-span	Tensile bottom chord at mid-span (928.4 kN)	2.14	Strain, deflection	3
Load case 2	Maximum positive bending moment at side-span	Tensile bottom chord at side-span mid-span (627.4 kN)	1.90	Strain, deflection	2
Load case 3	Maximum negative bending moment at pier top	Compressive top chord at pier top (−496.0 kN)	3.05	Strain	7
	Maximum shear force load case	Compressive brace member E7A7 at pier top (−439.8 kN)			

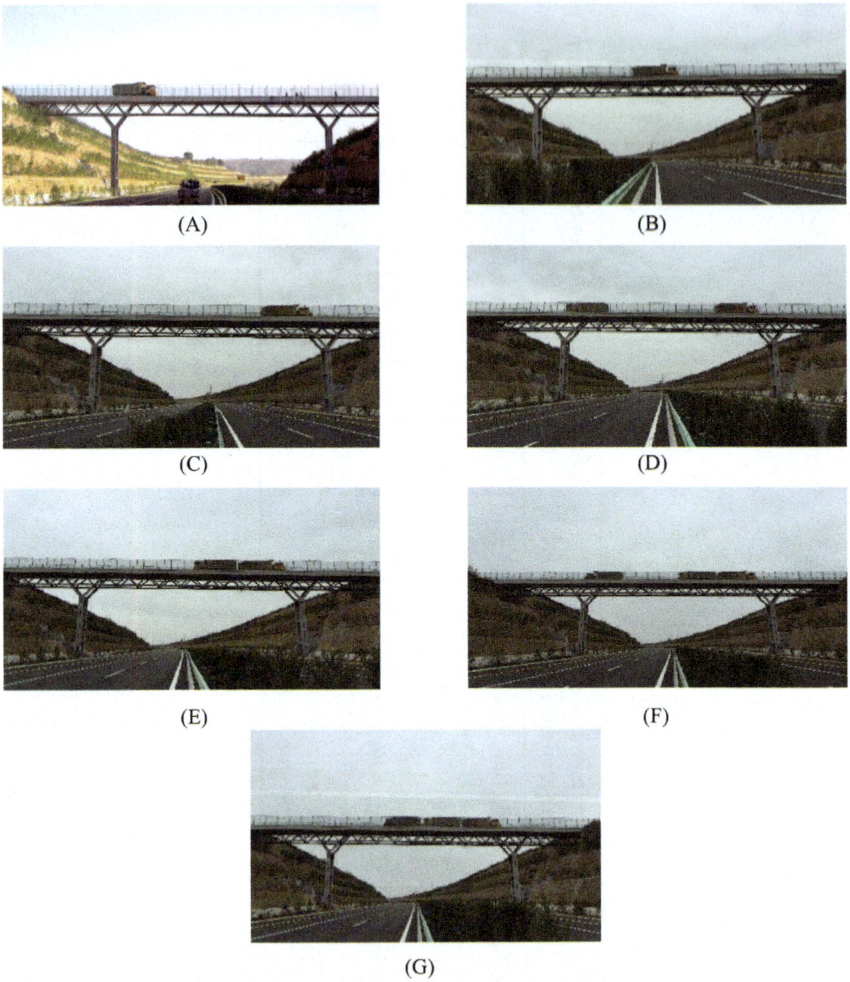

Figure 6.18 Photos of loading for case 3. (A) Step 1 loading. (B) Step 2 loading. (C) Step 3 loading. (D) Step 4 loading. (E) Step 5 loading. (F) Step 6 loading. (G) Step 7 loading.

the bottom surface of the deck at section I–I (node A8), section II–II (quarter-point of the panel), and section III–III (mid-panel). At each section, five strain gauges were arranged along the longitudinal direction of the bridge, covering areas above the top chord and on the deck bottom, as shown in Fig. 6.22E. Site limitations prevented the installation of strain gauges on the cantilever ends of the deck.

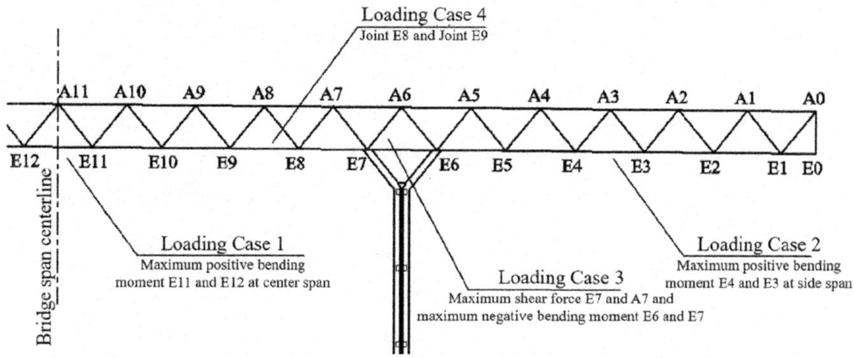

Figure 6.19 Schematic diagram of load test cases and control members.

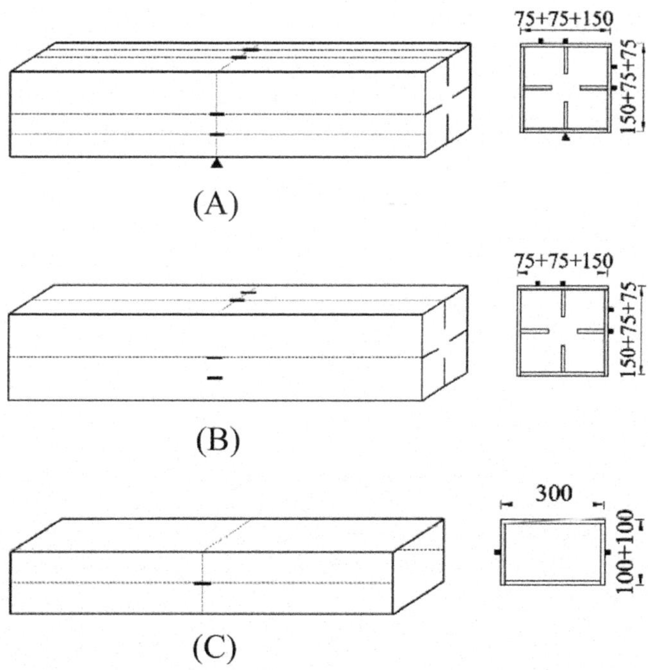

Figure 6.20 Sensor arrangement for case 1 (unit: mm). (A) Cross-section at mid-span of bottom chord E11E12. (B) Cross-section at mid-span of bottom chord E9E10. (C) Cross-sections at mid-span of braces E9A8 and E9A9.

Figure 6.21 Sensor arrangement at the mid-span section of member E3E4 for case 2 (unit: mm).

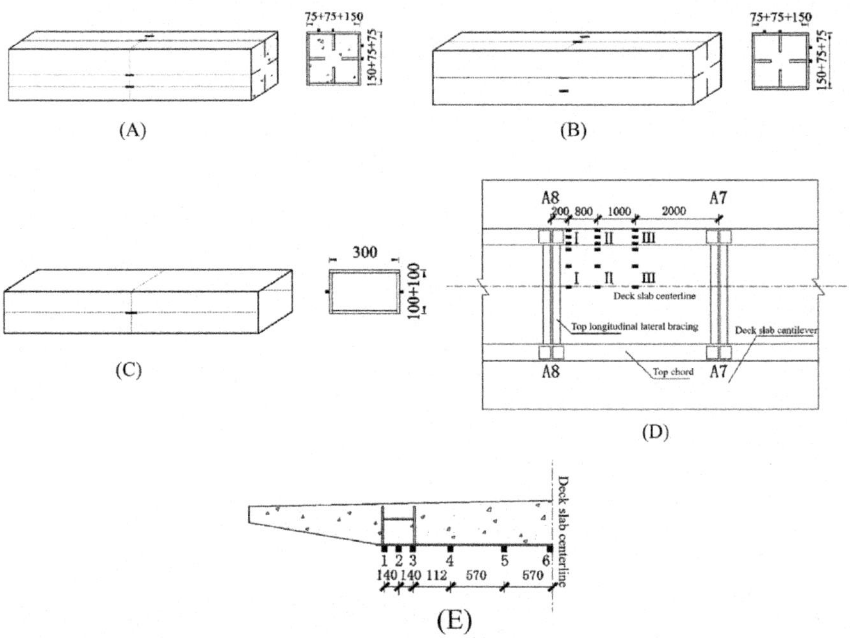

Figure 6.22 Sensor arrangement for case 3 (unit: mm). (A) Cross-sections at mid-span of bottom chords E6E7 and E7E8. (B) Cross-section at mid-span of bottom chord E8E9. (C) Cross-sections at mid-span of braces E7A6, E7A7, E8A7, and E8A8. (D) Plan layout of measurement points on deck slab. (E) Arrangement of measurement points on cross-section of deck slab.

6.3.2 Global mechanical performance

The average measured strain from all sensors on the test section of a control member for each loading step was taken as the member's axial strain. The corresponding axial force in the member was obtained from the calculation results of the beam-element finite element model. This allowed for the creation of scatter plots showing the calculated axial force versus the measured average sectional strain for the control members under different loading steps of each test case, as shown in Fig. 6.23.

Based on the principle of equivalent mid-span displacement, the vehicle loads in case 1 and case 2 were converted into equivalent concentrated loads applied at the mid-span of the bridge's center span. Combining this with the measured deflection data for the corresponding cases, it yielded the scatter plots of equivalent mid-span load versus measured displacement shown in Fig. 6.24. Finally, linear fitting was performed on each scatter plot.

Fig. 6.23 shows that the scatter plots of member axial force versus strain for all cases exhibit a significant linear relationship.

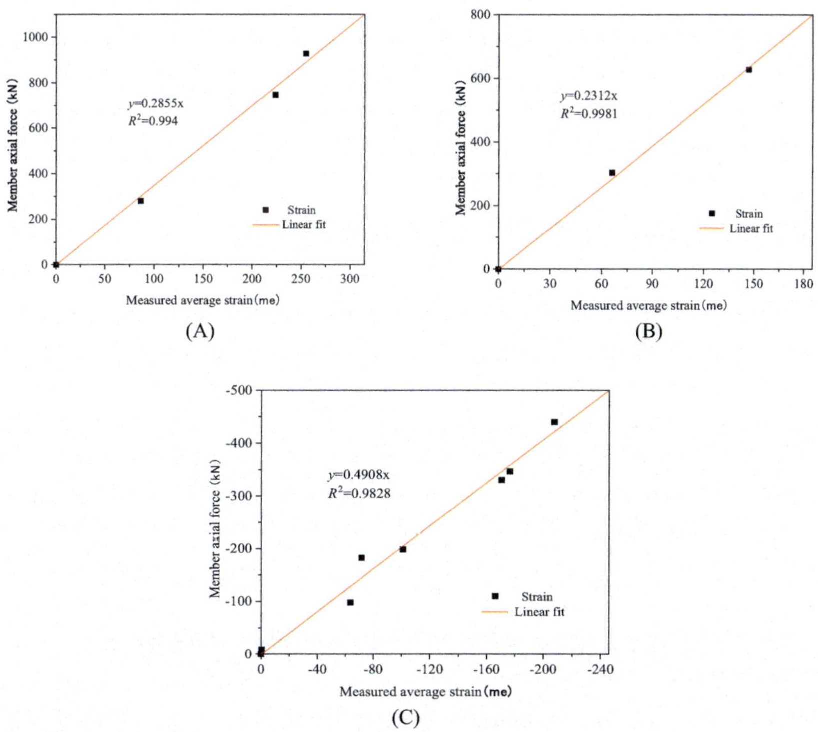

Figure 6.23 Axial force versus measured strain curves for control members in each test case. (A) E11E12 Bottom chord E11E12 under load case 1. (B) Bottom chord E3E4 under load case 2. (C) Bottom chord E6E7 under load case 3.

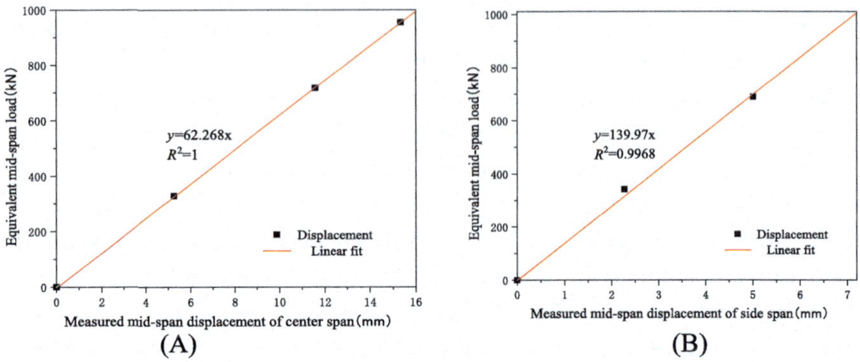

Figure 6.24 Equivalent mid-span load versus displacement curves for case 1 and case 2. (A) E11E12 Bottom chord E11E12 under load case 1. (B) Bottom chord E3E4 under load case 2.

Fig. 6.24 indicates a significant linear relationship for the equivalent mid-span load versus displacement plots in cases 1 and 2. This demonstrates that the test bridge remained in a well-behaved elastic state even under the overload conditions with load efficiencies ranging from 1.90 to 3.05.

6.3.3 Mechanical performance of members

The compactness of the concrete infill inside the steel tubes is closely related to its mechanical performance and is one of the most concerned issues in the industry. Through the full-scale bridge test, a quantitative analysis was conducted on the mechanical behavior of the CFST members in the actual structure during the elastic stage. The study investigated the load-sharing ratio between the steel tube and the infill concrete, determined the axial compression and axial tension stiffness of the CFST members, and deduced the effective cross-sectional area of the infill concrete.

6.3.3.1 Load distribution ratio in concrete-filled steel tube compression members

C50 micro-expansive concrete was poured into the bottom chord members of the main truss near the pier supports of the test bridge. Node E8 is located near the transition point between positive and negative bending moments in the steel truss girder under dead load. Therefore an end diaphragm was installed within the bottom chord member E8E9 connected to this node. The portion of the bottom chord on the side of the diaphragm closer to the mid-span, which is in tension, remained as an empty steel tube. The portion on the other side, closer to the main pier and in compression, was filled with concrete, which is shown in Fig. 6.25.

The measured strain data from the seventh loading step of case 3 was selected. The average strain value from all measurement points on the instrumented cross-section was taken as the strain of the steel tube. The axial force in the steel tube (N_s) for each member connected to node E8 was calculated using Formula (6.1), based on the member's cross-sectional area and the material's elastic modulus. In the formula, As is the cross-sectional area of the steel tube, and E_s is the elastic modulus of steel, taken as 206,000 MPa.

Although no strain gauges were installed on the concrete infill of member E8E7, the total axial force N_{CFST} of the CFST member E8E7 at the instrumented section was calculated to be −248.34 kN based on the nodal force equilibrium condition. Using Eq. (6.2), the axial force carried by the concrete infill at that section was calculated to be −95.25 kN. The measured force state of node E8 under case 3 is shown in Fig. 6.25.

$$N_s = E_s A_s \varepsilon_s \tag{6.1}$$

$$N_c = N_{\mathrm{CFST}} - N_s \tag{6.2}$$

Based on the elastic modulus and cross-sectional area of the steel tube and the infill concrete in the test section of member E8E7, the theoretical axial stiffness values for both components were calculated, as shown in Table 6.5. The elastic moduli for the bottom chord steel tube and concrete were taken as 206,000 and 34,500 MPa, respectively. The cross-sectional areas of the steel tube and the infill concrete were calculated according to

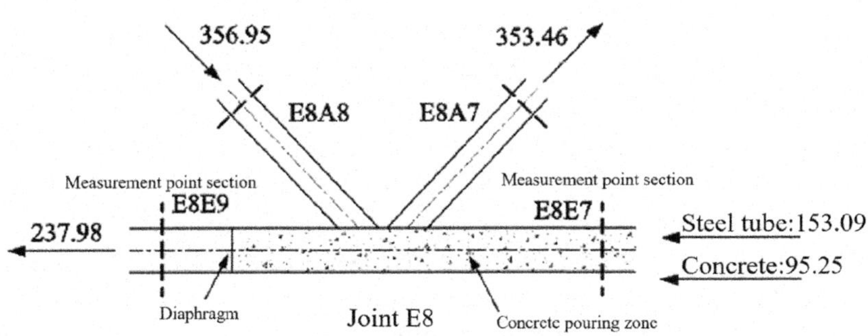

Figure 6.25 Measured member end axial forces at steel-concrete node E8 (unit kN).

Table 6.5 Load distribution ratio between steel tube and infill concrete for member E8E7.

Member section	Measured axial force/kN	Measured load distribution ratio	Theoretical stiffness/ kN/m	Theoretical load distribution ratio
Steel tube section	153.09	61.65%	3737664	60.12%
Concrete	95.25	38.35%	2479032	39.88%

the construction drawings, with the sectional area of the internal PBL stiffeners included in the steel tube's area.

The full-scale bridge test remained entirely within the elastic stage. Therefore it can be assumed that the ratio of the load shared by the steel tube to the load shared by the infill concrete in a CFST member is the same as the ratio of their axial stiffnesses. A comparison between the measured and theoretical load distribution ratios for the steel tube and the infill concrete in member E8E7 is shown in Fig. 6.26. The figure indicates that the measured load distribution ratio is quite close to theoretical value.

6.3.3.2 Stiffness of main truss bottom chord members

Based on their stress states under different load cases, the bottom chord members of the test bridge can be categorized into three main types: CFST compression members, CFST tension members, and hollow steel tube tension members. Using the measured data from various loading steps across the cases and the corresponding member axial forces calculated by the finite element model for those steps, a series of scatter plots of calculated axial force versus measured strain were generated for these three member types. Linear fitting was then performed on the scatter plots for each type, resulting in the three fitted axial force-strain curves shown in Fig. 6.27. To facilitate comparison between tension and compression members, the absolute values of the measured strain are used in the figure.

As given by Eq. (6.1), the slope of a member's axial force-strain curve represents its axial stiffness. Fig. 6.27 shows that the CFST compression member has the highest axial stiffness (5,619,800 kN/m), followed by the CFST tension member (4,523,400 kN/m), with the hollow steel tube tension member having the lowest (3,573,800 kN/m). The ratio of the fitted curve slopes indicates an axial stiffness ratio of 1:0.805:0.636 for the

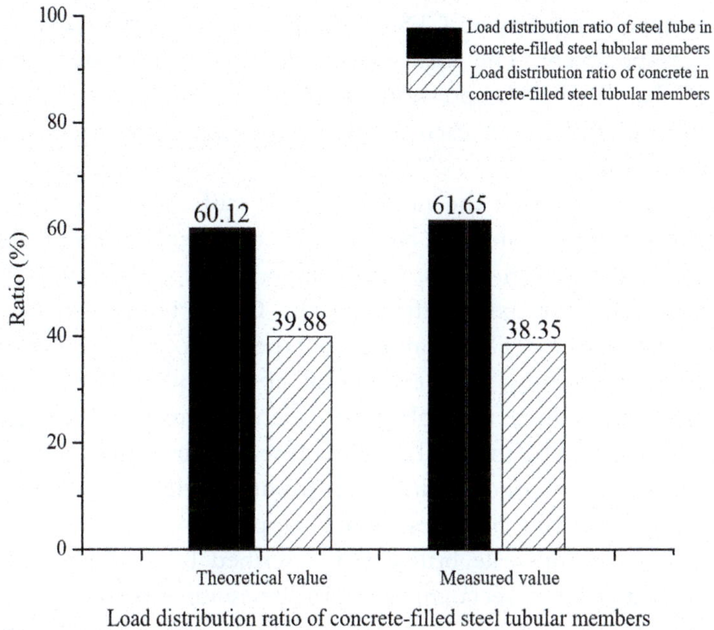

Figure 6.26 Load distribution ratio for CFST member E8E7.

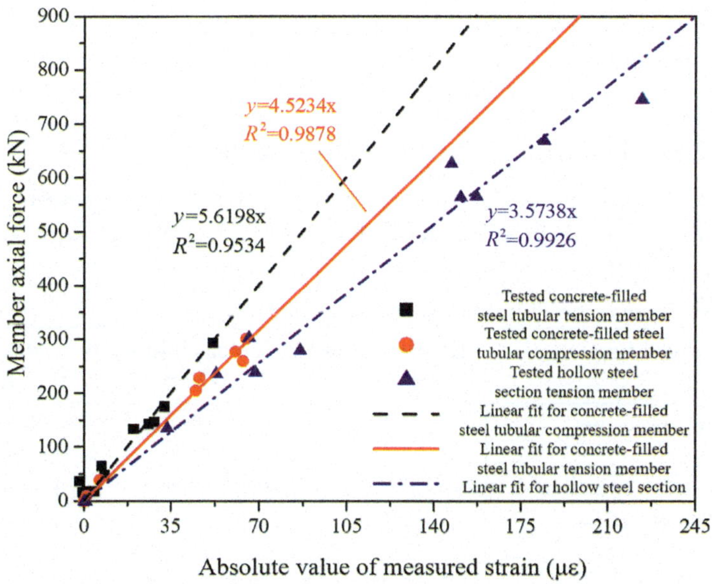

Figure 6.27 Comparison of axial stiffness for different member types.

CFST compression member, CFST tension member, and hollow steel tube member, respectively. The axial stiffness of the CFST tension member is 80.5% of that of the compression member. A comparison between the measured axial stiffness of each member type and its theoretical value is presented in Table 6.6.

Table 6.6 shows that the measured axial stiffness of both the CFST compression members and the hollow steel tube tension members agrees well with the theoretical values, with relative errors within 10%. Based on these test results, the axial stiffness of the CFST tension members falls between that of the CFST compression members and the hollow steel tubes and is significantly greater than that of the hollow steel tubes.

Relevant literature, through finite element analysis, has proposed formulas for calculating the axial tensile stiffness of circular CFST members. It suggests that the concrete infill contributes little to the axial tensile stiffness and primarily provides lateral support to the steel tube wall. However, in the full-scale bridge test described here, the tensile CFST bottom chords feature rectangular steel tubes with longitudinally welded PBL stiffeners. The concrete dowels formed through the holes in the PBL ribs enhance the composite action between the steel and concrete. Consequently, the contribution of concrete infill to the axial tensile stiffness is increased, which is reflected in the measured axial stiffness of the CFST tension members being substantially greater than that of the hollow steel tubes.

6.3.4 Effective width of the bridge deck

Current design codes in various countries provide specifications for the effective width of bridge decks. However, the applicability of these

Table 6.6 Comparison of bottom chord axial stiffness.

Bottom chord type	Theoretical value (kN/m)	Measured value (kN/m)	Relative error
CFST compression member	6,216,696	5,619,800	−9.6%
CFST tension member	–	4,523,400	–
Hollow steel tubular tension member	3,737,664	3,573,800	−4.38%

provisions to composite truss bridges with rectangular tubular chords has not been reported in the literature or codes. In this full-scale bridge test, strain measurements were taken on the deck slab between the two main trusses at both the joint locations and the panel mid-points, allowing for the calculation of the effective width coefficient of the deck slab.

The measured strain data from the sixth and seventh loading steps of case 3 were selected for comparative analysis, treating these two datasets as independent tests. Three strain gauges on the cross-section were installed on the bottom plate of the top chord; the average strain from these three points was taken as the measured strain of the top chord's bottom plate. The two gauges closest to the centerline of the deck were located on the bottom surface of the concrete deck slab.

The effective width, be, used to account for shear lag effects, was calculated using Eq. (6.3). Where E_c is the elastic modulus of concrete, taken as 34,500 MPa,ε_x is the measured longitudinal concrete strain, and $\varepsilon_{\max}$ is the maximum absolute value of the measured longitudinal concrete strain. The effective width coefficient ψ, was calculated using Eq. (6.4), where b is the actual width of the deck slab.

Due to constraints in the full-scale test setup, no strain gauges were installed on the cantilever ends of the cross-section. Therefore the effective width was calculated only for the deck slab between the two main trusses. The calculation results are shown in Fig. 6.29.

$$b_e = \frac{\int_{-b}^{b} E_c \varepsilon_x \, dx}{E_c \varepsilon_{\max}} \tag{6.3}$$

$$\psi = \frac{b_e}{b} \tag{6.4}$$

Fig. 6.28 shows that the average effective width coefficient for the deck slab between the main trusses at the joint section (I–I) is 0.63. At the quarter-point panel section (II–II), the average coefficient is 0.91, and at the mid-panel section (III–III), it is 0.79. Under the sixth and seventh loading steps of case 3, the transverse strain distribution patterns across the deck slab were similar. Significant shear lag effects were observed at the I–I section near the pier top and the joint. In contrast, the shear lag effect was less pronounced at the II–II and III–III sections located within the panel.

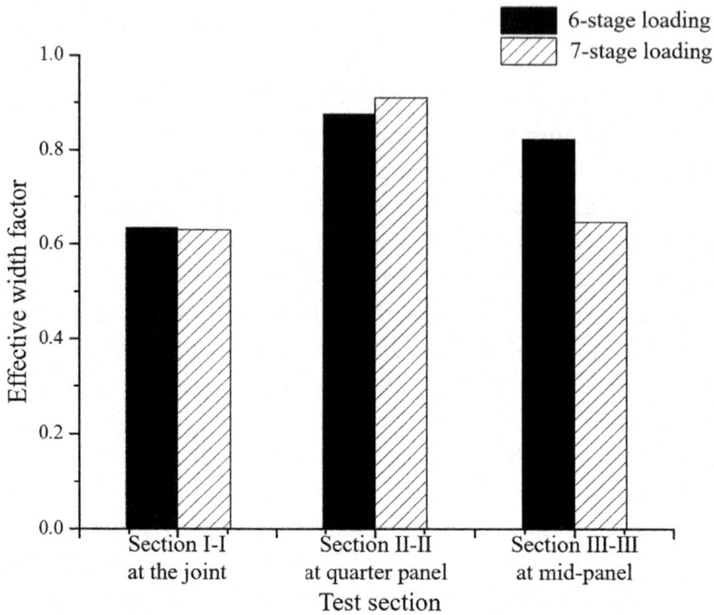

Figure 6.28 Effective width coefficient of the deck slab between main trusses.

6.4 Simplified calculation of flexural capacity for concrete-filled steel tube composite truss girders

6.4.1 Failure mode determination and bearing capacity calculation

When the concrete deck slab and the connected truss chord of a composite truss girder are subjected to tension, the concrete slab cracks at an early stage. In this scenario, the load-bearing capacity is provided by the tensile chord and the longitudinal reinforcement within the deck slab. Therefore the capacity of the composite truss girder can still be calculated by referring to the method for steel-concrete trusses provided in Section 6.3.2 of this book.

When the deck slab and the connected chord are in compression, the composite action is fully utilized. In this case, the concrete slab of the composite truss girder can be considered as part of the truss chord. The distribution ratio of the axial load between the deck slab and the truss chord can be determined according to the ratio of their axial stiffnesses.

Based on the characteristics of CFST composite truss girders, the following calculation assumptions are proposed: (1) The stiffness and bearing capacity of the connectors between the concrete slab and the top chord are infinite, meaning the influence of bond-slip at the interface between the deck and the main truss is neglected. (2) The vertical loads acting on the concrete slab of the composite truss girder can all be equivalently represented as a series of concentrated loads applied at the nodes.

Based on the assumptions, a method for determining the failure modes and a simplified calculation method for the flexural capacity of CFST composite truss girders can be proposed. Taking a simply supported triangular truss girder (without vertical brace members) as an example, the failure modes and corresponding flexural capacity under positive bending moments with different loading patterns are derived. This calculation method is also applicable to other types of truss structures.

When a concentrated load P acts solely on the concrete slab of the composite truss girder, its simplified calculation model is shown in Fig. 6.29. According to structural mechanics principles, the maximum internal forces in the compressed concrete slab, the compressed chord, the tension chord, and the brace members under nodal loading can be calculated using Eqs. (6.5) to (6.8), respectively.

$$N_{\mathrm{dc}} = \frac{k_1 Pna}{H}\left(\frac{N - n - 0.5}{N}\right) \tag{6.5}$$

$$N_c = \frac{k_2 Pna}{H}\left(\frac{N - n - 0.5}{N}\right) \tag{6.6}$$

$$N_t = \frac{Pna}{H}\left(\frac{N - n}{N}\right) \tag{6.7}$$

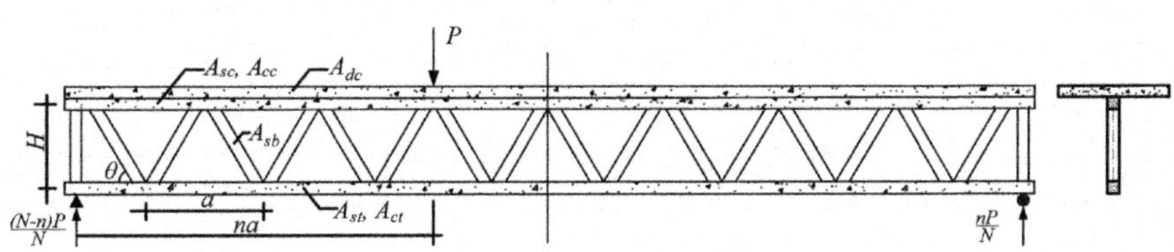

Figure 6.29 Simplified calculation model for the CFST composite truss girder under nodal loading.

$$N_b = \frac{(N-n)P}{N \sin \theta} \tag{6.8}$$

In the equations, k_1 and k_2 are the load distribution ratios for the compressed deck slab and the compressed chord, respectively. For a CFST truss, k_1 can be taken as 0 and k_2 as 1.

The sectional resistance of the concrete slab, the compression chord, the tension chord, and the brace members can be calculated using Eqs. (6.9–6.12) respectively.

$$R_{\mathrm{dc}} = A_{\mathrm{dc}} f_c \tag{6.9}$$

$$R_c = \chi\left(A_{\mathrm{sc}} f_y + A_{\mathrm{cc}} f_c\right) \tag{6.10}$$

$$R_t = A_{\mathrm{st}} f_y \tag{6.11}$$

$$R_b = \eta_{\mathrm{eff}} A_{\mathrm{sb}} f_y \tag{6.12}$$

In the equations, A_{dc} is the cross-sectional area of the compressed concrete deck slab; A_{cc} is the cross-sectional area of the concrete infill in the compressed chord; the meanings of the remaining parameters are the same as those in the formulas found in Chapter 3 of this book.

Under load, the efficiency coefficient S for each truss member and the concrete slab can be calculated using Eq. (6.13):

$$S_i = \frac{N_i}{R_i} \tag{6.13}$$

In the equation, the subscript i can represent dc, c, t, or b.

In practical engineering, the cross-sectional area of the concrete slab in a composite truss girder is typically much larger than that of the chords to meet detailing requirements. When the sectional resistance of the compressed concrete slab, R_{dc}, is significantly greater than that of the compressed chord, R_c and k_1 can be taken as 1.

When the composite truss girder fails in bending due to the tension chord reaching its ultimate state first, its efficiency coefficient S_t should be higher than the efficiency coefficients of the concrete slab S_{dc} and the brace members S_b:

$$S_t > S_{\mathrm{dc}} \tag{6.14}$$

$$S_t > S_{\mathrm{sb}} \tag{6.15}$$

Substituting Eq. (6.13) into Eqs. (6.14) and (6.15) yields:

$$\frac{A_{dc}}{A_{st}} > \frac{f_y}{f_c}\left(\frac{N - n - 0.5}{N - n}\right) \tag{6.16}$$

$$\frac{\eta_{eff} A_{sb}}{A_{st}} > \frac{1}{\lambda \sin \theta} \tag{6.17}$$

In the equations, λ is the shear-span ratio, $\lambda = na/H$.

Therefore when the cross-sectional areas of the composite truss girder members simultaneously satisfy Eqs. (6.16) and (6.17), the failure mode is bending failure induced by yielding of the bottom chord. Similarly, the criteria for a bending failure mode caused by crushing of the concrete slab in compression, and for a shear failure mode caused by failure of the brace members or joints, can be derived.

The maximum bending moment in the composite truss girder under a concentrated load P is $Pna(N\text{-}n)/N$. When the composite truss girder reaches its ultimate state, the magnitude of the corresponding external load P can be calculated based on the resistance of the members. This allows for the further determination of the ultimate flexural capacity corresponding to different truss failure modes:

1. When the tension chord of the composite truss girder fails first: $M_u = A_{st} f_y H$.
2. When the brace member of the composite truss girder fails first: $M_u = \eta_{eff} A_{sb} f_y na \sin \theta$.
3. When the compressed concrete slab of the composite truss girder fails first: $M_u = \chi A_{dc} f_c H$.

The derived results are summarized in Table 6.7, which presents the failure mode criteria and bearing capacity calculation formulas for CFST composite truss girders under both deck-in-compression and deck-in-tension loading modes.

When the deck slab of the composite truss girder is located in the plane of the tension chord, the concrete cracks and ceases to contribute at an early stage due to its low tensile strength. However, the truss structure remains intact after the deck cracks and can continue to carry load. In this case, the tensile capacity provided by the deck slab and the tension chord is the sum of the tensile capacities of the steel tube and the longitudinal reinforcement in the deck slab. The tensile capacity provided by the longitudinal reinforcement in the deck slab can be equivalently incorporated into the chord's cross-sectional area, as shown in Eq. (6.18):

Table 6.7 Failure mode criteria and corresponding flexural capacity under positive bending moment (deck slab in compression).

Member cross-section dimension ratio	Failure mode	Flexural capacity
$\begin{cases} \dfrac{A_{st}}{\chi A_{dc}} < \dfrac{f_c}{f_y}\left(\dfrac{N-n}{N-n-0.5}\right) \\ \dfrac{A_{st}}{\varphi A_{sb}} < \lambda \sin\theta \end{cases}$	Failure of bottom chord	$M_u = A_{st} f_y H$
$\begin{cases} \dfrac{\chi A_{dc}}{A_{sb}} < \dfrac{f_y}{f_c}\left(\dfrac{N-n-0.5}{N-n}\right) \\ \dfrac{\chi A_{dc}}{\eta_{eff} A_{sb}} < 2cos\theta \dfrac{f_y}{f_c}\dfrac{n(N-n-0.5)}{N-n} \end{cases}$	Failure of deck slab	$M_u = \chi A_{dc} f_c H$
$\begin{cases} \dfrac{\eta_{eff} A_{sv}}{\chi A_{dc}} < \dfrac{f_c}{f_y}\dfrac{N-n}{n(N-n-0.5)}\dfrac{1}{2cos\theta} \\ \dfrac{\eta_{eff} A_{sb}}{A_{st}} < \dfrac{1}{\lambda \sin\theta} \end{cases}$	Failure of brace member	$M_u = \eta_{eff} A_{sb} f_y na \sin\theta$

$$k_{rt} = \frac{A_{st} f_y + A_{rt} f_{y_r}}{A_{st} f_y} \tag{6.18}$$

In the equation, A_{rt} is the cross-sectional area of the longitudinal reinforcement in the deck slab, f_{yr} is the yield strength of the longitudinal reinforcement, k_{rt} is the modification factor for the cross-sectional area of the tension chord, accounting for the influence of the longitudinal reinforcement in the deck slab.

It is evident that in this scenario, the failure mode of the composite truss girder is similar to that of a truss. Consequently, the failure mode criteria and bearing capacity calculation formulas for the CFST composite truss girder with the deck slab in tension are finalized and presented in Table 6.8.

6.4.2 Formula verification and comparison

A total of 24 test specimens were collected, comprising 17 CFST truss tests and 7 CFST composite truss girder tests. The statistical results of the test data are shown in Table 6.9, where the reference citations correspond to the references listed at the end of this chapter.

Based on the proposed formulas in this book for calculating the bearing capacity of rectangular steel tube and CFST joints, and the calculation methods provided in Table 6.7 and Table 6.8, a comparison between the

Table 6.8 Failure mode criteria and corresponding flexural capacity under negative bending moment (deck slab in tension).

Member cross-section dimension ratio	Failure mode	Flexural capacity
$\begin{cases} \dfrac{A_{sc}}{k_{rt}A_{st}} > \left(\dfrac{N-n-0.5}{N-n}\right)\left(\dfrac{\xi}{1+\xi}\right) \\ \dfrac{\eta_{eff}A_{sb}}{A_{st}} > \dfrac{1}{\lambda \sin\theta} \end{cases}$	Failure of tensile chord	$M_u = k_{rt} A_{st} f_y H$
$\begin{cases} \dfrac{\eta_{eff}A_{sb}}{k_{rt}A_{st}} < \dfrac{1}{\lambda \sin\theta} \\ \dfrac{\eta_{eff}A_{sb}}{A_{sc}} < \dfrac{(1+1/\xi)}{\lambda \sin\theta}\dfrac{(N-n)}{(N-n-0.5)} \end{cases}$	Failure of brace member	$M_u = \eta_{eff} A_{sb} f_y na \sin\theta$
$\begin{cases} \dfrac{A_{sc}}{k_{rt}A_{st}} < \left(\dfrac{N-n-0.5}{N-n}\right)\left(\dfrac{\xi}{1+\xi}\right) \\ \dfrac{\eta_{eff}A_{sb}}{A_{sc}} > \dfrac{(1+1/\xi)}{\lambda \sin\theta}\dfrac{(N-n)}{(N-n-0.5)} \end{cases}$	Failure of compressive chord	$M_u = \left(A_{sc} f_y + A_{cc} f_c\right) H \dfrac{(N-n)}{(N-n-0.5)}$

Table 6.9 Summary of test data for CFST trusses and composite truss girders.

No.	Literature source	Truss type	Loading mode	Number of specimens
1	Huang et al.	Four-chord circular CFST truss	Two-point symmetric loading	1
2	Han et al.He Shanhu	Three-chord circular CFST truss	Two-point symmetric loading	6
		Three-chord circular CFST composite truss girder	Two-point symmetric loading	4
3	Liu Yongjian et al.	Planar rectangular CFST truss	Mid-span loading	3
		Planar circular CFST truss	Mid-span loading	3
4	Gao Yimin et al.	Planar rectangular CFST truss	Mid-span loading	4
5	Chen Rong et al.	Three-chord circular CFST composite truss girder	Two-point symmetric loading	1
6	Liu Yongjian et al.	Planar rectangular CFST composite truss girder	Two-point symmetric loading	2

calculated values and the measured test values is presented in Table 6.10. Specimens numbered 1 to 17 are CFST trusses, and specimens numbered 18 to 24 are CFST composite truss girders. The specimen identifiers correspond to those used in the original references.

For some trusses whose failure mode involved chord failure and where the corresponding member stress reached the ultimate strength of the steel, the yield strength f_y in the flexural capacity formulas in Table 6.7 was replaced by the measured ultimate strength f_u from the tests cited in the literature.

Table 6.10 shows that the failure modes of the CFST trusses and composite truss girders calculated using the formulas proposed in this book are basically consistent with the failure modes observed in the tests. Furthermore, the calculated ultimate flexural capacities are quite close to the measured values (for all specimens, the average relative error $\mu = 1.27\%$, with a standard deviation $\sigma = 0.128$).

Based on the comparison results in Table 6.10, Fig. 6.30 presents a comparison between the calculated and measured flexural capacities for the CFST trusses and composite truss girders. The figure indicates that for the CFST truss tests, the measured capacities under the three different failure modes all agree well with the values calculated by the formulas proposed in this chapter. The average relative error between calculated and measured values is $\mu = 0.23\%$, with a standard deviation of $\sigma = 0.123$. The calculated flexural capacities for the CFST composite truss girders also show good agreement with the measured values, and the accuracy is even higher than that for the CFST trusses ($\mu = -1.50\%$, standard deviation $\sigma = 0.029$).

6.4.3 Design recommendations

Based on the formulas proposed in this chapter and the comparative analysis of test specimen results, the following design recommendations are provided for CFST trusses and composite truss girders:

1. During the design of CFST trusses and composite truss girders, the formulas proposed in this chapter can be used to verify the cross-sectional size ratios of various members. This ensures that the flexural capacities under different failure modes are relatively close, avoiding material waste and improving economic efficiency. It is also essential to ensure that the bending failure mode occurs before the shear failure mode.

Table 6.10 Comparison of failure modes and flexural capacity for CFST trusses and composite truss girders.

No.	Specimen No.	Specimen type	Comparison of failure modes		Comparison of flexural capacity (kN)		
			Calculated failure mode	Measured failure mode	Calculated value M_{uc}	Measured value M_{ue}	$\frac{M_{uc}-M_{ue}}{M_{ue}}$
1	R0	I-1	Joint failure	Failure of top chord joint	141.09	154.44	−8.64%
2	R1	I-2	Joint failure	Failure of bottom chord joint	141.09	158.73	−11.11%
3	R2	I-3	Joint failure	Failure of bottom chord joint	141.09	177.61	−20.56%
4	RCFST	I-3	Joint failure	Failure of bottom chord joint	388.35	357.05	8.77%
5	PBL-1	I-3	Joint failure	Failure of bottom chord joint	488.50	453.55	7.70%
6	PBL-2	I-3	Joint failure	Failure of bottom chord joint	604.79	525.93	14.99%
7	PBL-3	I-3	Joint failure	Failure of bottom chord joint	604.24	564.53	7.03%
8	W2	I-2	Brace member failure	Failure of compressive brace member	121.51	90.50	34.26%
9	C0	I-1	Joint failure	Failure of top chord joint	202.69	180.02	12.59%
10	C1	I-2	Joint failure	Failure of top chord joint	202.69	199.07	1.82%
11	C2	I-3	Joint failure	Failure of top chord joint	313.84	268.31	16.97%
12	T8s-1	I-3	Brace member failure	Failure of compressive brace member	157.57	175.20	−10.06%
13	T8s-2	I-3	Brace member failure	Failure of compressive brace member	157.57	170.10	−7.37%

14	T8–1	I-3	Bottom chord failure	Bottom chord failure	305.55	279.00	9.52%
15	T8–2	I-3	Bottom chord failure	Bottom chord failure	305.55	279.00	9.52%
16	T6–1	I-3	Bottom chord failure	Bottom chord failure	305.55	377.60	−19.08%
17	T6–2	I-3	Bottom chord failure	Bottom chord failure	305.55	333.60	−8.41%
18	T8b-1	II-3	Bottom chord failure	Bottom chord failure	321.85	333.00	−3.35%
19	T8b-2	II-3	Bottom chord failure	Bottom chord failure	321.85	313.20	2.76%
20	T8b-180–1	II-1	Bottom chord failure	Bottom chord failure	332.98	327.60	−1.64%
21	T8b-180–2	II-1	Bottom chord failure	Bottom chord failure	332.98	335.40	−0.72%
22	CR	II-3	Bottom chord failure	Bottom chord failure	237.33	255.20	−7.00%
23	CTB1	II-3	Top chord failure	Top chord failure	899.72	914.66	−1.63%
24	CTB2	II-3	Top chord failure	Top chord failure	849.09	839.96	+1.09%

In the specimen types, I-1, I-2, and I-3 represent a steel tube truss, a truss with only the top chord being concrete-filled, and a truss with both top and bottom chords being concrete-filled, respectively. II-1 and II-3 represent a composite truss girder with hollow chords and a composite truss girder with both chords concrete-filled, respectively.

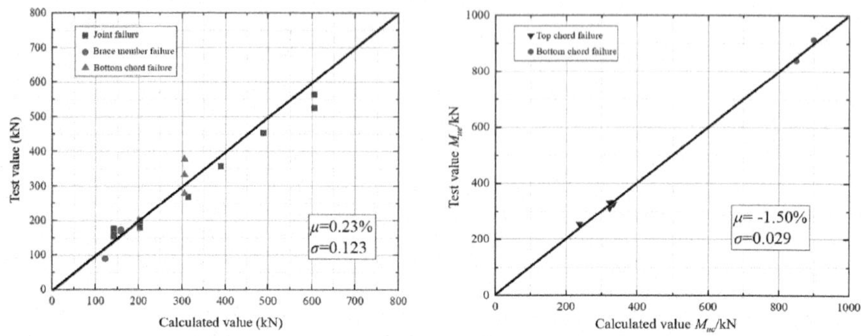

(A) (B)

Figure 6.30 Comparison of calculated and measured flexural capacity for CFST trusses and composite truss girders. (A) CFST truss. (B) CFST composite truss girder.

2. Strive to avoid shear failure in composite truss girders caused by the failure of brace members or joints: (1) Structural measures can be adopted to enhance joint capacity and prevent joint failure preceding member failure. These measures include using joints where the brace and chord have equal widths, filling the chord at the joint with concrete, and installing PBL stiffeners. (2) Control the product of the shear-span ratio (λ) and the sine of the brace member inclination angle ($\sin\theta$) under the most unfavorable loading position to avoid shear failure induced by brace member failure.
3. For the positive moment region of a composite truss girder where the deck slab is in compression, the compression chord and the concrete slab, integrated through shear connectors, form a robust composite section. This makes the tension chord relatively weaker. Therefore the size of the compression chord can be appropriately reduced, meeting only the requirements of the construction stage, while the cross-sectional area of the bottom (tension) chord should be adequately increased to prevent its premature failure.

6.5 Summary

Experiments were conducted on rectangular CFST composite truss girders subjected to negative bending moments, investigating the composite action when the deck slab is in tension. The flexural performance of the composite truss girders utilizing PBL-stiffened rectangular CFST equal-width joints was fully demonstrated.

A full-scale test was performed on a rectangular CFST composite truss girder used in an actual engineering project. The results indicated that the concrete infill within the tubes was in good working condition. The incorporation of PBL stiffeners in the rectangular CFST chords enhanced the axial stiffness of the members, and the engineering application achieved the expected performance level.

A calculation method for the flexural capacity of CFST composite truss girders, which accounts for the composite action between the deck and the truss, was proposed. This method is characterized by clear physical concepts, simplicity, and practicality. It shows good agreement with existing test results and can provide effective guidance for the design of CFST composite truss girders.

Extended reading

Chen R, Chen JB, Yao JF, et al. Experimental study on bending performances of spatial pipe truss concrete composite beams[J]. Journal of Henan University Of Urban Construction 2014;23(02):23–7.

Gao YM, Liu YJ, Jiang L, et al. Experiment on flexural behaviour of rectangular concrete filled steel tubular truss stiffened with PBL[J]. Journal of Architecture and Civil Engineering 2017;34(05):171–80.

Han L, Wu X, He S, et al. Flexural behaviour of CFST chord to hollow tubular brace truss: experiments[J]. Journal of Constructional Steel Research 2015;105(15):137–51.

He SH. Flexural performance of three-main components CFST chord - steel tubular brace truss[D]. Beijing: Tsinghua University; 2012.

Huang W, Fenu L, Chen B, et al. Experimental study on joint resistance and failure modes of CFST truss girders[J]. Journal of Constructional Steel Research 2018;141:241–50.

Liu B, Liu YJ, Jiang L, et al. Flexural behavior of concrete-filled rectangular steel tubular composite truss beams in the negative moment region[J]. Engineering Structures 2020;216:110738.

Liu YJ, Liu JP, Yang GJ, et al. Experimental research on mechanical behavior of RHS trusses with concrete-filled in chord[J]. Journal of Building Structures 2009;30(06):107–12.

Liu YJ, Liu JP, Zhang JG. Experimental research on RHS and CHS truss with concrete-filled chord[J]. Journal of Building Structures 2010;31(04):86–93.

Liu YJ, Ma YP, Tian ZJ, et al. Field test of rectangular concrete filled steel tubular composite truss bridge with continuous rigid system[J]. China Journal of Highway and Transport 2018;31(05):53–62.

Liu YJ, Wang KN, Liu B, et al. Experimental research on mechanical behavior of RCFST composite truss beam under negative bending[J]. Journal of Building Structures 2019;40(09):74–83.

Liu YJ, Xiong ZH, Luo YL, et al. Double-composite rectangular truss bridge and its joint analysis [J]. Journal of Traffic and Transportation Engineering (English Edition 2015;2(4):249–57.

Ma YP, Liu YJ, Long X, et al. Simplified calculation method for flexural strength of CFST composite truss girder[J]. Journal of Building Structures 2020;41(05):76–84.

CHAPTER 7

Calculation methods for concrete-filled rectangular steel tubular composite truss bridges

7.1 Overview

This chapter summarizes existing experimental and theoretical research findings, proposing detailing requirements, structural internal force analysis methods, and bearing capacity calculation methods applicable to concrete-filled rectangular steel tubular (CFRST) composite truss bridges. These calculation methods can provide reference and basis for determining structural parameters and optimizing the design of such bridges.

7.2 Detailing requirements for concrete-filled rectangular steel tubular composite truss girders

7.2.1 Joint detailing requirements

7.2.1.1 T-type joint

To enhance joint bearing capacity, for long-span simply supported trusses, concrete infill can be placed only within the chords near the supports where shear forces are high, as shown in Fig. 7.1. For trusses with smaller spans, concrete may be filled along the entire length of the chords for construction convenience. For continuous trusses, both the top and bottom chords can utilize CFRST sections (Fig. 7.2). This approach not only increases the bearing capacity of the joints but also enhances the load-bearing capacity and stiffness of the compression chords, allowing for a reduction in truss depth and saving steel material. To ensure joint capacity and the quality of concrete placement and compaction, the chord members (main tubes) of the CFRST truss should remain continuous, and the brace members should not penetrate through the chords.

Fig. 7.3 shows the joint details for the CFRST truss illustrated in Fig. 7.2. Concentrated loads are applied at top chord joints 1, 3, and 5 and at bottom chord support joint 2, where steel bearing plates are provided.

Composite Truss Bridges With Concrete-filled Rectangular Steel Tube Members
DOI: https://doi.org/10.1016/B978-0-443-44011-3.00008-1

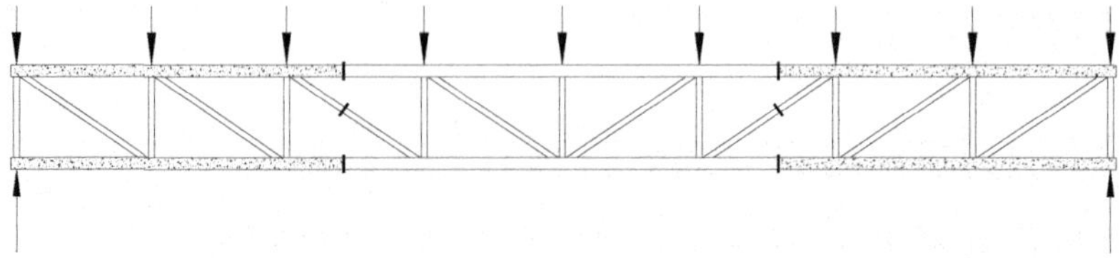

Figure 7.1 Simply supported truss with concrete-filled rectangular steel tubular partially filled with concrete near the supports.

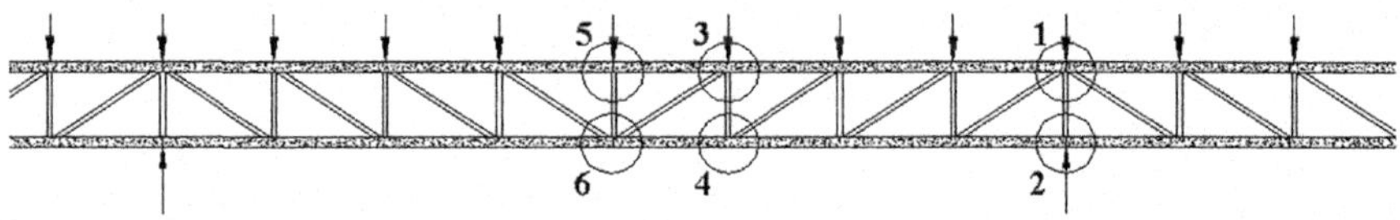

Figure 7.2 Continuous truss with concrete-filled rectangular steel tubular (chords filled with concrete).

Based on the force transfer mechanism, joint 2 and joint 5 are classified as X-type compression joints. For joints 1 and 3, part of the internal force in the vertical brace member balances the concentrated load, while another part balances the vertical component of the diagonal brace member's force. Therefore joints 1 and 3 can be considered as combined K-type and X-type joints. Bottom chord joint 4 is a K-type joint, while joint 6 should be treated as a pair of K-type joints.

For joints in CFRST trusses where concrete is filled locally near the joints, the concrete fill length should satisfy $L_c \geq \frac{h_1}{\sin\theta_1} + 4h_0$, and the concrete must be ensured to be compacted densely.

The detailed requirements for welds in CFRST joints should comply with the current code "*Standard for design of steel structures*" (GB 50017-2017).

7.2.1.2 Joint dimensions

For welded tubular joints where the eccentricity of the connection between the brace member and the chord satisfies the limit specified in Eq. (7.1), the influence of the bending moment induced by the eccentricity may be ignored when calculating the bearing capacity of the joint and the tension chord. However, this eccentric moment, calculated as $M = \Delta N \times e$ (where ΔN is the difference in axial force between the

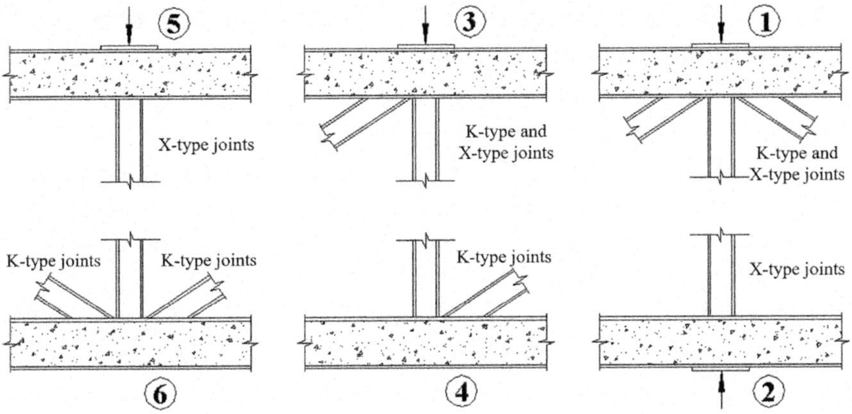

Figure 7.3 Continuous truss with concrete-filled rectangular steel tubular (chords filled with concrete).

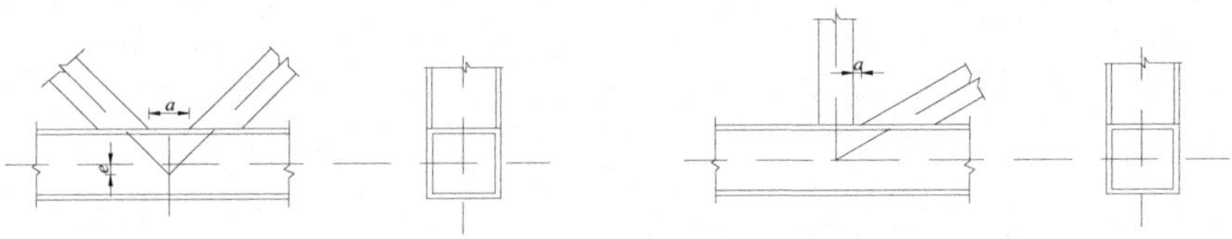

(A) (B)

Figure 7.4 Eccentricity and gap in K-type and N-type tubular joints. (A) Gapped K-type joint ($e > 0$). (B) Gapped N-type joint (e=0).

two sides of the chord at the joint), must be considered for the compression chord. The joint gap is illustrated in Fig. 7.4.

$$-0.55 \leq e/h \leq 0.25 \tag{7.1}$$

The width ratio β between the brace member and the chord should fall within the following ranges: for T-type, Y-type, or X-type joints, $0.25 \leq \beta \leq 1$, and for K-type or N-type gap joints, $0.35 \leq \beta \leq 1$. Joints with $\beta = 1$ (equal-width joints) exhibit superior mechanical performance and are recommended as the preferred choice.

$$0.5[b_0 - (b_1 + b_2)/2] \leq g \leq 1.5[b_0 - (b_1 + b_2)/2] \,\&\, g \geq t_1 + t_2 \tag{7.2}$$

7.2.2 Detailing requirements for concrete-filled rectangular steel tubular members

The following detailed recommendations are proposed for CFRST members:

1. The wall thickness of the chord should not be less than that of the brace. The width-to-thickness or depth-to-thickness ratio of the steel tube for the chord and tension braces should not be greater than $35\sqrt{235/f_y}$, nor less than $15\sqrt{235/f_y}$.
2. The width-to-thickness or depth-to-thickness ratio of the steel tube for compression braces should not be greater than $\sqrt{E_S/f_y}$.
3. The width ratio between the brace and chord (β): for T-type, Y-type, or X-type joints, β should be $0.25 \le \beta \le 1.0$. For K-type joints, β should be $0.35 \le \beta \le 1.0$.
4. The inclination angle of the brace members should generally not be less than 45 degrees. A reasonable angle is about 50 degrees for trusses with vertical members and about 60 degrees for trusses without vertical members.
5. The minimum side dimension of a CFRST member's cross section should not be less than 100 mm. The steel tube wall thickness should not be less than 4 mm. The aspect ratio (depth to width) of the section should not exceed 2.
6. When Perforated Bond Link (PBL) stiffeners are installed inside a CFRST chord, the thickness of the PBL plate should not be less than the wall thickness of the steel tube. The height of the PBL plate is recommended to be 1/4 to 1/3 of the tube's depth. The diameter of the perforations is recommended to be 1/2 to 3/4 of the PBL plate height. The spacing between the centers of the perforations is also recommended to be equal to the PBL plate height. PBL stiffeners can be arranged at the midpoint or third points of the side and top plates of the rectangular steel tube and should run continuously along the length.

7.3 Calculation of internal forces in concrete-filled steel tubular composite truss girders

7.3.1 Simplified calculation method

When the linear stiffness (the ratio of sectional flexural rigidity to member length) of the truss members is relatively small, or when the sectional depth is less than 1/15 of the member length, the internal forces in the structure can be calculated based on a pin-connected truss model.

7.3.2 Finite element method using beam elements

In the finite element model using beam elements for concrete-filled steel tubular (CFST) structures, the members of the CFST truss are simulated using beam elements. The truss members are connected at the joints using shared nodes. The bridge deck is also modeled using beam elements. When the transverse width of the deck is large, multiple beam elements can be used, accompanied by virtual beam elements (considering stiffness only, not weight) to simulate its transverse stiffness.

When the deck is connected to the main truss by being embedded, the shared node method can be used for simulation. When shear connectors are used for the connection, elastic links can be employed, with the tangential and normal spring stiffnesses calculated based on the cross-sectional dimensions of the shear connectors.

For CFST members, when it is necessary to simulate the construction stage involving concrete pouring into the tubes, it is recommended to use a staged composite section for simulation. If only the completed bridge stage is analyzed or simulating the concrete pouring process is not required, the dual-element model can be used. For CFST members in tension, a conservative approach is to simulate them using only the steel tube element. Taking the CFRST composite truss bridge described in Chapter 8 as the engineering background, the full-bridge finite element model established using beam elements is shown in Fig. 7.5.

The beam-element finite element model is suitable for analyzing the global mechanical behavior of the entire bridge. Compared to solid-element finite element models, it offers advantages such as faster modeling and higher analysis efficiency. However, beam-element models cannot accurately simulate stress in regions with complex force conditions, such as truss joints.

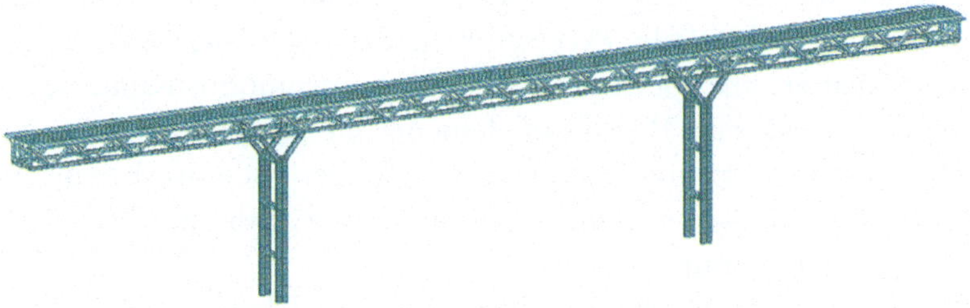

Figure 7.5 Full-bridge finite element modeling using beam elements.

7.3.3 Multiscale finite element calculation method

7.3.3.1 Characteristics of multiscale finite element models

Depending on the scale of the object analysis, finite element analysis of CFST composite truss bridges can be primarily divided into two types: beam-element models for analyzing the global mechanical performance of the entire bridge and shell/solid-element models for detailed analysis of local structures like joints. However, the former cannot analyze the detailed stress state in complex structural regions, while the latter involves high computational costs. A real bridge structure exhibits mechanical behavior at both the global structural scale and the local member scale, and these interact and are closely linked. A major limitation of traditional numerical models is the use of a single-scale model to simulate a multiscale real structure.

Under these circumstances, full-bridge multiscale finite element models have emerged. Their advantages are mainly twofold: Firstly, they allow for the creation of refined finite element models in regions of interest and simplified models in noncritical areas. This enables the analysis of mechanical performance at different scales within a single multiscale model, offering high computational efficiency. Secondly, full-bridge multiscale models can accurately simulate the boundary conditions and loading patterns that members and the structure experience in the real bridge, leading to highly precise finite element calculation results.

7.3.3.2 Establishment of the multiscale finite element model

Again, using the CFRST composite truss bridge from Chapter 8 of this book as the engineering background and based on the full-scale bridge test described in Section 6.3, a full-bridge multiscale finite element model was established using Abaqus, as shown in Fig. 7.6. This model incorporates three scales: the full bridge, the bridge deck, and the truss members.

At the full-bridge scale, except for the deck slab and the instrumented members that are the focus of interest, the remaining truss members and piers are modeled using B31 beam elements. CFST members are simulated by converting them into equivalent concrete sections based on the principle of stiffness equivalence. The mesh size at the full-bridge scale is approximately 600 mm.

At the bridge deck scale, the deck slab is modeled using C3D8R solid elements, with mesh refinement applied at the instrumented sections of the deck. To accurately simulate the stress state of the deck, a reinforcement

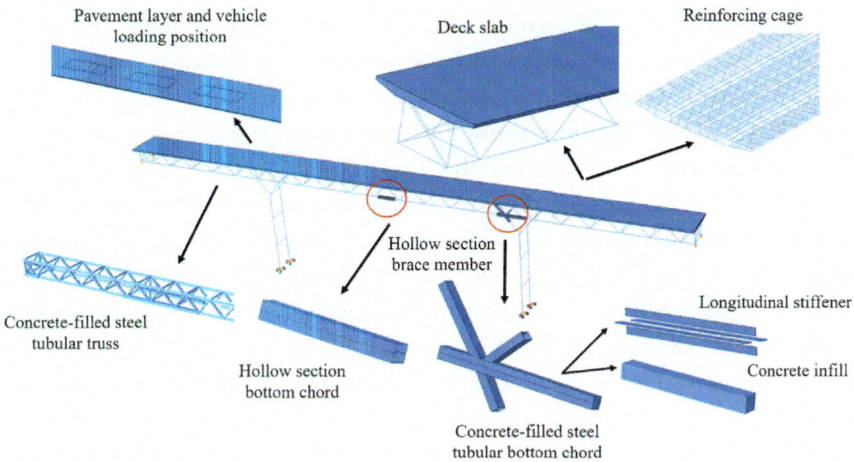

Figure 7.6 Full-bridge multiscale finite element modeling.

skeleton is modeled using B31 beam elements, and the wearing surface is modeled using C3D8R solid elements. The wearing surface is connected to the deck slab using the "Tie" constraint. The reinforcement skeleton is embedded within the deck slab elements using the "Embedded" constraint. The deck solid elements and the top chord B31 beam elements at the full-bridge scale are also connected using the "Embedded" constraint. The mesh size for the deck slab and wear surface ranges from 100 mm to 300 mm.

At the member scale, for Case 1 and Case 3, axial strain measurements were taken on the hollow steel bottom chord at mid-span, the CFST bottom chord at the pier top, and the hollow steel compression brace member at the pier top. The steel tubes and internal stiffeners of these three members are modeled using S4R shell elements. The concrete infill in the pier top bottom chord is modeled using C3D8R solid elements.

The internal stiffeners are integrated with the concrete infill elements using the "Embedded" constraint. The concrete infill is connected to the inner surface of the steel tube wall using a contact pair: hard contact in the normal direction and a Coulomb friction model in the tangential direction, with a critical bond strength of 0.462 MPa and an interface friction coefficient of 0.6.

The members of interest are connected to the other members modeled with beam elements using a "Coupling" constraint to eliminate redundant degrees of freedom. To avoid Saint-Venant's principle effects, the model length of each member is appropriately extended so that the distance from the test section to the coupling location is greater than three times the chord's cross-sectional dimension. The mesh size for the steel members and the concrete infill is approximately 25 mm.

The vehicle loads are applied to the wearing surface elements according to their positions in the full-scale test. The model ignores the soil-pile interaction at the pier bases, which are modeled as fixed supports. Sliding bearings are provided at both ends of the bridge; therefore only the vertical displacement is constrained at these ends in the model. In the initial model, the elastic modulus for both the infill concrete and the deck concrete is taken as 3.5×10^4Mpa with a Poisson's ratio of 0.2. For the steel, the initial model uses an elastic modulus of 2.06×10^5Mpa and a Poisson's ratio of 0.3.

To conduct a comparative analysis of the computational efficiency and accuracy of the multiscale finite element model, a conventional finite element analysis model (hereinafter referred to as the solid model) composed of solid and shell elements with similar mesh density was also established. A comparison of the number of elements after meshing and computational efficiency under identical software and hardware environments is shown in Table 7.1. A comparison of the errors between the finite element analysis results of the two models and the measured results from the full-scale bridge test described in Section 6.3 of this book is shown in Fig. 7.7.

As can be seen from Table 7.1 and Fig. 7.7, the multiscale finite element model significantly reduces the number of elements through simplification, thereby improving the efficiency of the finite element analysis. At the same

Table 7.1 Comparison of computational efficiency between the multiscale model and the solid model.

Finite element model type	Number of elements			Computation time
	Shell/plate element	Solid element	Beam element	
Multiscale model	20,752	30,655	19,640	12 min
Solid-element model	34,462	497,215	15,764	152 min

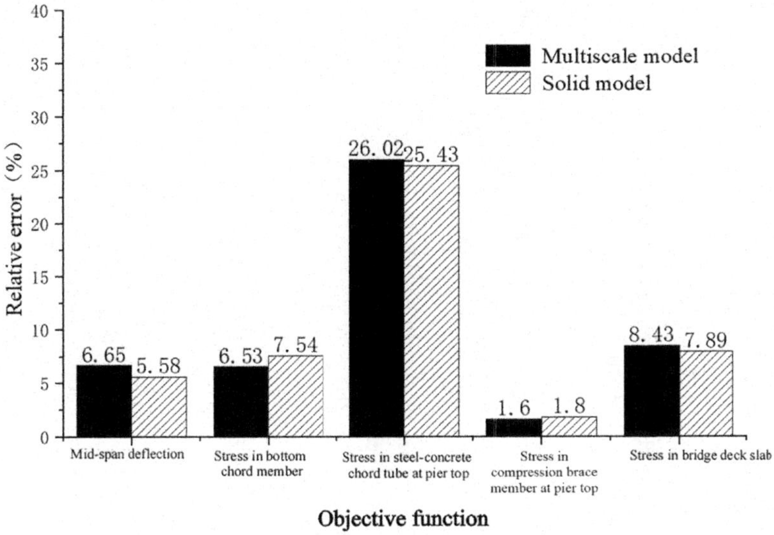

Figure 7.7 Comparison of calculation results between the multiscale model and the solid model.

time, it ensures the accuracy of the finite element calculation results for the regions of interest. Therefore the multiscale full-bridge finite element model can simultaneously analyze both the global and local mechanical behavior of bridge structure, characterized by high efficiency and high precision.

7.4 Calculation method for bearing capacity of concrete-filled rectangular steel tubular composite truss girders

The calculation method for the bearing capacity of CFRST members can be referred to the "*Technical specification for structures with CFRST members*" CECS 159:2004. Based on existing research findings, this section provides calculation methods for the bearing capacity of CFRST joints and composite truss girders (Tables 7.2 and 7.3).

7.4.1 Calculation method for bearing capacity of concrete-filled rectangular steel tubular joints

The joint types in CFRST composite truss bridges mainly include rectangular hollow section joints and CFRST joints. This section

Table 7.2 Calculation formulas for bearing capacity of rectangular hollow section joints.

Joint type		Applicable conditions	Parametric strength formula
Axially loaded brace	T-type, Y-type, and X-type joints	$\beta \leq 0.85$	$N_{iu} = \frac{f_{y0} t_0^2}{(1-\beta)\sin\theta_i}\left[\frac{2h_i}{b_0 \sin\theta_i} + 4(1-\beta)^{0.5}\right] f(n)$ Where $f(n)$ is the influence factor accounting for the effect of chord stress on joint strength, $n = \frac{N_0}{A_0 f_{yo}} + \frac{M_0}{W_0 f_{y0}}$; When $n \geq 0$(chord in tension) $f(n) = 1.0$; When $n < 0$(chord in compression) $f(n) = 1.3 + \frac{0.4}{\beta} n$ and $f(n) \leq 1.0$; Where N_0 and M_0 are the axial force and bending moment in the chord, respectively.
		$\beta \leq 0.85$	$N_{iu} = f_{yi} t_i (2h_i - 4t_i + 2b_e)$ $b_e = \frac{10}{b_0/t_0} \frac{f_{y0} t_0}{f_{yi} t_i} b_i \leq b_i$
		$0.85 \leq \beta \leq (1 - 1/\gamma)$	$N_{iu} = \frac{f_{y0} t_0}{\sqrt{3}\sin\theta_i}\left(\frac{2h_i}{\sin\theta_i} + 2b_{ep}\right)$ $b_{ep} = \frac{10}{b_0/t_0} b_i \leq b_i$
		β=1.0	$N_{1u} = \frac{f_{k0} t_0}{\sin\theta_1}\left(\frac{2h_1}{\sin\theta_1} + 10t_0\right)$ Where: T- and Y-type joints: $f_{k0}=\phi f_{y0}$; X-type joint: $f_{k0}=0.8\sin\theta_1(\phi f_{y0})$; Where ϕ is the equivalent column buckling reduction factor, determined based on the slenderness ratio $\lambda = 3.46\left(\frac{h_0}{t_0} - 2\right)\left(\frac{1}{\sin\theta_1}\right)^{0.5}$

	K-type joint	$\beta \le 0.85$	$N_2^u = \min\left\{ 8.9 \frac{f_{y0} t_0^2}{\sin\theta_i} \left(\frac{b_1 + b_2 + h_1 + h_2}{4b_0} \right) \gamma^{0.5} f(n), \; f_{yi} t_i (2h_i - 4t_i + b_i + b_e) \right\}$
		$0.85 \le \beta \le (1 - 1/\gamma)$	$N_{iu} = \frac{f_{y0} t_0}{\sqrt{3} \sin\theta_i} \left(\frac{2h_i}{\sin\theta_i} + b_i + b_{ep} \right)$
Brace subjected to bending moment		$\beta \le 0.85$	$M_u = f_{y1} [Z_1 - (1 - b_e/b_1) b_1 t_1 (h_1 - t_1)]$

Table 7.3 Calculation formulas for bearing capacity of concrete-filled rectangular steel tubular joints.

Joint in tension	Joint type	Applicable conditions	Parametric strength formula
	T-, Y-, and X-type joints	$\beta < 0.7$	$N_2^u = \frac{ft_0^2}{(1-\beta)\sin\theta_2}\left[\frac{2h_2}{b_0 \sin\theta_2} + 4(1-\beta)^{0.5}\right]$
		$1 \le \beta < 0.7$	$N_2^u = \min\left\{\begin{array}{l} f_2 t_2(2h_2 - 4t_2 + b_2 + b_e), \\ \frac{\sqrt{3}}{3} ft_0(2h_2/\sin\theta_2 + b_2 + b_{ep})/\sin\theta_2 \end{array}\right\}$ $b_e = \frac{10}{b_0/t_0}\frac{f_{y0}t_0}{f_{y2}t_2}b_2 \le b_2$
	K-type joint	$\beta < 0.7$ and $g \ge 0.5b_0\sqrt{1-\beta}$	$N_2^u = \frac{ft_0^2}{(1-\beta)\sin\theta_2}\left[\frac{2h_2}{b_0 \sin\theta_2} + 4(1-\beta)^{0.5}\right]$
		$\beta < 0.7$ and $g < 0.5b_0\sqrt{1-\beta}$	$N_2^u = \min\left\{\begin{array}{l} f_2 t_2(2h_2 - 4t_2 + b_2 + b_e), \\ \frac{\sqrt{3}}{3} ft_0(2h_2/\sin\theta_2 + b_2 + b_{ep})/\sin\theta_2 \end{array}\right\}$
		$1 \le \beta < 0.7$	$N_2^u = \min\left\{\begin{array}{l} f_2 t_2(2h_2 - 4t_2 + b_2 + b_e), \\ \frac{\sqrt{3}}{3} ft_0(2h_2/\sin\theta_2 + b_2 + b_{ep})/\sin\theta_2 \end{array}\right\}$
Joint in compression	$N_1^u = (1 + k_s)k_c f_c A_b$ Where: $k_s = \frac{3f_{y0}}{\nu_c f_{ck}}\frac{(k_c h_1/h_0)^2}{(h_0/t_0)^2}k_c = \sqrt{A_d/(b_1 h_1/\sin\theta_1)}$; When the concrete grade inside the tube exceeds C40: $k_c = \frac{190 - f_{cu}}{150}\sqrt{A_d/A_b}$; $A_b = b_1 h_1/\sin\theta_1$; When $3b_1 < b_0$: $A_d = 3b_1(h_1/\sin\theta_1 + 2b_1)$; When $3b_1 \ge b_0$: $A_d = b_0 h_1(1 + 2\alpha)/\sin\theta_1, \alpha = b_1\sin\theta_1/h_1 \le 1.0$		

Joint subjected to bending moment	$M_u = f_i (Z_i - (1 - b_e/b_i) b_i t_i (h_i - t_i))$ Where: $Z_i = b_i h_i t_i + \frac{t_i (h_i - 2t_i)^2}{2}$; $b_e = \frac{10}{b_0 / t_0} \frac{f_{y0} t_0}{f_{yi} t_i} b_i \leq b_i$

summarizes the calculation methods for the bearing capacity of these two types of joints.

7.4.2 Calculation method for flexural capacity of concrete-filled steel tubular composite truss girders

The derivation and verification of the flexural capacity calculation formulas for CFST trusses and composite truss girders have been presented in Sections 3.1 and 6.4. Here, only the verified calculation methods are summarized and listed. When the deck slab of the composite truss girder is in compression under nodal loading, the simplified calculation model is shown in Fig. 7.8. In this case, the failure modes and flexural capacity of the composite truss girder can be calculated according to Table 7.4. When the deck slab is in tension, the failure modes and flexural capacity can be calculated according to Table 7.5.

For calculations under complex loading conditions, the complex loads should first be simplified into a combination of a series of nodal loads. Then, linear superposition can be performed based on the two loading modes described above (Tables 7.6 and 7.7).

7.4.3 Fatigue calculation method for joints of concrete-filled rectangular steel tubular trusses

7.4.3.1 Fatigue load

Fatigue Load Model I uses an equivalent lane load, consisting of a concentrated load of $0.7P_k$ and a uniformly distributed load of $0.3q_k$·P_k and q_k shall be taken according to the current "*General Specifications for Design of Highway Bridges and Culverts*" (JTG D60-2015). The fatigue load

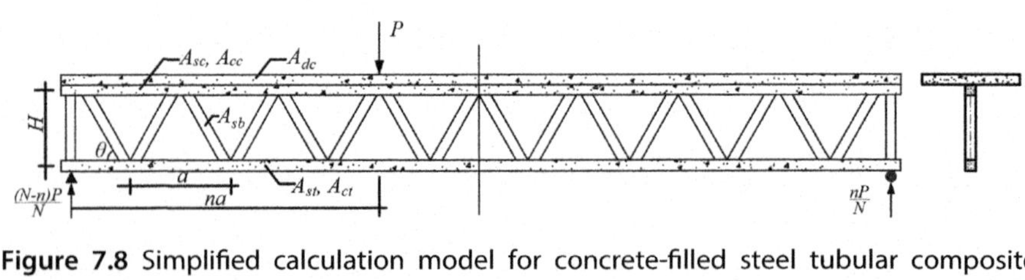

Figure 7.8 Simplified calculation model for concrete-filled steel tubular composite truss girders.

Table 7.4 Failure mode criteria and corresponding flexural capacity under nodal loading (deck slab in compression).

Member cross-section dimension ratio	Failure pattern	Flexural capacity
$\begin{cases} \dfrac{A_{st}}{\chi A_{dc}} < \dfrac{f_c}{f_y}\left(\dfrac{N-n}{N-n-0.5}\right) \\ \dfrac{A_{st}}{\varphi A_{sb}} < \lambda \sin\theta \end{cases}$	Failure of bottom chord	$M_u = A_{st} f_y H$
$\begin{cases} \dfrac{\chi A_{dc}}{A_{sb}} < \dfrac{f_y}{f_c}\left(\dfrac{N-n-0.5}{N-n}\right) \\ \dfrac{\chi A_{dc}}{\eta_{eff} A_{sb}} < 2cos\theta \dfrac{f_y}{f_c} \dfrac{n(N-n-0.5)}{N-n} \end{cases}$	Failure of concrete slab	$M_u = \chi A_{dc} f_c H$
$\begin{cases} \dfrac{\eta_{eff} A_{sv}}{\chi A_{dc}} < \dfrac{f_c}{f_y} \dfrac{N-n}{n(N-n-0.5)} \dfrac{1}{2cos\theta} \\ \dfrac{\eta_{eff} A_{sb}}{A_{st}} < \dfrac{1}{\lambda \sin\theta} \end{cases}$	Failure of brace member	$M_u = \eta_{eff} A_{sb} f_y na \sin\theta$

shall be applied in the most unfavorable position, and the impact factor for the fatigue load shall be calculated as specified (Tables 7.8 and 7.9).

Fatigue Load Model II adopts a two-vehicle model. The two model vehicles have the same axle spacing and axle weights. The axle weight and spacing configuration for a single vehicle are shown in Fig. 7.9. When loading, the distance between the centers of the two model vehicles shall not be less than 40 m.

7.4.3.2 *Hot-spot stress calculation*

When using the hot-spot stress method to assess joint fatigue life, a three-dimensional solid model can be established to directly obtain the hot-spot stress through extrapolation. For joint types with sufficient research foundation, the hot-spot stress range can be obtained indirectly by multiplying the nominal stress range in the member by the corresponding hot-spot stress concentration factor (SCF). For K-type joints commonly used in bridge structures, they are typically divided into three basic load cases: balanced loading in the braces (tension/compression), chord axial force, and chord bending moment, as shown in Fig. 7.10. The total hot-spot stress range is obtained by the linear superposition of the hot-spot stress ranges from each load case, expressed as:

$$S_h = SCF_I \cdot \sigma_I + SCF_{II} \cdot \sigma_{II} + SCF_{III} \cdot \sigma_{III} \tag{7.3}$$

Table 7.5 Failure mode criteria and corresponding flexural capacity under nodal loading (deck slab in tension).

Member cross-section dimension ratio	Failure pattern	Flexural capacity
$\begin{cases} \dfrac{A_{sc}}{k_{rt}A_{st}} > \left(\dfrac{N-n-0.5}{N-n}\right)\left(\dfrac{\xi}{1+\xi}\right) \\ \dfrac{\eta_{eff}A_{sb}}{A_{st}} > \dfrac{1}{\lambda \sin\theta} \end{cases}$	Failure of tensile chord	$M_u = k_{rt}A_{st}f_y H$
$\begin{cases} \dfrac{\eta_{eff}A_{sb}}{k_{rt}A_{st}} < \dfrac{1}{\lambda \sin\theta} \\ \dfrac{\eta_{eff}A_{sb}}{A_{sc}} < \dfrac{(1+1/\xi)}{\lambda \sin\theta}\dfrac{(N-n)}{(N-n-0.5)} \end{cases}$	Failure of brace member	$M_u = \eta_{eff}A_{sb}f_y\, na \sin\theta$
$\begin{cases} \dfrac{A_{sc}}{k_{rt}A_{st}} < \left(\dfrac{N-n-0.5}{N-n}\right)\left(\dfrac{\xi}{1+\xi}\right) \\ \dfrac{\eta_{eff}A_{sb}}{A_{sc}} > \dfrac{(1+1/\xi)}{\lambda \sin\theta}\dfrac{(N-n)}{(N-n-0.5)} \end{cases}$	Failure of compressive chord	$M_u = \left(A_{sc}f_y + A_{cc}f_c\right)H\dfrac{(N-n)}{(N-n-0.5)}$

Notes: η_{eff} in the formula is the reduction factor of brace member resistance, which can be calculated according to Table 7.2, Table 7.3, and *Code for Design of Highway Steel Structure Bridges* (JTG d64-2015).

Table 7.6 Parameter formulas for SCF_{max} of concrete-filled rectangular steel tubular K-type joints.

Location	SCF_{max} formula
Load case (I): tension-compression balanced load of brace members	
Chord	$SCF_{max} = (0.437 + 0.121\beta + 0.046\beta^2)\cdot(2\gamma)^{0.626}\cdot\tau^{0.311}\cdot(g')^{-0.00005}\cdot(\sin\theta)^{0.793}$
Brace member	$SCF_{max} = (0.529 + 0.646\beta + 0.131\beta^2)\cdot(2\gamma)^{0.509}\cdot\tau^{0.162}\cdot(g')^{-0.00005}\cdot(\sin\theta)^{0.420}$
Load case (II): axial force of chord members	
Chord	$SCF_{max} = (1.170 + 0.116\beta - 0.341\beta^2)\cdot(2\gamma)^{0.139}\cdot\tau^{-0.692}\cdot(g')^{-0.006}\cdot(\sin\theta)^{0.194}$
Brace member	$SCF_{max} = 0$
Load case (III): bending moment of the chord	
Chord	$SCF_{max} = (2.048 + 0.495\beta - 0.852\beta^2)\cdot(2\gamma)^{0.047}\cdot\tau^{-0.537}\cdot(g')^{-0.003}\cdot(\sin\theta)^{-0.068}$
Brace member	$SCF_{max} = 0$
Application scope: $10 \le 2\gamma \le 35$;$0.25 \le \tau \le 1.0$;$0.4 \le \beta \le 1.0$;$30° \le \theta \le 60°$	

Note: The calculated stress concentration factor *(SCF)* is not less than 1.

Table 7.7 Parameter formulas for SCF_{max} of Perforated Bond Link–stiffened concrete-filled rectangular steel tubular K-type joints with gusset plates (θ = 30°).

Location	SCF_{max} formula
Load case (I): tension-compression balanced load of brace members	
Chord	$SCF_{max} = 0.211 \cdot 2\gamma^{0.512} \cdot \tau^{0.608}$
Brace member	$SCF_{max} = 0.013 \cdot 2\gamma^{0.629} \cdot -8.257\tau^2 + 12.408\tau + 0.005^{1.705}$
Load case (II): axial force of chord members	
Chord	$SCF_{max} = 1.962 \cdot 2\gamma^{0.202} \cdot 24.596\tau^2 - 30.637\tau + 14.996^{-0.456}$
Brace member	$SCF_{max} = 0$
Load case (III): bending moment of the chord	
Chord	$SCF_{max} = 1.555 \cdot 2\gamma^{0.666} \cdot 39.114\tau^2 - 54.023\tau + 35.629^{-0.815}$
Brace member	$SCF_{max} = 0$
Application scope: $20 \leq 2\gamma \leq 50; 0.5 \leq \tau \leq 1.0$	

Table 7.8 Parameter formulas for SCF_{max} of Perforated Bond Link–stiffened concrete-filled rectangular steel tubular K-type joints with gusset plates (θ = 45°).

Location	SCF_{max} formula
Load case (I): tension-compression balanced load of brace members	
Chord	$SCF_{max} = 0.443 \cdot 2\gamma^{0.497} \cdot \tau^{0.444}$
Brace member	$SCF_{max} = 0.008 \cdot 2\gamma^{0.934} \cdot -4.882\tau^2 + 7.759\tau + 0.004^{2.026}$
Load case (II): axial force of chord members	
Chord	$SCF_{max} = 2.216 \cdot 2\gamma^{0.192} \cdot 28.491\tau^2 - 36.520\tau + 19.358^{-0.505}$
Brace member	$SCF_{max} = 0$
Load case (III): bending moment of the chord	
Chord	$SCF_{max} = 1.726 \cdot 2\gamma^{0.518} \cdot 40.238\tau^2 - 51.690\tau + 33.462^{-0.578}$
Brace member	$SCF_{max} = 0$
Application scope: $20 \leq 2\gamma \leq 50; 0.5 \leq \tau \leq 1.0$	

Table 7.9 Parameter formulas for SCF_{max} of Perforated Bond Link–stiffened concrete-filled rectangular steel tubular K-type joints with gusset plates ($\theta = 60°$).

Location	SCF_{max} formula
Load case (I): tension-compression balanced load of brace members	
Chord	$SCF_{max} = 0.197 \cdot 2\gamma^{0.360} \cdot \tau^{0.535}$
Brace member	$SCF_{max} = 0.009 \cdot 2\gamma^{0.913} \cdot -5.357\tau^2 + 8.164\tau + 0.004^{2.204}$
Load case (II): axial force of chord members	
Chord	$SCF_{max} = 2.256 \cdot 2\gamma^{0.186} \cdot 35.112\tau^2 - 45.735\tau + 22.688^{-0.515}$
Brace member	$SCF_{max} = 0$
Load case (III): bending moment of the chord	
Chord	$SCF_{max} = 1.805 \cdot 2\gamma^{0.485} \cdot 45.826\tau^2 - 62.713\tau + 35.690^{-0.579}$
Brace member	$SCF_{max} = 0$
Application scope: $20 \le 2\gamma \le 50; 0.5 \le \tau \le 1.0$	

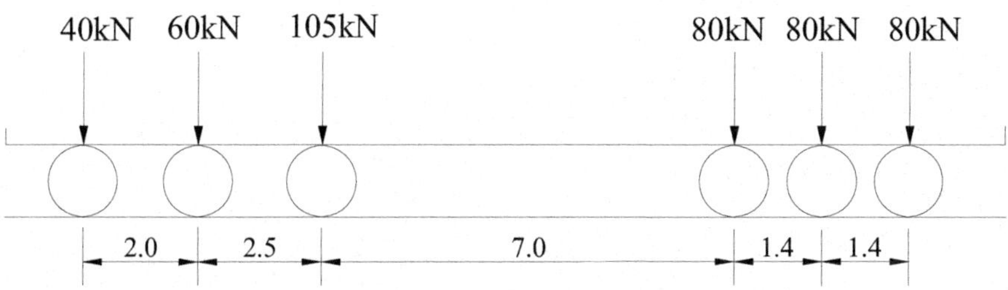

Figure 7.9 Fatigue load calculation model II (dimension unit: m).

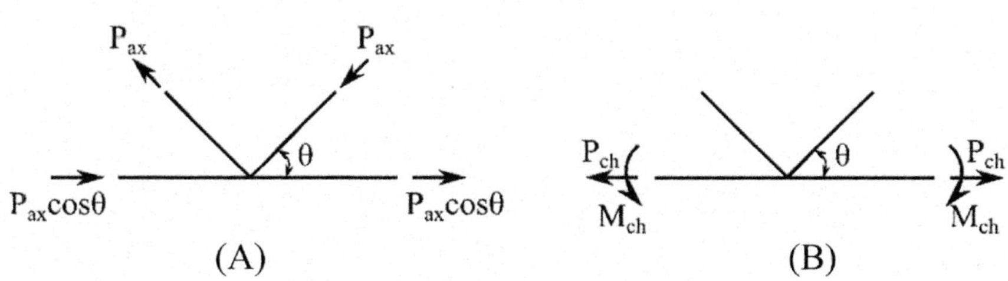

Figure 7.10 Basic load cases for K-type joints. (A) Load case I: self-equilibrating brace load. (B) Load cases II and III: chord load (axial force or bending moment).

In the equation, σ_{I}, σ_{II}, and σ_{III} are the nominal stress ranges under the three basic load cases: balanced brace loading (tension/compression), chord axial force, and chord bending moment, respectively. SCF_{I}, SCF_{II}, and SCF_{III} are the corresponding hot-spot stress concentration factors for the joint under these three basic load cases, respectively.

The calculation formulas for the SCF of CFRST K-type joints are shown in the table below:

The calculation formulas for the SCF of PBL-stiffened CFRST K-type joints with gusset plates are shown in the table below. Linear interpolation should be used when a value cannot be directly calculated, and the resulting SCF should not be less than 1.

7.4.3.3 Fatigue resistance S-N *curve*

The procedure for fatigue life assessment based on the hot-spot stress method is similar to that of the nominal stress method. However, the fatigue resistance *S-N* curve for the hot-spot stress method needs to consider the plate thickness correction factor. The fatigue resistance *S-N* curve for CFRST joints is shown in Fig. 7.11, and its expression is as follows:

When $10^3 < N < 5 \times 10^6$:

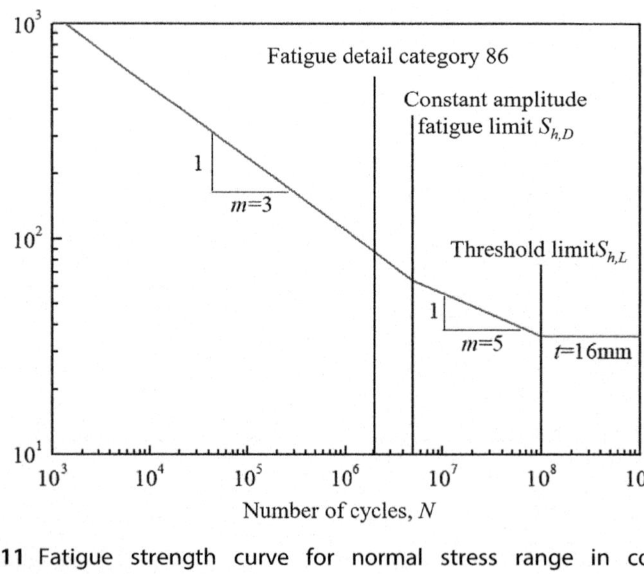

Figure 7.11 Fatigue strength curve for normal stress range in concrete-filled rectangular steel tubular joints.

$$\lg S_h = \frac{1}{3}(12.103 - \lg N) + 0.06 \lg N \cdot \lg\left(\frac{16}{t}\right) \tag{7.4}$$

When $5 \times 10^6 < N < 10^8$:

$$\lg S_h = \frac{1}{5}(15.706 - \lg N) + 0.402 \lg\left(\frac{16}{t}\right) \tag{7.5}$$

The fatigue resistance *S-N* curve for CFRST joints with gusset plates is shown in Fig. 7.12, and its expression is as follows:

When $10^3 < N < 5 \times 10^6$:

$$\lg S_h = \frac{1}{3}(12.328 - \lg N) + 0.06 \lg N \cdot \lg\left(\frac{16}{t}\right) \tag{7.6}$$

When $5 \times 10^6 < N < 10^8$:

$$\lg S_h = \frac{1}{5}(16.081 - \lg N) + 0.402 \lg\left(\frac{16}{t}\right) \tag{7.7}$$

7.4.3.4 Joint fatigue checking

When members and connections do not satisfy the checking requirements using Fatigue Load Model I, the checking should be performed using

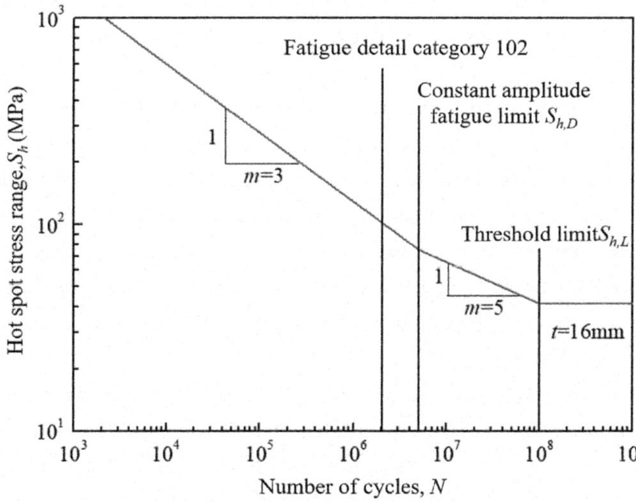

Figure 7.12 Fatigue strength curve for normal stress range in concrete-filled rectangular steel tubular joints with gusset plates.

Model II. When using Fatigue Load Calculation Model I, the following formula shall be used for verification:

$$\gamma_{\mathrm{Mf}}\Delta S_h \leq \Delta S_{h,D} \tag{7.8}$$

In the equation, γ_{Mf} is the partial factor for fatigue resistance, taken as 1.35 for primary members and 1.15 for secondary members; ΔS_h is the normal stress range calculated using Fatigue Load Calculation Model I; and $\Delta S_{h,D}$ is the constant amplitude fatigue limit for normal stress, adopted according to the corresponding joint detail category.

When using Fatigue Load Calculation Model II, the following formula shall be used for verification:

$$\gamma_{\mathrm{Mf}}\Delta S_h \leq \Delta S_{h,C} \tag{7.9}$$

In the equation, $\Delta S_{h,c}$ is the fatigue detail category, representing the fatigue stress intensity corresponding to 2×10^6 constant amplitude fatigue cycles.

7.5 Summary

Based on the theoretical and experimental research findings presented in this book, a calculation methodology suitable for CFRST composite truss bridges has been summarized and proposed. Regarding the preliminary size of the structure, detailing requirements for the members and joints of CFRST trusses have been put forward.

After determining the structural dimensions and layout of the bridge type, simplified calculation methods can be used for a rapid estimation of internal forces in composite truss bridges with simple structural forms. For composite truss bridges with more complex structural forms, finite element models using beam elements can be employed to analyze the global mechanical performance. Utilizing a multiscale finite element model allows for the simultaneous simulation of both the global and local mechanical behavior of the bridge.

Once the internal forces are obtained, structural verification can be performed according to the calculation methods provided in this chapter for the bearing capacity of rectangular hollow section and CFST joints, the bearing capacity of composite truss girders, and the fatigue assessment of joints.

Extended reading

CECS: 280-2010, Technical specification for structures with steel hollow sections[S]. Beijing: China Planning Press, 2010.

CIDECT Design Guide No.3, Design guide for rectangular hollow section joints under predominantly static loading[S]. Cologne: Verlag Tuv Rheinland Press, 2009.

GB 50017-2017, Standard for design of steel structures[S]. Beijing: China Architecture & Building Press, 2017.

Jiang L. Research on fatigue behaviour and calculation method of joints in concrete-filled rectangular hollow section truss bridge[D]. Xian City: Chang'an University, 2019.

JTG D60-2015, General specifications for design of highway bridges and culverts[S]. Beijing: China Communications Press Co., Ltd, 2015.

JTG D64-2015, Specifications for design of highway steel bridge[S]. Beijing: China Communications Press Co., Ltd, 2015.

Liu YJ. Experiments on ultimate bearing capacity and research on design method of joints of concrete-filled rectangular steel tube truss[D]. Changsha: Hunan University, 2003.

Ma YP, Liu YJ, Liu J. Multi-scale finite element model updating of CFST composite truss bridge based on response surface method[J]. China Journal of Highway and Transport 2019;32(11):51–61.

Ma YP, Liu YJ, Long X, et al. Simplified calculation method for flexural strength of CFST composite truss girder[J]. Journal of Building Structures 2020;41(05):76–84.

CHAPTER 8

Continuous rigid frame bridge with concrete-filled rectangular steel tubular composite truss girders

8.1 Overview

As an illustrative example, this chapter presents the scheme comparison and final design of China's first continuous rigid frame bridge utilizing concrete-filled rectangular steel tubular (CFRST) composite truss girders. Applying the calculation methods for such bridges proposed in this book, a structural verification was conducted for this bridge. Finally, an economic analysis was performed by comparing its material consumption with that of a concrete bridge of equivalent span.

8.2 Project overview

At Chainage K15+644 of the Huangyan Expressway, there is a cutting where the ramp overpasses the main highway line. The bridge location plan is shown in Fig. 8.1. The subgrade width is 5.5 m, configured as a single lane. The clearance height at the Huangyan Bridge site is approximately 17 m. The engineering geology primarily consists of loess and paleosol (Figs. 8.2 and 8.3).

The design of the Huangyan Bridge adheres to the fundamental principles of "Safety, Serviceability, Economy, Aesthetics, and Environmental Friendliness" while particularly emphasizing the characteristics of the CFRST composite truss bridge. The design focuses on ensuring structural safety, serviceability, and economic efficiency. Safety and serviceability are primarily reflected in meeting the requirements for flexural and shear capacity of the main girder under the ultimate limit state and satisfying the requirements for girder deflection and deck slab crack resistance under the serviceability limit state. Economic performance is mainly demonstrated through material consumption and construction

Composite Truss Bridges With Concrete-filled Rectangular Steel Tube Members
DOI: https://doi.org/10.1016/B978-0-443-44011-3.00009-3

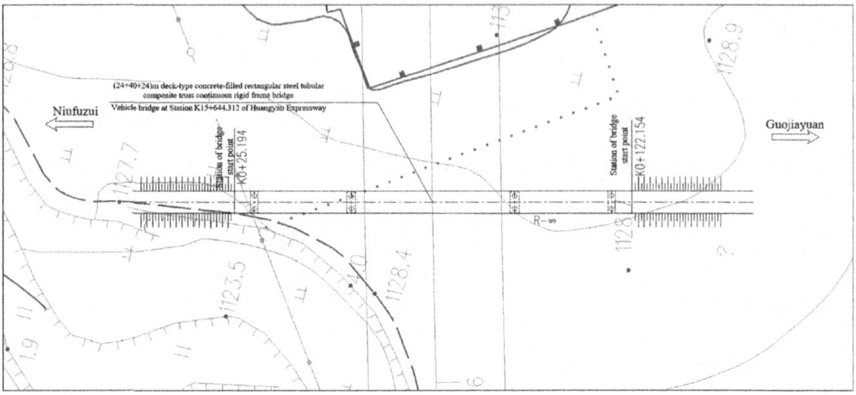

Figure 8.1 Bridge location plan.

convenience, with key material indicators being steel tonnage and concrete volume (Figs. 8.4 and 8.5).

The main design technical standards are as follows:

1. Structural form: Deck-type continuous rigid frame bridge with CFRST composite truss girders
2. Span arrangement: 24 + 40 + 24 m
3. Overall bridge width: 5.5 m, with a bidirectional 2% cross slope
4. Cross-section layout: 0.5 m (barrier) + 4.5 m (traffic lane) + 0.5 m (barrier)
5. Design loading: Highway-Class II
6. Seismic peak ground acceleration coefficient: 0.05 g

8.3 Structural design

8.3.1 Scheme comparison and selection

Bridge scheme 1: A 24 + 40 +24 m continuous rigid frame bridge with CFRST composite trusses. The deck width is 5.5 m. The elevation features a triangular brace member system with a panel length of 4 m. The girder depth is 2.8 m, resulting in a depth-to-span ratio of 1/14.3. The cross-section form is an inverted triangular spatial section. The bottom chords and the compression brace members at the supports are Perforated Bond Link (PBL)-stiffened CFRST members. The top chords utilize PBL perforated embedded steel plates, and the deck slab is a C40 reinforced concrete slab. The main piers are "Y"-shaped PBL-stiffened CFRST structures, supported by monolithic reinforced concrete pile foundations integrating the pile and column.

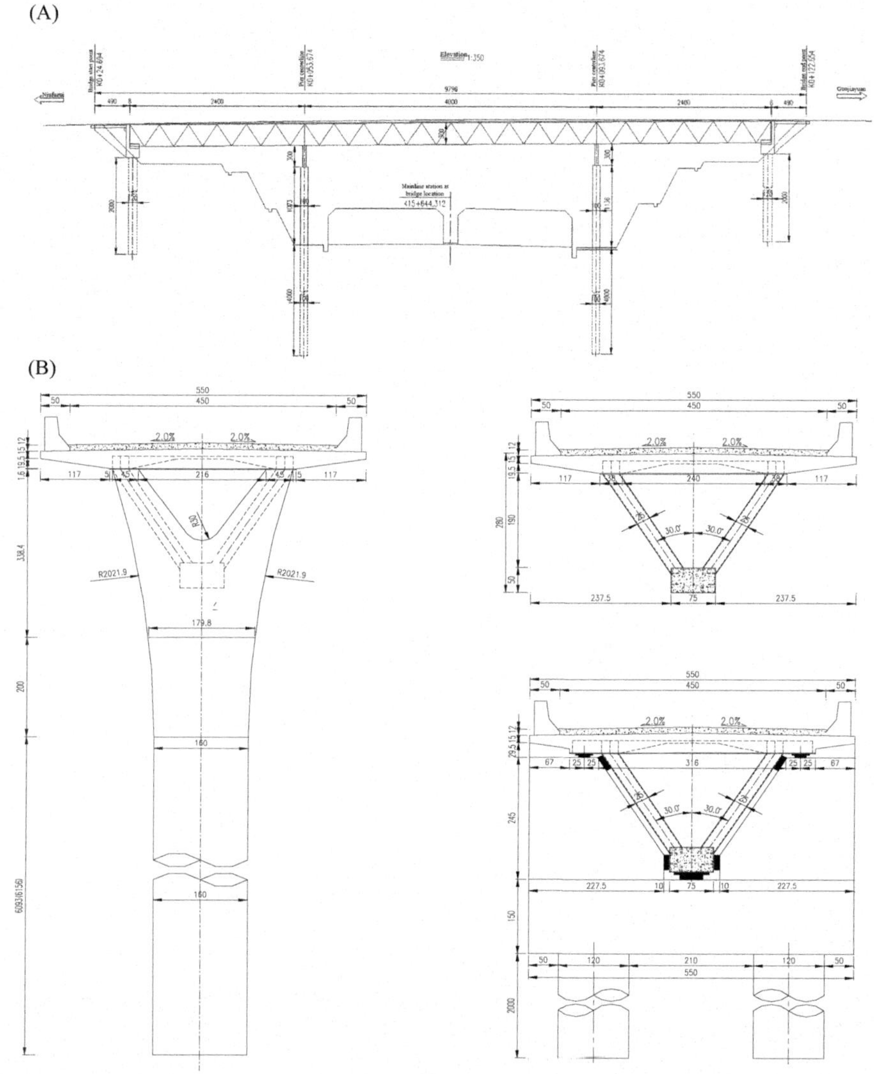

Figure 8.2 General layout of bridge scheme 1 (unit: cm). (A) Elevation layout. (B) Cross-section layout.

Bridge scheme 2: A 24 + 40 + 24 m continuous rigid frame bridge with CFRST composite trusses. The deck width is 5.5 m. The elevation features a triangular brace member system with a panel length of 4 m. The girder depth is 3.0 m, resulting in a depth-to-span ratio of 1/13.3. The cross-

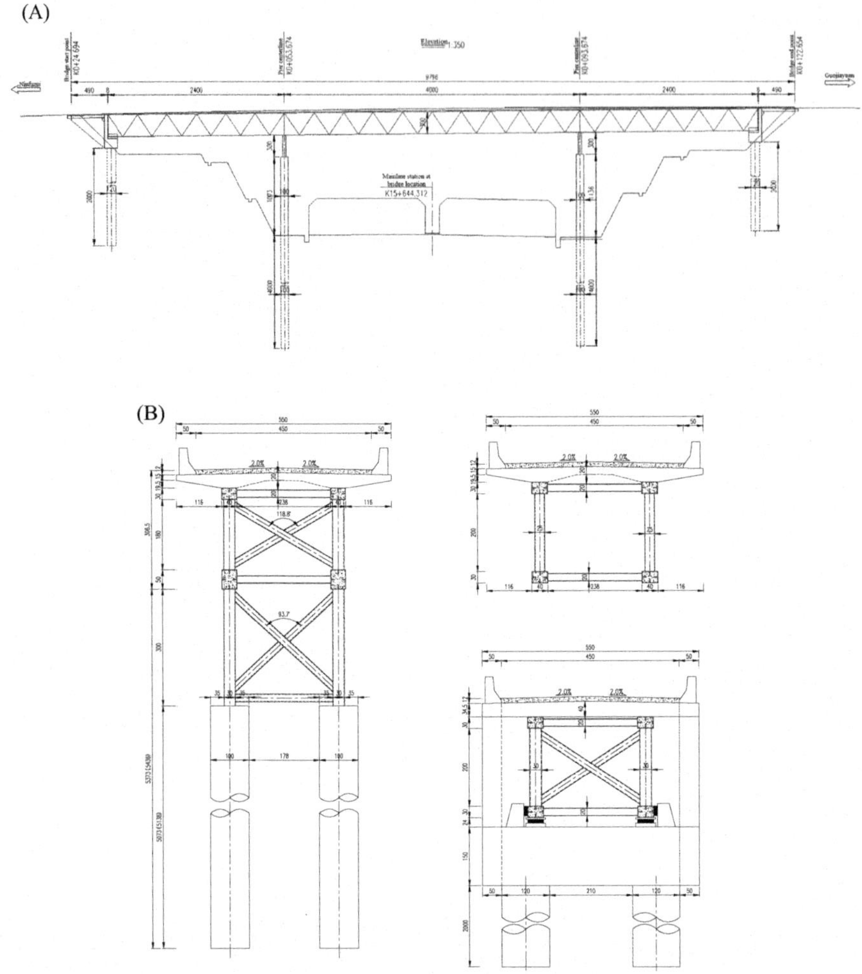

Figure 8.3 General layout of bridge scheme 2 (unit: cm). (A) Elevation layout. (B) Cross-section layout.

section form is a four-limb–type section. The bottom chords, top chords, and the brace members at the supports are PBL-stiffened CFRST members. The deck slab is a C50 prestressed concrete slab, designed as a Class B prestressed concrete member. The main piers are PBL-stiffened CFRST structures, supported by monolithic reinforced concrete pile foundations integrating the pile and column.

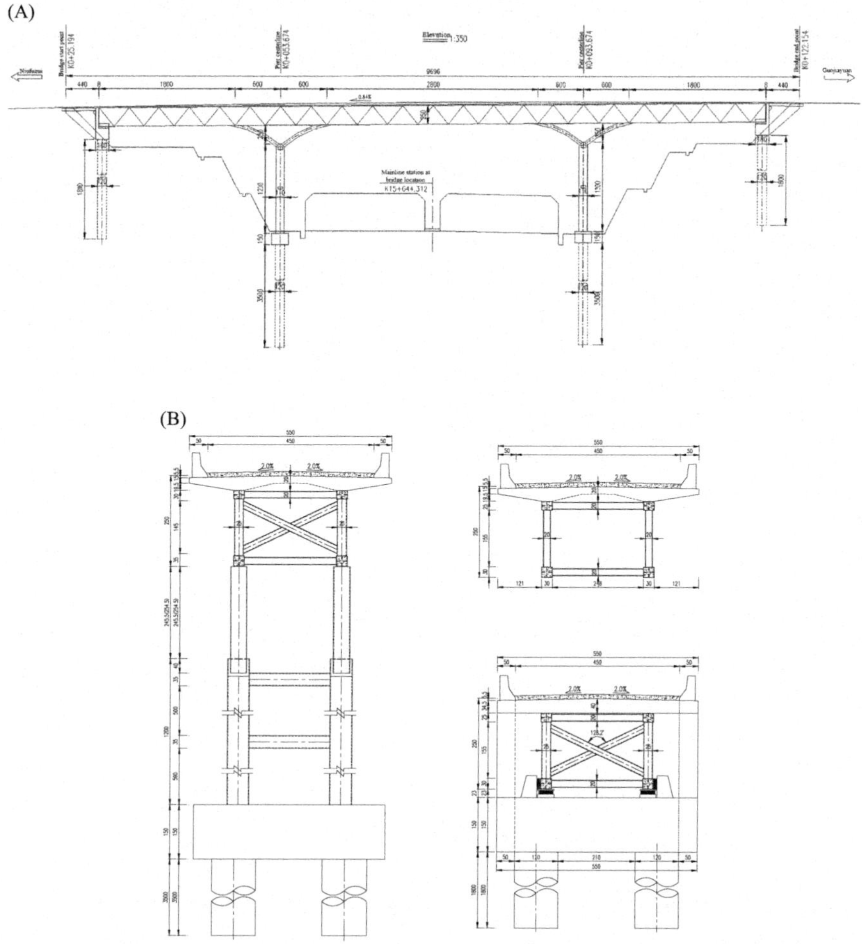

Figure 8.4 General layout of bridge scheme 3 (unit: cm). (A) Elevation layout. (B) Cross-section layout.

Bridge scheme 3: A 24 + 40 + 24 m continuous rigid frame bridge with CFRST composite trusses. The deck width is 5.5 m. The elevation features a triangular brace member system with a panel length of 4 m. The girder depth is 2.5 m, resulting in a depth-to-span ratio of 1/16. The cross-section form is a four-limb–type section. The bottom chords, top chords, and the brace members at the supports are PBL-stiffened CFRST members. The deck slab is a C50 precast prestressed concrete slab, designed as a Class B

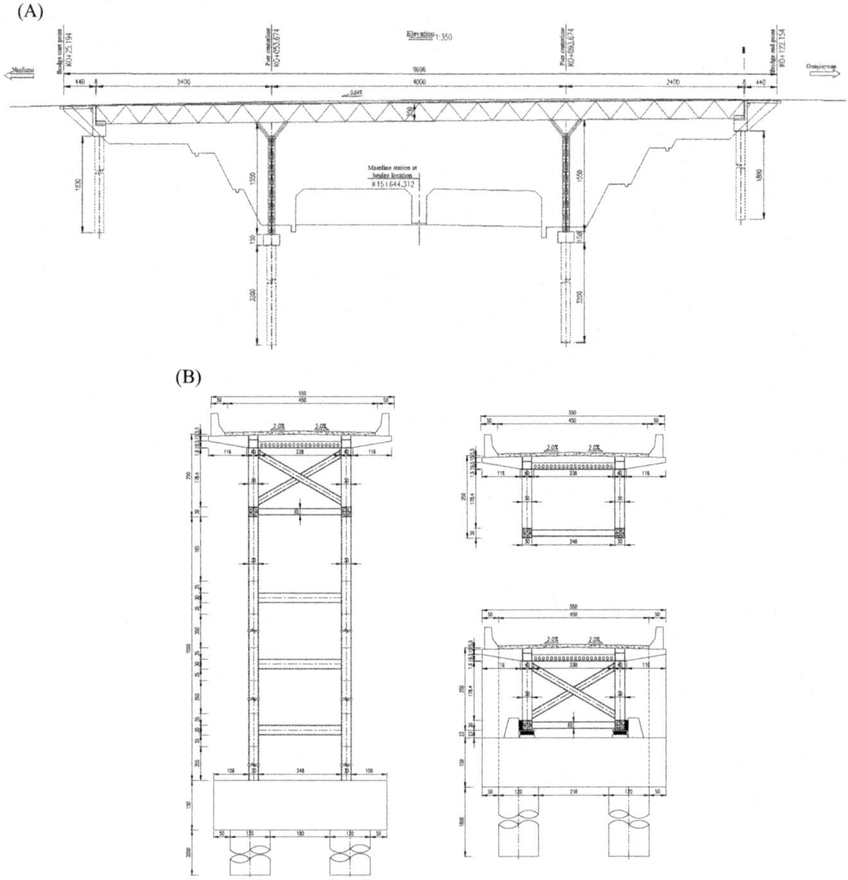

Figure 8.5 General layout of bridge scheme 4 (unit: cm). (A) Elevation layout. (B) Cross-section layout.

prestressed concrete member. The main piers are arc-shaped PBL-stiffened CFRST structures, supported by column-type PBL-stiffened CFST piers.

Bridge scheme 4: A 24 + 40 + 24 m continuous rigid frame bridge with CFRST composite trusses. The deck width is 5.5 m. The elevation features a triangular brace member system with a panel length of 4 m. The girder depth is 2.5 m, resulting in a depth-to-span ratio of 1/16. The cross-section form is a four-limb–type section. The bottom chords are PBL-stiffened CFRST members, the brace members are hollow steel tubes, and the top chords utilize PBL closed-form embedded steel plates. The chords and brace members of the composite truss are arranged with equal width.

The deck slab is a C40 reinforced concrete slab, with anticracking prestressed tendons provided in the 10-m segments of the slab on each side of the pier tops. The main piers are double-legged "Y"-shaped PBL-stiffened CFST piers.

Table 8.1 lists the material consumption for each scheme of this bridge. The table shows that the steel consumption ranges from 146.8 to 210.0 kg/m^2, indicating a significant variation. This highlights the importance of rationally optimizing the force distribution in the main truss for cost reduction. The configurations of the concrete deck slabs are quite similar across the four schemes, resulting in comparable concrete volumes. Comparing schemes 3 and 4 with schemes 1 and 2 reveals a substantial reduction in steel consumption, demonstrating that optimizing the form of the main piers has a significant impact on the material usage of the main truss. A direct comparison between scheme 3 and scheme 4 shows that the pier-girder connection design in scheme 4 is more rational. Scheme 4 employs a CFRST composite truss girder with chords and brace members of equal width. Although its steel consumption is slightly higher than that of scheme 3, the mechanical performance of its joints and the truss itself is significantly improved.

Considering all the factors mentioned above, scheme 4 was selected as the final design. The following sections provide a detailed description of the specific structural design of scheme 4.

8.3.1.1 Cross-section layout

The cross-section layout of the main truss is shown in Fig. 8.6. The bottom chord, located in the tension zone under dead load at mid-span, uses 300×300×12 mm rectangular steel tubes. In the compression zones at the pier tops and supports, PBL-stiffened CFRST sections with the same dimensions are used. The brace members have a cross section of 200 mm (8 mm thick) × 300 mm (12 mm thick). The chords and brace members of the main truss are arranged with equal width (i.e., $\beta = 1$). The top chord utilizes a closed-form PBL perforated embedded steel plate. The top lateral bracing adopts an inverted "T"-shaped section. The bottom lateral bracing uses a 200×200×8/12 "H"-shaped steel. At the support points, cross frames are installed to enhance the transverse connection between the two trusses. The cross-section layout is shown in the figure below.

Table 8.1 Material consumption of the main truss for different design schemes.

No.	Material	Alternative 1	Alternative 2	Alternative 3	Alternative 4
1	Steel quantity (kg)	92,404.9	101,664.0	71,072.2	79,884.6
	Steel quantity per unit area (kg/m^2)	190.9	210.0	146.8	165.1
2	Concrete quantity (m^3)	173.3	173.1	161.6	176.2
	Concrete quantity per unit area (m^3/m^2)	0.36	0.36	0.34	0.37
3	Ordinary reinforcement quantity (kg)	43,503.8	29,980.0	32,589.5	46,715.0
	Ordinary reinforcement quantity per unit area (kg/m^2)	89.9	61.9	67.3	96.5
4	Prestressing steel strand quantity (kg)	/	8547.5	6783.8	1359.6
	Prestressing steel strand quantity per unit area (kg/m^2)	/	17.7	14.0	2.8

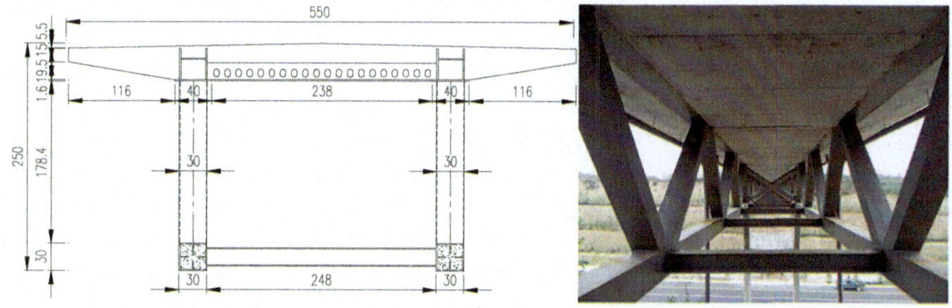

Figure 8.6 Cross-section layout (unit: cm).

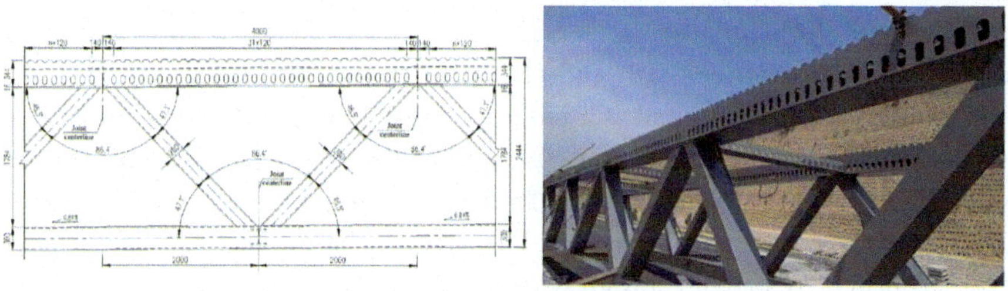

Figure 8.7 Elevation layout of the main truss segment (unit: cm).

8.3.1.2 Elevation layout

Fig. 8.7 shows the detailed drawing of the main truss members in a standard panel. The eccentricity (e) between the intersection point of the brace member axes and the chord axis is 22.8 mm, satisfying the code requirement for positive eccentricity of $e/h_0 = 22.08/300 \leq 0.25$. This allows the internal forces in the truss members to be calculated using a pin-connected model. The minimum angle between the brace member axes and the chord axis at the bridge ends is 38.3 degrees, meeting the code requirement of being not less than 30 degrees. The joints are designed as K-type joints. The elevation layout of the main truss segments is shown in the figure below.

8.3.1.3 Deck slab structural design and calculation principles

Given that the bridge width is only 5.5 m and considering the simplicity and convenience of setting up formwork for cast-in-place concrete, the deck slab configuration shown in Fig. 8.8 is adopted. The top chord is

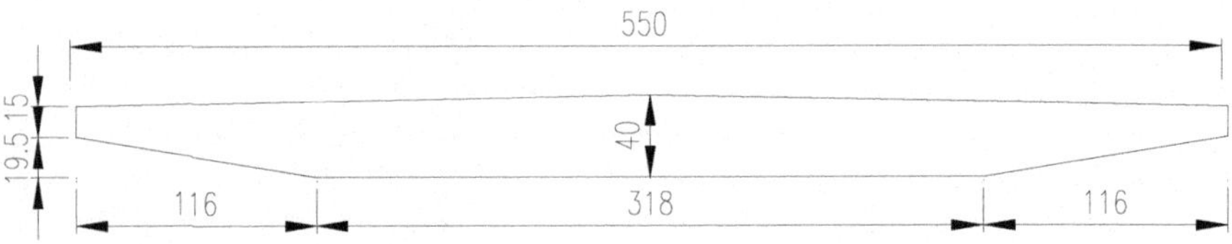

Figure 8.8 Deck slab configuration.

embedded within the concrete deck. The shear connection utilizes a PBL perforated steel plate form, consisting of a base plate and two vertically welded, perforated steel plates running longitudinally along the girder. The concrete dowels formed in the plate holes resist the longitudinal shear and uplift forces between the steel girder and the concrete. The connection detail between the deck slab and the main truss is shown in Fig. 8.9.

The deck slab is designed as a reinforced concrete member, with the design controlled by crack width limitations. Within the 10-meter segments on each side of the pier tops, six bundles of 15-5 prestressing tendons for crack control are provided, as shown in Fig. 8.10. The segments at the pier tops are designed and calculated as partially prestressed concrete Class B members. Under the short-term effect combination of loads considering the influence of long-term effects, the crack width limit is 0.10 mm. Other segments are designed and calculated as reinforced concrete members, with a crack width limit of 0.20 mm.

8.3.1.4 Pier configuration and pier-girder connection form

As shown in Fig. 8.11, the main piers adopt a double-legged "Y"-shaped PBL-stiffened CFST design. The cross-sectional dimensions are 400×300×16 mm. The two legs are spaced 10 cm apart, and three transverse tie beams, which are 600×300×16 mm rectangular steel tubes, are installed between them. The double-legged "Y"-shaped pier design optimizes the force distribution in the main truss and deck slab near the pier top, allowing for a reduction in the main truss depth and a decrease in steel consumption. The bifurcation point of the "Y" shape uses a curved transition, which is visually more appealing.

The connection between the main truss bottom chord and the pier column is a welded joint, utilizing a Grade I grooved full penetration weld. This ensures that the flexural, shear, and torsional strength of the weld is not less than that of the base metal. The Y-shaped pier and the main truss bottom chord are arranged with equal width. The connection at the pier and girder employs an integral gusset plate design, as shown in Fig. 8.12. During construction, the structural components at the pier-girder

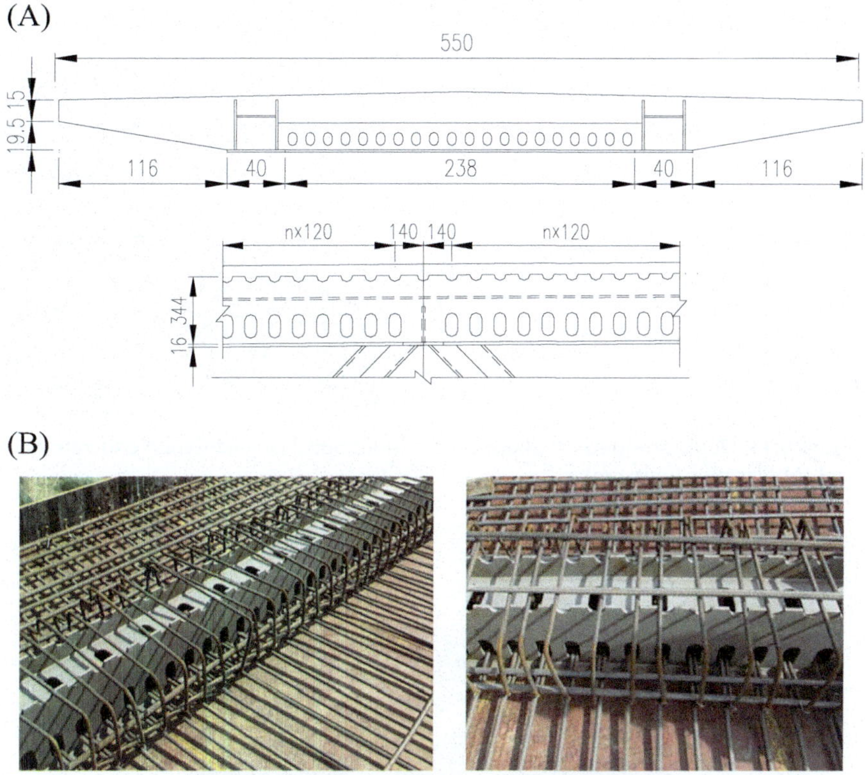

Figure 8.9 Connection detail between the deck slab and the main truss (unit: cm). (A) Connection detailing between the deck slab and the main truss (drawing). (B) Actual bridge connection detail between the deck slab and the main truss.

Figure 8.10 Prestressing tendons in the deck slab.

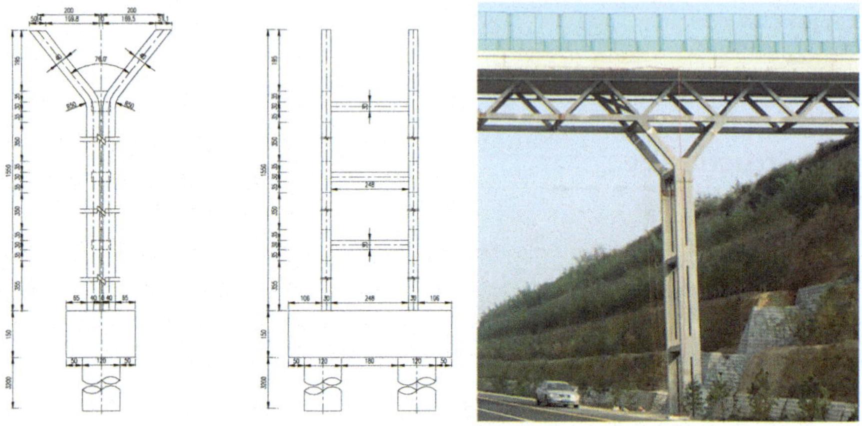

Figure 8.11 Double-legged "Y"-shaped Perforated Bond Link–stiffened concrete-filled steel tubular pier (unit: cm).

Figure 8.12 Joint design at the pier-girder connection.

connection are first welded into an integral unit on the ground and then hoisted into place for final assembly at the pier site. This process ensures the reliability of the pier-girder connection joint.

8.4 Structural verification

8.4.1 Structural verification during construction stages

The segments of the Huangyan Overpass Bridge are relatively lightweight. To improve construction efficiency, segmental lifting and assembly construction method was adopted. The construction stages are divided as follows:

CS0: Main truss fabrication in the factory.
CS1: Erect falsework and construct the hollow steel tube piers.
CS2: Lift and install the main truss segments near the intermediate supports (10 + 12 m segments) and complete the pier-girder rigid connection.
CS3: Lift and install the 18 m main truss segment at the mid-span of the center span and the 13 m segments near the end supports of the side spans.
CS4: Pour concrete into the piers and the main truss connection segments.
CS5: Remove the falsework.
CS6: Install bridge deck reinforcement and cast the deck slab (wet weight stage).
CS7: Deck slab reaches design strength.
CS8: Tension the prestressing tendons near the pier tops.
CS9: Apply deck pavement, install ancillary facilities, and complete the bridge.

The main truss is divided into segments of 13 + 12 + 10 + 18 + 10 + 12 + 13 m. The segmentation is based on lifting and transportation capacities, and the division is planned to avoid joint locations. The hollow steel tubes of the piers are erected first. After the main truss segments are lifted into place and integrated with the piers to form the overall joints, concrete is poured into the piers. The deck slab within the 10-m range on each side of the pier tops is designed as a partially prestressed concrete Class A member, while the remaining segments are designed as reinforced concrete members.

Analysis of forces during construction stages: As shown in Table 8.2 and Fig. 8.13, the most critical stage during construction is the cast-in-place deck slab stage. The maximum tensile stress in the top chord at the pier top is 158.7 MPa. Since the top chord has a perforated cross section, its stiffness is reduced in the calculation, specifically by reducing its moment of inertia. During the pouring of the deck slab concrete, before the concrete reaches

Table 8.2 Mechanical performance during construction stages.

Construction phase	Maximum stress (MPa)		Design strength (MPa)	Maximum deflection (mm)	Stability safety coefficient
	Compressive stress	Tensile stress			
CS1	−0.9	1.7	275	0.03	715.5
CS2	−8.7	11.9	275	1.6	614.6
CS3	−20.0	25.0	275	4.5	85.0
CS4	−21.0	25.2	275	4.7	85.0
CS5	−17.0	26.9	275	6.3	69.2
CS6	−107.8	158.7	275	38.0	14.3
CS7	−109.4	155.4	275	37.3	34.8
CS8	−105.0	134.6	275	32.5	33.7
CS9	−111.3	141.6	275	39.6	86.7

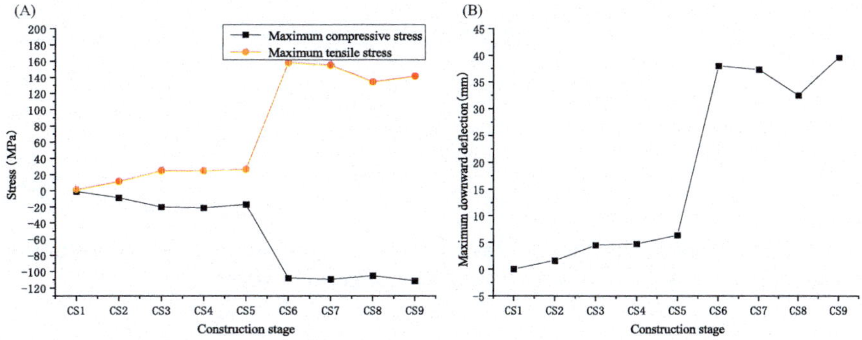

Figure 8.13 Mechanical analysis of the main truss during construction stages. (A) Stress envelope diagram during construction phases (unit: MPa). (B) Deflection envelope diagram during construction phases (unit: mm).

its strength, the stiffness of the deck slab is not included, and its self-weight is applied as a uniformly distributed load on the top chord. After the concrete reaches its strength, its stiffness contribution is included. The calculated result for CS6 is greater than that for CS7 because the influence of construction machinery is considered in the CS6 stage. After the application of the secondary dead load (pavement, etc.), the maximum deflection at the mid-span of the main truss is 38 mm, indicating that the main truss has relatively high stiffness. The minimum stability safety factor occurs during the cast-in-place deck slab stage, which is 14.3, meeting the code requirements.

Strength calculation of the main truss during construction stages: The finite element model simulates the concrete pouring process for the bottom chord at the connection segments using the "staged composite section" function, to account for concrete shrinkage and creep. The results are shown in Fig. 8.14.

Fig. 8.14 shows that the stresses are relatively high in the areas to the left and right of the double-legged "Y"-shaped pier tops and at the mid-span of the center span. During the calculation, the strength capacity and stability capacity of the sections at the pier tops are calculated, considering them as concrete-filled steel tubes. Table 8.3 lists the bearing capacity verification results for the critical sections of the bottom chord. The sections to the left and right of Pier #2 top are symmetrical to those of Pier #1 top. The member efficiency coefficient is calculated according to the definition in

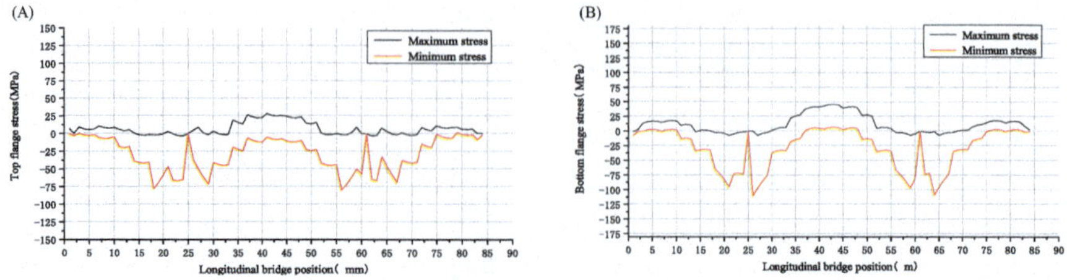

Figure 8.14 Envelope diagram of stresses at the top and bottom edges of the bottom chord (unit: MPa). (A) Upper edge stress. (B) Lower edge stress.

Eq. (3–8) from Chapter 3 (the ratio of a member's axial force to its bearing capacity) (Table 8.4 and 8.5).

Fig. 8.15 shows that the top chord is in tension near the pier top, with a maximum tensile stress of 158.7 MPa. The remaining segments are in compression, with a maximum compressive stress of −131.4 MPa at the mid-span of the center span. All these values are less than the design strength of 275 MPa. Here, considering that the top chord is embedded within the concrete deck slab, the stability reduction factor φ may be disregarded.

Verification of main truss brace members: The brace members at and near the pier locations are subjected to unfavorable forces. All brace members are hollow steel tubes with constant cross-sectional dimensions. Therefore it is only necessary to perform strength and overall stability checks for the members experiencing the most critical tension and compression forces, respectively (Table 8.6).

Stiffness calculation and analysis of the main truss during construction stages: The deflection curve variations for each construction stage are shown in Fig. 8.16.

The deflection curve variation diagram during the construction stages shows that the deflection of the main truss is relatively small before the deck slab is cast, with a maximum of only 6.3 mm at the mid-span of the center span. The casting of the deck slab causes a significant increase in the main truss deflection, raising it to 38.0 mm at mid-span. With the application of the secondary deadload (pavement, etc.), the deflection increases further, reaching a maximum of 39.6 mm at mid-span. The deflection calculations demonstrate that the composite truss girder possesses substantial flexural stiffness.

Table 8.3 Bearing capacity verification table for the bottom chord during construction stages.

Control section	Axial force (kN)	Bending moment (kN.m)	Member capacity	Member efficiency factor
Left side of pier 1 top	−2105.5	−76.2	−6250.1	0.34
Right side of pier 1 top	−2182.5	−72.9	−6250.1	0.35
Mid-span of central span	1073.8	21.2	6227.3	0.17

Table 8.4 Strength verification table for brace members during construction stages.

Critical member	Stress (MPa)	Design strength (MPa)	Axial force (kN)	Bending moment (kN.m)	Member capacity (kN)	Member efficiency factor
Tensile brace member	101.9	275	/	/	/	0.36
Compressive brace member	/	/	−793.4	−10.1	−2915.5	0.32

Table 8.5 Bearing capacity verification table for the bottom chord.

Control section	Axial force (kN)	Bending moment (kN.m)	Member bearing capacity	Member efficiency factor
Pier 1 top left	−2067.3	−87.8	−6250.1	0.33
Pier 1 top right	−2053.0	−80.5	−6250.1	0.33
Midpoint of mid-span	1812.1	24.7	5624.6	0.32

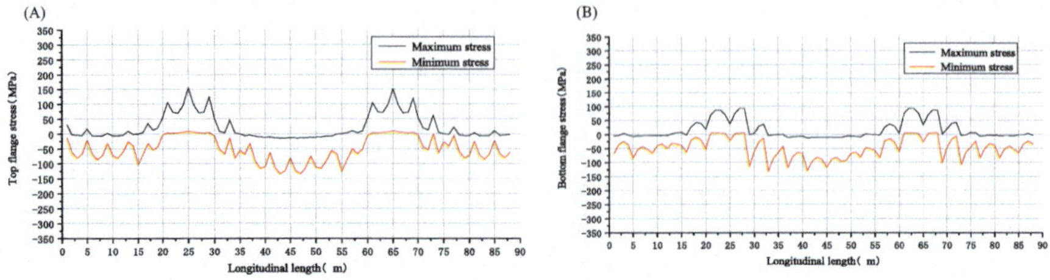

Figure 8.15 Envelope diagram of stresses at the top and bottom edges of the top chord (unit: MPa). (A) Upper edge stress. (B) Lower edge stress.

8.4.2 Structural verification of the completed bridge stage

8.4.2.1 Strength verification of the main truss

The structure in the completed bridge stage is calculated considering the most unfavorable combination under the ultimate limit state. The loads include dead load, live load (vehicle), prestressing load, secondary effects of shrinkage and creep, uniform temperature rise/drop, gradient temperature, and foundation displacement (Fig. 8.17).

Calculation of bottom chord bearing capacity: The critical sections for the bottom chord calculation are still selected as the most critical compression member (near the pier top) and the most critical tension member (mid-span of the center span) for verification.

The top chord is in tension near the pier tops, with a maximum tensile stress of 128.6 MPa. The remaining segments are in compression, with a maximum compressive stress of −175.7 MPa at the mid-span of the center span. All these stress values are less than the design strength of 275 MPa.

The stress envelope for the brace members under the ultimate limit state is shown in Fig. 8.18. Since the brace members use a uniform cross section, it is only necessary to perform strength and overall stability checks for the members subjected to the most critical tension and compression forces, respectively.

Calculations show that the maximum slenderness ratio for the compression brace members is 29.1, which is less than the allowable slenderness ratio of 100. It can be observed that the efficiency coefficient of the brace members is relatively low, indicating a significant stress reserve.

8.4.2.2 Calculation and analysis of deck slab bearing capacity and crack resistance

The deck slab is a cast-in-place concrete member. Within the 20-m range on each side of the pier tops, six bundles of 15-5 prestressing tendons are

Table 8.6 Bearing capacity verification table for brace members.

Critical member	Stress (MPa)	Design strength (MPa)	Axial force (kN)	Bending moment (kN.m)	Member capacity (kN)	Efficiency factor
Tensile brace member	151.2	275	/	/	/	0.55
Compressive brace member	/	/	−1206.6	−37.8	−2915.5	0.41

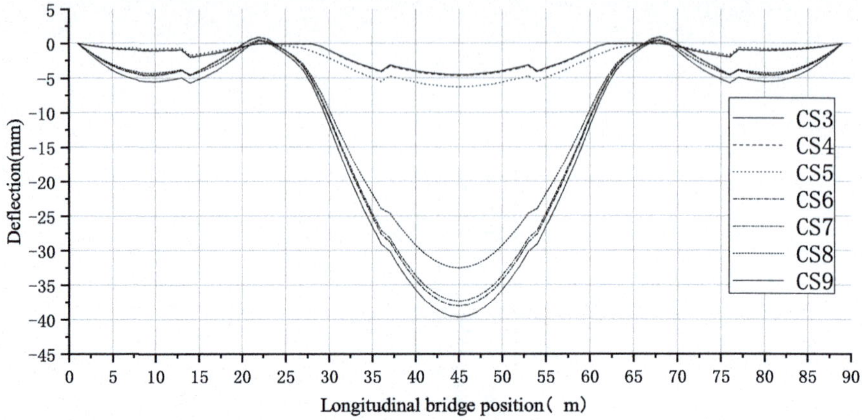

Figure 8.16 Deflection curve of the main truss during construction stages.

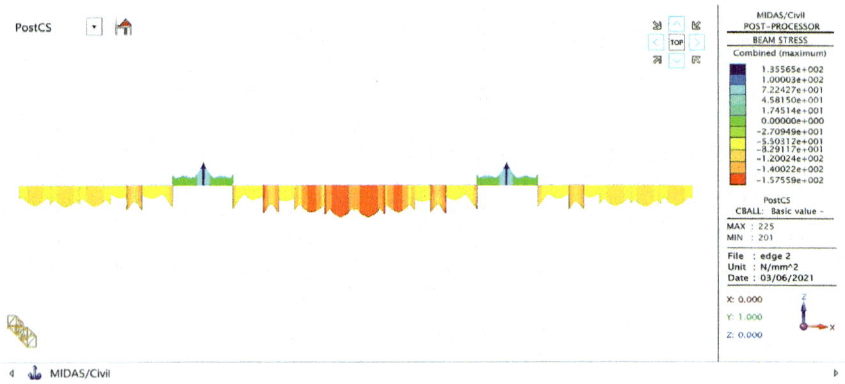

Figure 8.17 Stress envelope diagram for the top chord.

Figure 8.18 Stress envelope diagram for brace members.

provided. The verification is conducted according to the "*Specifications for Design of Highway Reinforced Concrete and Prestressed Concrete Bridges and Culverts*" (JTG 3362-2018). Critical sections are selected for verification under both the ultimate limit state and the serviceability limit state.

As can be seen from Table 8.7 and Table 8.8, under the ultimate limit state, the design values for the deck slab are all less than the structural resistance. Under the short-term effect combination, and considering the prestressing effect, the crack width in the deck slab at the pier top is 0.018 mm. Compared to the 0.152 mm crack width calculated without considering prestressing, this is a reduction of 0.134 mm, demonstrating that the prestressing tendons effectively prevent cracking in the deck slab at the pier top. The crack width is less than the code-specified limit of 0.10 mm for Class B prestressed concrete members; thus the crack resistance requirement is satisfied.

8.4.2.3 Calculation and analysis of global stiffness of the main truss

According to Article 1.1.5 of the "Specifications for Design of Highway Steel Bridge" (JTG D64-2015), the vertical deflection caused by vehicular live load (excluding impact) shall not exceed L/500 for simply supported or continuous trusses.

Fig. 8.19 shows that the maximum sum of the absolute values of positive and negative deflections in the main truss at the mid-span of the center span under vehicular live load is 9.8 mm. This is less than the code-specified limit of L/500 = 80 mm.

8.4.2.4 Joint bearing capacity verification

The internal forces in the brace members of different truss joints under the ultimate limit state were extracted. The bearing capacity of the truss joints for the project under study was verified according to the calculation methods provided in Chapter 7 of this book. The bearing capacity of the rectangular hollow section and CFST joints was calculated using the formulas given in Chapter 7. The joint efficiency coefficient was calculated according to the definition in Eq. (3–15) from Chapter 3 (Joint Bearing Capacity/Member Bearing Capacity). The calculation results are shown in Table 8.9.

In Table 8.9, serial numbers 1 to 4 represent CFRST K-type joints, differing in the inclination angle of the brace members. Serial number 5 represents a rectangular hollow section K-type joint. The member internal forces were obtained by extraction from the finite element model. The

Table 8.7 Internal force verification of the deck slab under ultimate limit state.

Design internal force section		Basic effect combination for ultimate limit state			Resistance		Verification result
		N (kN)	Q (kN)	M (kN·m)	N (kN)	M (kN·m)	
Mid-span of side span	Maximum M	1046	492.4	424.3	4420	2180	Pass
	Maximum N	2071.5	506.2	−223.4	12,400	2180	Pass
2m left of pier 1 top	Maximum M	−1914.1	387.5	486.8	4580	2180	Pass
	Maximum N	−1914.1	387.5	486.8	14,600	2180	Pass
Pier 1 top	Maximum M	−1901.3	−537.8	372.5	5520	2180	Pass
	Maximum N	−1901.3	−537.8	372.5	11,400	2180	Pass
2m right of pier 1 top	Maximum M	−1947.6	−393.5	484.8	4270	2180	Pass
	Maximum N	−1947.6	−393.5	484.8	14,400	2180	Pass
Mid-span of central span	Maximum M	−1275.6	−481.8	434.8	3950	2180	Pass
	Maximum N	2000.6	−480.9	−198.0	−9060	2180	Pass

Table 8.8 Crack resistance verification of the deck slab under short-term effect combination of the serviceability limit state.

Design internal force section		Short-term action combination for serviceability limit state			Crack width (mm)	Allowable value (mm)	Verification result
		N (kN)	Q (kN)	M (kN·m)			
Mid-span of side span	Maximum M	1372.5	262.6	280.5	0.036	0.1	Pass
	Maximum N	1790.1	270.4	−154.8	0.046	0.1	Pass
2m left of pier 1 top	Maximum M	−1357.7	203.8	315.4	0.019	0.1	Pass
	Maximum N	−1357.7	203.8	315.4	0.019	0.1	Pass
Pier 1 top	Maximum M	−1375.3	−286.5	252.9	0.018	0.1	Pass
	Maximum N	−1375.3	−286.5	252.9	0.018	0.1	Pass
2m right of pier 1 top	Maximum M	−1398.3	−208.9	313.7	0.021	0.1	Pass
	Maximum N	−1398.3	−208.9	313.7	0.021	0.1	Pass
Mid-span of central span	Maximum M	−1094.5	−251.8	281.1	0.015	0.1	Pass
	Maximum N	1856.7	−251.0	−140.8	0.048	0.1	Pass

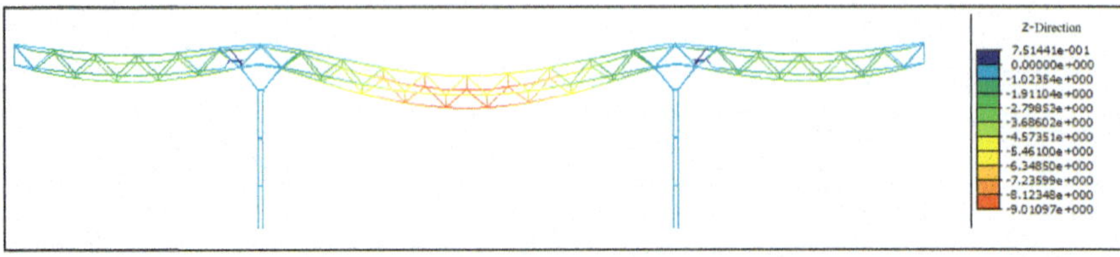

Figure 8.19 Structural deflection under lane load (unit: mm).

joint efficiency coefficient was determined by the ratio of the joint bearing capacity to the corresponding brace member bearing capacity. Table 8.9 shows that all joint bearing capacity verifications are passed. The joint efficiency coefficients range between 0.7 and 0.83, indicating that the joint design is reasonable and the material utilization is relatively high.

8.4.2.5 Verification of joint fatigue resistance

The fatigue resistance of the joints in the project under study was verified according to the calculation method provided in Chapter 7 of this book. The internal forces in the brace members of the truss joints under the Fatigue Load Model II combination were extracted, as shown in Fig. 8.20. The complex internal force distribution in the joint shown in Fig. 8.20 was then decomposed into a combination of internal forces under the three basic load cases shown in Fig. 8.21. From Fig. 8.20, Brace 1 is subjected to tension, and Chord 2 is subjected to greater tension compared to Chord 1. Therefore Brace 1 and Chord 2 should be selected as the fatigue-critical members. The nominal stress ranges for the fatigue-critical members under each basic load case are:

- **I.** For load case I: The nominal stress range in the brace due to axial force, $\Delta\sigma_{\mathrm{brace,P}}$, is 8.2 MPa, the nominal stress range in the brace due to bending moment, $\Delta\sigma_{\mathrm{brace,M}}$, is 1.8 MPa.
- **II.** For load case II: The nominal stress range in the chord due to axial force, $\Delta\sigma_{\mathrm{chord,P}}$, is 15.1 MPa.
- **III.** For load case III: The nominal stress range in the chord due to bending moment, $\Delta\sigma_{\mathrm{chord,M}}$, is 0.8 MPa.

It is worth noting that the bending moment in the chord induces compressive stress on the chord's top plate, which partially alleviates the tensile stress caused by the chord's axial force.

Calculate the $\mathrm{SCF}_{\max}$ on the brace and chord surfaces under each basic load case. First, the geometric dimensionless parameters for the CFRST

Table 8.9 Joint bearing capacity verification.

Joint no.	Chord dimensions $b_0 \times h_0 \times t_0$/mm	Brace dimensions $b_{1,2} \times h_{1,2} \times t_{1,2}$/mm	Brace member inclination angle		Axial force in joint brace/kN		Joint capacity/kN		Brace capacity/kN		Verification result	Joint capacity efficiency factor
			Tension	Compression	Tension	Compression	Tension	Compression	Tension	Compression		
1	300×300×20	300(12)×200(8)	46.5°	51.8°	526	−550	2252	−4212	3142	−3142	Pass	0.72
2	300×300×20	300(12)×200(8)	47.1°	51.2°	1257	−1467	2282	−4190	3142	−3142	Pass	0.73
3	300×300×20	300(12)×200(8)	46.5°	47.1°	1257	−1467	2548	−4190	3142	−3142	Pass	0.81
4	300×300×20	300(12)×200(8)	46.5°	47.1°	526.1	−549.8	2548	−4190	3142	−3142	Pass	0.81
5	300×300×20	300(12)×200(8)	46.5°	47.1°	830	−831	3055	−2194	2871	−3141.6	Pass	0.70

Notes: Node numbers 1–4 are CFRST K-type joints, and node number 5 is a rectangular hollow section K-type joint.

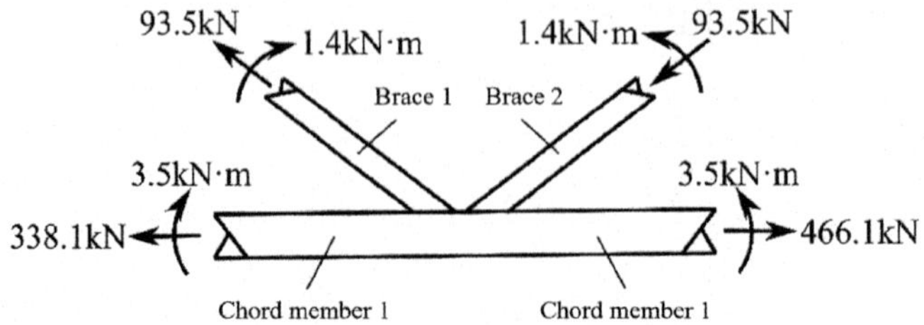

Figure 8.20 Internal force distribution in the joint.

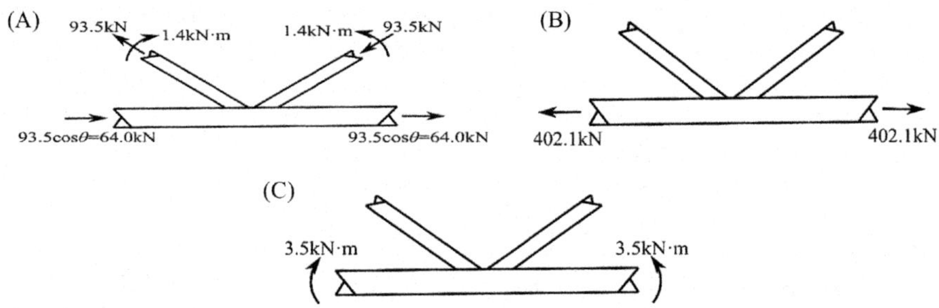

Figure 8.21 Decomposition of joint internal forces. (A) Self-equilibrating brace load. (B) Axial force in the chord. (C) Bending moment in the chord.

joint are calculated as $\beta = 1$, $2\gamma = 25$, $\tau = 1$, $g' = 4.23$, and $\sin\theta = 0.73$. Then, using the formulas from Chapter 7 of this book, the following are obtained:

I. For load case I: The stress concentration factor on the brace surface, $SCF_{brace,I}$, is 5.89, and the stress concentration factor on the chord surface, $SCF_{chord,I}$, is 3.53.

II. For load case II: The stress concentration factor on the brace surface, $SCF_{brace,II}$, is 0, and the stress concentration factor on the chord surface, $SCF_{chord,II}$, is 1.38.

III. For load case III: The stress concentration factor on the brace surface, $SCF_{brace,III}$, is 0, and the stress concentration factor on the chord surface, $SCF_{chord,III}$, is 2.00.

By linearly superimposing the hot-spot stress ranges generated under the three basic load cases, the maximum hot-spot stress ranges on the brace and chord surfaces under the complex loading $S_{h,brace}$ and $S_{h,chord}$ can be calculated. The calculation results are shown in Table 8.10.

Table 8.10 Joint fatigue verification.

Joint location		Normal stress $\triangle\sigma_P$ (MPa)	Bending stress $\triangle\sigma_M$ (MPa)	Hot-spot stress concentration factor				Hot-spot stress range S_h (MPa)	Fatigue stress range limit $\triangle\sigma_C/\gamma_{Mf}$ (MPa)	Verification result
				$SCF_{brace,I}$	$SCF_{chord,I}$	$SCF_{chord,II}$	$SCF_{chord,III}$			
Mid-span hollow tube joint	Chord	15.1	0.8	5.89	3.53	1.38	2.00	54.09	71	Pass
	Brace	8.2	1.8					58.9	71	Pass

The calculated maximum hot-spot stress ranges on the brace and chord surfaces are 58.9 and 54.09 MPa, respectively. Both values are less than the allowable fatigue strength $\Delta\sigma_c$ of 71 MPa, indicating that the joints of this CFRST truss bridge satisfy the fatigue design requirements.

8.4.3 Modal and stability analysis of the completed bridge stage

Based on the bridge modal analysis results, the first vertical vibration frequency of the bridge is 3.73 Hz, indicating that the bridge possesses significant vertical stiffness (Fig. 8.22).

Principle of Buckling Analysis: Constant Load + Buckling Factor × Variable Load=Buckling Load. The constant loads considered in the calculation include self-weight, secondary dead load, and prestressing load. The variable load is Highway-Class II load, which is equivalently represented as a uniformly distributed load (7.875 kN/m) plus a concentrated force (240 kN). The analysis model also accounts for the full-bridge eccentric loading effect. As shown in Fig. 8.23, the first-order buckling factor k for bridge scheme 4 is 168.6, corresponding to overall longitudinal horizontal vibration buckling, and stability meets the code requirements.

Photos of the completed Huangyan Overpass Bridge and the full-scale bridge test are shown in Figs. 8.24 and 8.25, respectively.

8.5 Economic analysis

A prestressed concrete box-girder bridge with a span and deck width similar to the project under study was selected. The material consumption of the superstructure and substructure for both bridges is compared and listed in Table 8.11. The table shows that, compared to the prestressed concrete

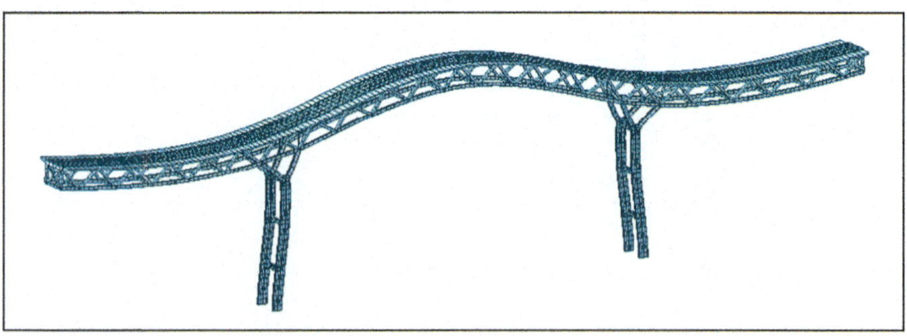

Figure 8.22 Vertical bending vibration (3.73 Hz).

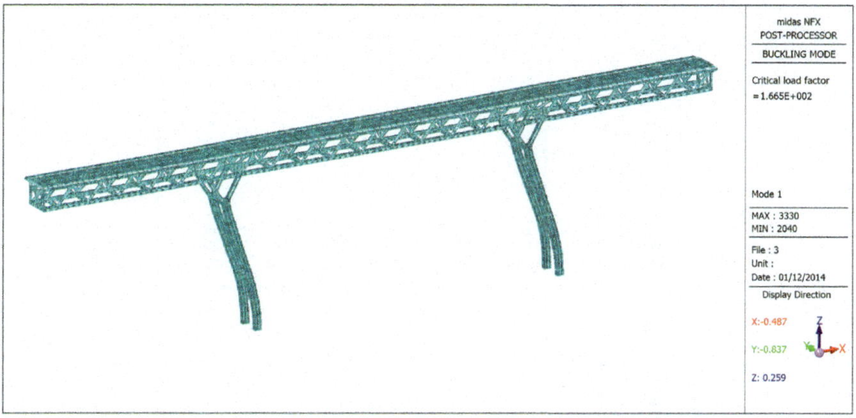

Figure 8.23 First-order buckling mode.

Figure 8.24 Photo of the completed bridge.

Figure 8.25 Photo of the full-scale bridge test.

Table 8.11 Material consumption comparison.

Bridge type category		Huangyan bridge	Alternative scheme	Material quantity ratio
		Concrete-filled rectangular steel tubular composite truss girder bridge	**Prestressed concrete box-girder bridge**	
Superstructure	Concrete (m^3/m^2)	0.528	0.929	0.57
	Steel (kg/m^2)	308.9	238.2	1.30
Substructure and foundation	Concrete (m^3/m^2)	1.135	1.433	0.79
	Steel (kg/m^2)	78.5	219.7	0.36
Total	Concrete (m^3/m^2)	1.663	2.361	0.70
	Steel (kg/m^2)	387.5	457.9	0.85

box-girder bridge, the CFRST composite truss bridge uses 43% less concrete in the superstructure but 30% more steel. However, the reduction in superstructure self-weight significantly reduces the material consumption for the substructure. Considering the total material usage for the superstructure, substructure, and foundations, the composite truss bridge uses 30% less concrete and 15% less steel overall. This result indicates that the CFRST composite truss bridge offers significant economic advantages.

8.6 Summary

The Huangyan Overpass Bridge adopts a continuous rigid frame system with a composite deck-and-truss structure. Both the main truss and the piers utilize CFRST members, resulting in a bridge with high overall stiffness. As an experimental bridge, it possesses a high safety margin while maintaining a lightweight and aesthetically pleasing appearance.

Compared to a concrete girder bridge of similar span, the Huangyan Overpass Bridge uses 30% less concrete and 15% less steel, demonstrating significant economic advantages. As a test bridge, the chords, brace members, and deck slab were conservatively designed with uniform cross sections based on the most critical sections. This indicates that there is still potential for further optimization of its economy.

Extended reading

Liu Y.J., Cheng G., Long G., et al. A steel truss-concrete slab composite beam based on shear connection structure: China, 201310744155. 1[P]. 2014-03-19.

Liu Y.J., Liu B., Dai L.X., et al. Fabricated rectangular concrete-filled steel tube composite truss bridge and construction method[P]. Shaanxi： CN106223183A [P]. 2016-12-14.

Liu YJ, Ma YP, Tian ZJ, et al. Field test of rectangular concrete filled steel tubular composite truss bridge with continuous rigid system[J]. China Journal of Highway and Transport 2018;31(05):53–62.

Liu YJ, Xiong ZH, Luo YL, et al. Double-composite rectangular truss bridge and its joint analysis[J]. Journal of Traffic and Transportation Engineering (English Edition 2015;2(4):249–57.

Liu Y.J., Yang Y.H., Yuan Z.Y., et al. A PBL stiffened rectangular concrete-filled steel tubular space truss: China, 201410032698. 5[P]. 2014-04-16.

Liu Y.J., Yuan Z.Y., Cheng G., et al. Continuous rigid frame bridge based on steel truss-concrete slab composite beam: China, 201310745945. 1[P]. 2014-03-26.

Liu Y.J., Yuan Z.Y., Xiong Z.H., et al. A Y-shaped pier for concrete-filled steel tube truss beam: China, 201310745942. 8[P]. 2014-03-26.

Tian ZJ, Liu YJ, Jiang L, et al. A review on application of composite truss bridges composed of hollow structural section members[J]. Journal of Traffic and Transportation Engineering (English Edition 2019;6(1):94–108.

CHAPTER 9

Prefabricated concrete-filled rectangular steel tubular composite truss bridge

9.1 Overview

This chapter adopts standardization and prefabrication as the core design concepts, fully leveraging the characteristics of the CFRST composite truss structure. General design drawings for prefabricated CFRST composite truss bridges with spans of 50 m, 60 m, 70 m, and 80 m are developed as examples. Similar to Chapter 8, relevant structural design details, structural verification, and techno-economic analyses are also provided.

9.2 Design overview

9.2.1 Main technical standards

(1) Structural type: Continuous upper-deck CFRST composite truss bridge.
(2) Span arrangement: Single-span lengths of 50 m, 60 m, 70 m, and 80 m (total bridge length consisting of 4 to 8 spans).
(3) Bridge deck width: 0.50 m (anti-collision barrier) + 11.75 m (carriageway) + 0.50 m (anti-collision barrier) = 12.75 m, with a transverse slope of 2.0%.
(4) Design load: Highway Class I.
(5) Highway classification: Expressway.

9.2.2 Material parameters

(1) Concrete: The chords are filled with C60 steel fiber-reinforced concrete; the precast slabs use C50 concrete; and the cast-in-place slabs use C50 micro-expansive concrete.
(2) Main truss steel: The primary load-bearing members of the truss adopt Q420qD steel, while other members use Q345qD steel.
(3) Ordinary reinforcement: HPB300 and HRB400 rebars.
(4) Prestressing steel: Strand1860 low-relaxation high-strength steel strands.

Composite Truss Bridges With Concrete-filled Rectangular Steel Tube Members
DOI: https://doi.org/10.1016/B978-0-443-44011-3.00010-X

9.2.3 Design specifications and standards

(1) *Technical Standard of Highway Engineering (JTG B01–2014)*
(2) *General Code for Design of Highway Bridges and Culverts (JTG D60–2015)*
(3) *Code for Design of Highway Reinforced Concrete and Prestressed Concrete Bridges and Culverts (JTG 3362-2018)*
(4) *Technical Specification for Rectangular Concrete-Filled Steel Tubular Structures (CECS 159:2004)*
(5) *Code for Design and Construction of Highway Steel–Concrete Composite Bridges (JTG/T D64-01-2015)*
(6) *Code for Design of Highway Steel Bridges (JTG D64–2015)*
(7) *Standard for Design of Steel Structures (GB 50017-2017)*
(8) *Code for Welding of Steel Structures (GB 50661-2011)*

9.3 Structural design

9.3.1 Overall layout and main truss selection

Based on the requirements of prefabricated bridge design for standardized span segments, the general design of the CFRST continuous composite truss bridge adopts an equal-span configureuration, with standard spans divided into two types: side spans and middle spans. The main truss elevation adopts a triangular truss layout, and the overall layout parameters are shown in Table 9.1. Here, the beam height refers to the distance from the bottom surface of the lower chord to the top surface of the bridge deck slab. The elevation layout of the composite truss girder is illustrated in Fig. 9.1 (taking the 80 m span as an example).

9.3.2 Cross-section, bridge deck selection, and plate-truss connection structure

After comparative analysis and optimization, a four-chord double main truss cross-section was finally adopted, as shown in Fig. 9.2. The total bridge width is 12.75 m, with a spacing of 6.4 m between the two main trusses. Local reinforcement measures are applied at the bridge supports as follows:

(1) End diaphragms are installed at the supports to increase the torsional stiffness of the section.
(2) The upper and lower chords and brace members near the supports are locally widened to enhance the bending and shear capacity of the section.

Table 9.1 General arrangement of 50~80 m continuous composite truss girder bridge with concrete-filled rectangular steel tubes.

Span/m	Bridge width/m	Girder height/m	Height-to-span ratio	Panel length/m	Brace member inclination angle/°
50	12.75	3.62	1/14	3.4+12×3.6+3.4=50	59.3
60		4.32	1/14	15×4=60	62.1
70		4.72	1/15	14×5=70	56.8
80		5.72	1/14	16×5=80	63.5

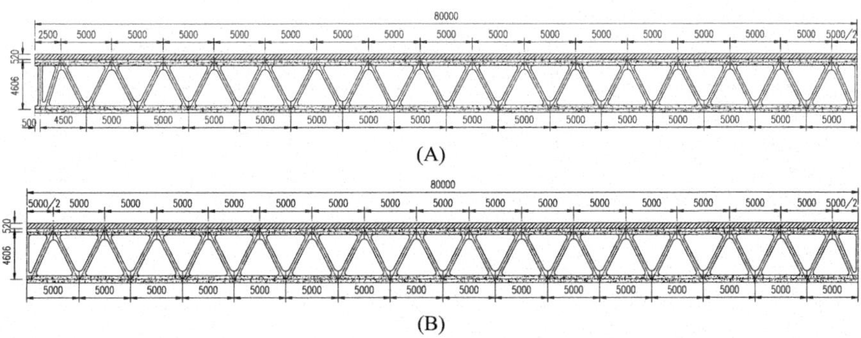

Figure 9.1 80 m Elevation layout of CFRST composite truss bridge (unit: mm). (A) Side span, (B) Mid-span.

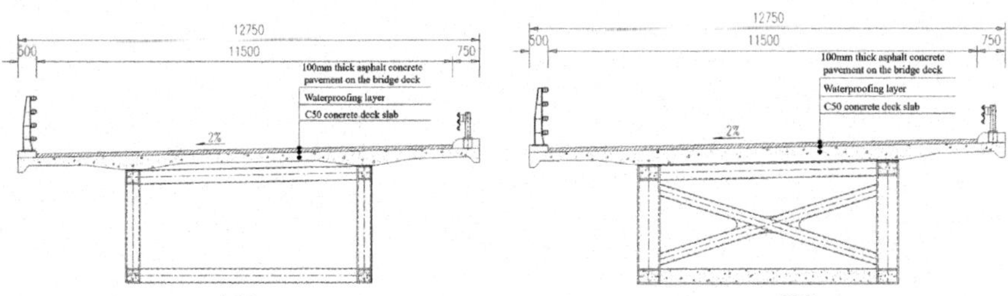

Figure 9.2 Cross-sectional layout (unit: mm). (A) Mid-span, (B) Support point.

(3) At the supports, cast-in-place concrete bottom slabs are provided for two panel lengths on each side of the pier top to assist the lower chords in compression.

(4) The concrete deck slab near the supports is locally thickened and densely reinforced, with longitudinal prestressing tendons arranged. Anti-uplift but non-shear-resisting headed studs are used to improve the prestress transfer efficiency in the deck slab, thereby further enhancing its crack resistance.

To accelerate deck construction and reduce secondary internal forces caused by concrete shrinkage and creep, prefabricated deck slabs are adopted in the mid-span regions in this general design. The deck slabs are fully prefabricated across the entire bridge width, with shear keys and wet joints reserved. At the pier tops, cast-in-place deck slabs are used to facilitate the application of prestressing, as shown in Fig. 9.3. The plan

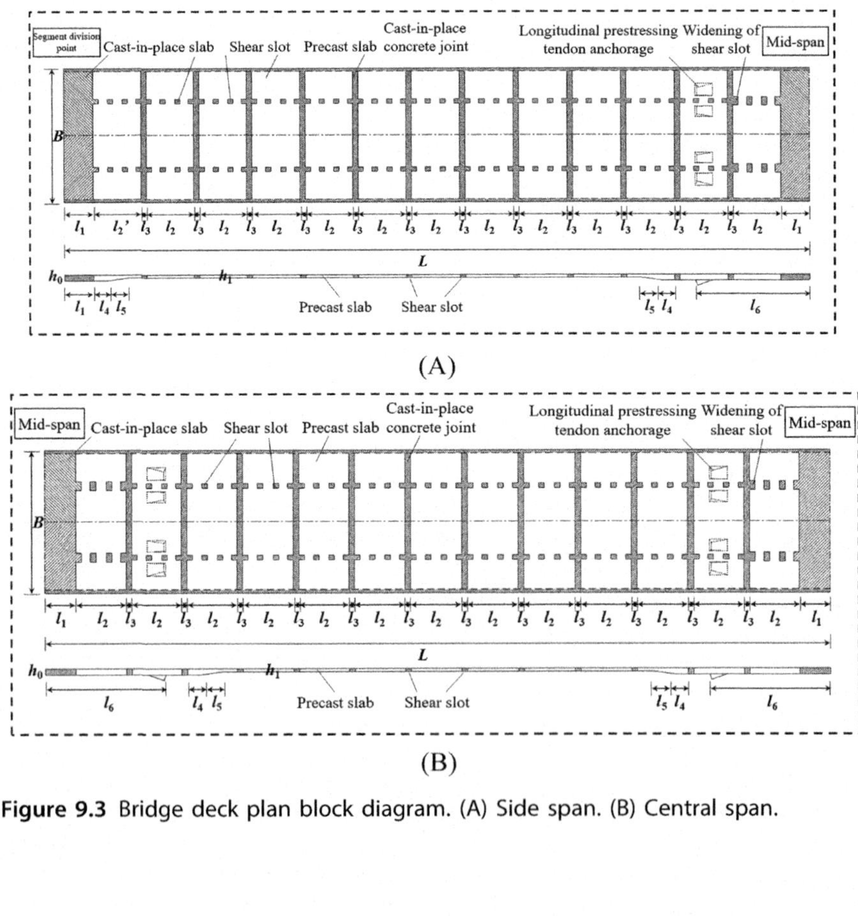

Figure 9.3 Bridge deck plan block diagram. (A) Side span. (B) Central span.

dimensions and longitudinal prestressing ranges of the deck slabs for spans of 50–80 m are given in Fig. 9.3 Bridge deck plan block diagram.

Table 9.2, and the longitudinal design of the deck slabs follows that of conventional reinforced concrete members.

Considering the 6.4 m spacing between the main trusses, transverse tensile stresses may develop at the bottom of the deck slab in the mid-span and at the top near the supports under wheel loads. To prevent transverse cracking of the deck slab, transverse prestressing tendons are arranged across the entire span. The deck slab is designed as a Class B prestressed concrete member in the transverse direction, with the transverse prestressing tendons passing through the shear keys or wet joints tensioned on site.

As shown in Fig. 9.4, the concrete deck slabs are classified into two types: standard deck slabs and locally thickened deck slabs. For spans of 50 m and 60 m, the slab thickness at the haunches is 400 mm, while for spans of 70 m and 80 m, the thickness at the haunches is 500 mm. The haunch width extends 100 mm beyond the top surface of the upper chord on both sides. To ensure a dense contact between the upper chord and the bottom surface of the deck slab, a 20 mm thick epoxy mortar layer is provided at the steel–concrete interface.

The connection between the concrete deck slab and the steel truss is achieved through shear studs welded on the top plate of the upper chord. The studs are headed cylindrical type made of ML15 steel and arranged in bundles. The number of studs is matched to the dimensions of the chord members. In negative moment regions, the chords are locally widened, and the number of studs is correspondingly increased. To improve the efficiency of prestress application and enhance the crack resistance of the concrete deck slab in negative moment regions, anti-uplift but non-shear-resisting headed studs are used in these areas, while standard headed studs are used elsewhere. The shapes and arrangements of ordinary shear studs and anti-uplift studs are shown in Figs. 9.5 and 9.6, respectively.

9.3.3 Main truss members, joint selection, and main truss connection structure

Based on the bending moment and shear force diagrams of the continuous beam (Fig. 9.7) and the force transfer mechanism of the truss (Fig. 9.8), it can be observed that the axial forces in the upper chords, lower chords, and brace members of the continuous composite truss bridge exhibit a stepped distribution pattern. Accordingly, a graded design can be adopted for the

Table 9.2 Bridge deck plane block size and prestressed steel bar range.

Span length L	Bridge width B	l_1	l_2	l_2'	l_3	l_4	l_5	l_6	h_0	h_1
50	12.75	1950	3100	3000	500	750	1500	7827	400	250
60		2250	3500	3500	500	1000	1500	8839	400	250
70		2750	4500	4500	500	1250	2000	10827	500	250
80		2750	4500	4500	500	1250	2000	12374	500	250

Note: Span and bridge width are recorded in m, and other parameters are recorded in mm.

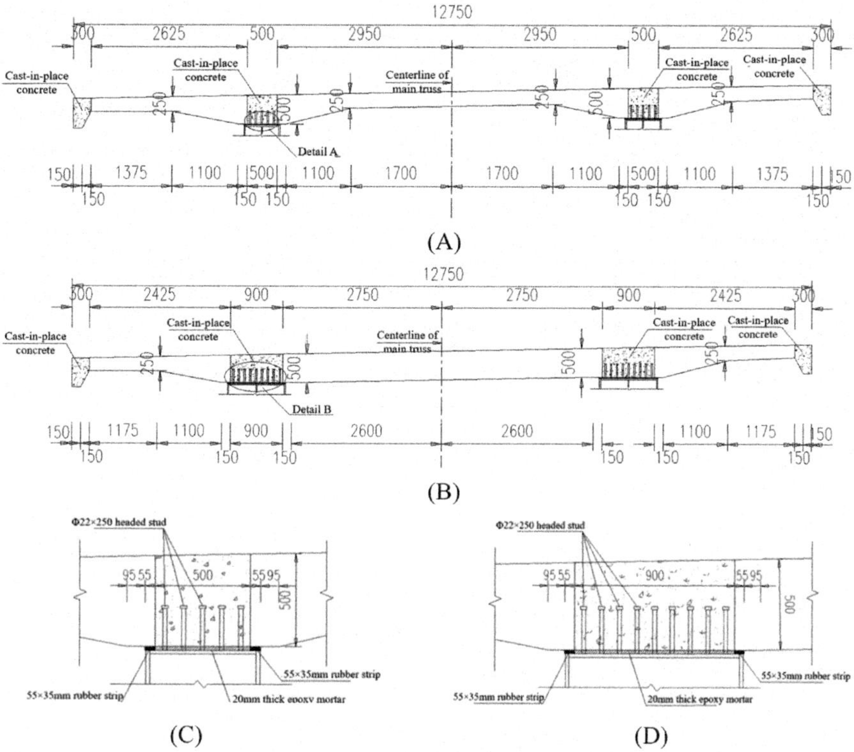

Figure 9.4 Transverse layout of bridge deck (80 m) (unit: mm). (A) Conventional deck slab. (B) Locally thickened deck slab. (C) Detail A. (D) Detail B.

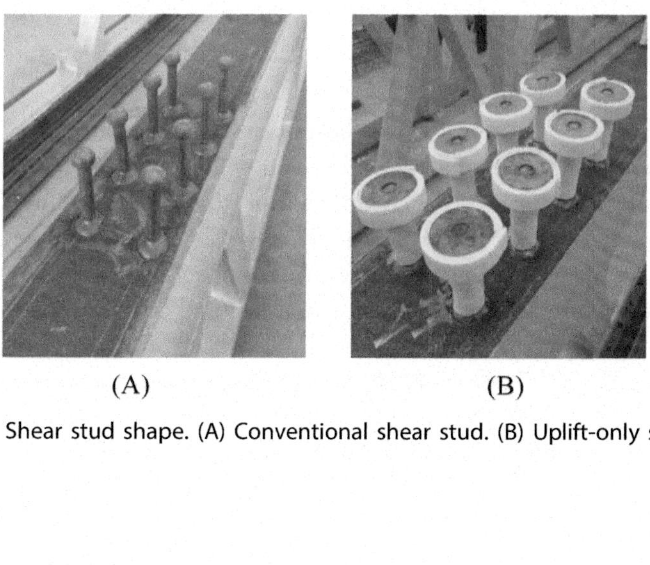

(A) (B)

Figure 9.5 Shear stud shape. (A) Conventional shear stud. (B) Uplift-only shear stud.

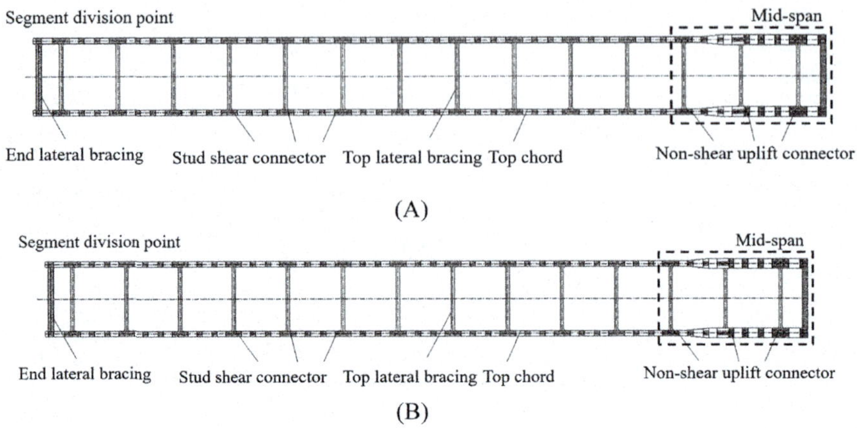

Figure 9.6 Shear stud arrangement. (A) Side span. (B) Central span.

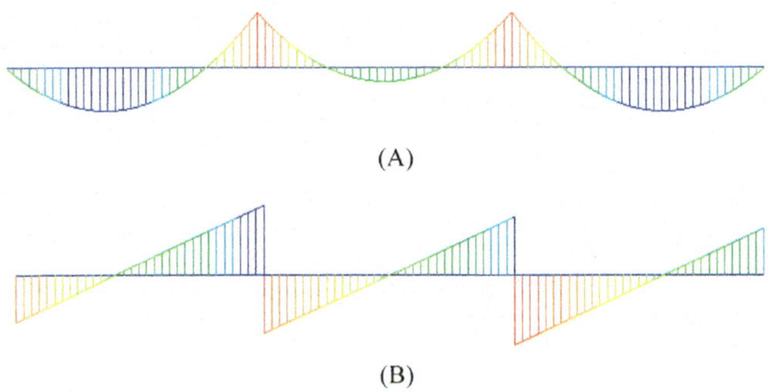

Figure 9.7 Bending moment and shear force distribution under self-weight for three-span continuous beam. (A) Bending moment. (B) Shear force.

rectangular steel tubes—under the condition of maintaining the same outer dimensions of the members—by varying the wall thickness of the steel tubes at different locations according to the magnitude of internal forces.

The chords adopt PBL-stiffened CFRST sections, while the brace members and end diaphragms use rectangular steel tubes, and the upper and lower lateral members adopt I-shaped sections. Following the concept of graded design and based on the internal force distribution of the continuous beam, a three-level design is applied to both the main truss chords and brace members, as shown in Fig. 9.9. Different numbers indicate corresponding members with different cross-sectional dimensions.

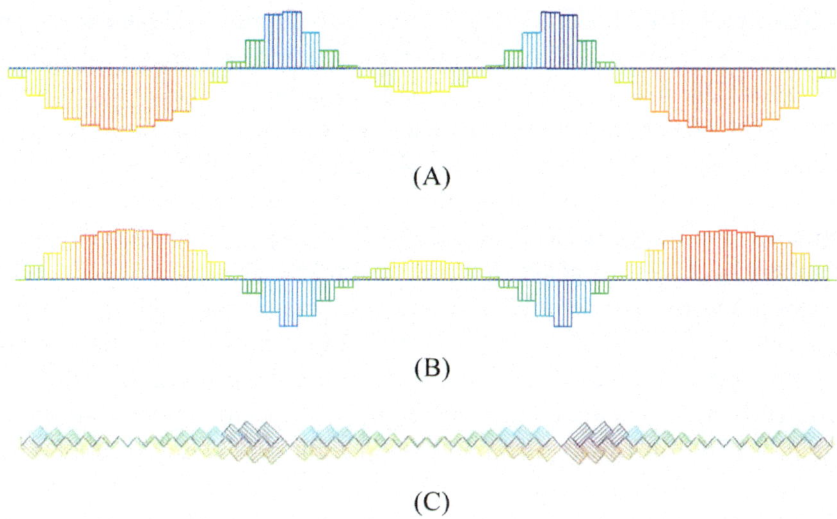

Figure 9.8 Internal force distribution of three-span CFRST truss beam under deadweight. (A) Axial force of upper chord. (B) Axial force of lower chord. (C) Axial force of brace members.

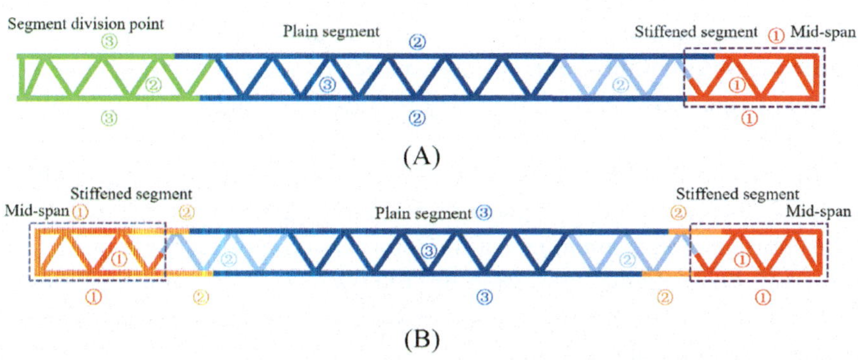

Figure 9.9 Truss grading design. (A) Side span. (B) Central span.

Due to the relatively large bending moments and shear forces at the pier tops, the member wall thickness is increased along with local widening of the sections, forming the strengthened segment regions of the main truss. As shown in Fig. 9.9, the side spans include one strengthened segment, while the middle spans include two symmetrical strengthened segments.

The brace plate thickness ranges from 12 to 36 mm, and the chord plate thickness ranges from 16 to 44 mm, with specific cross-sectional dimensions provided in Tables 9.3 and 9.4. Due to the equal-width design of the braces and chords, the braces in the strengthened segments may have width-to-height ratios greater than 2. To ensure compliance with the plate width-to-thickness ratio limits, the brace plates in the strengthened segments are designed with unequal thicknesses, while the remaining members have all four plates of equal thickness.

For smooth force transfer, the end diaphragms and upper and lower lateral members are designed to have the same height as the chords, with wall thicknesses primarily of 12 mm and 16 mm. Regarding the PBL keys, two main types are adopted in this design, with specific dimensions shown in Fig. 9.10 Form I, with a stiffening rib height of 130 mm, is used when the chord width or height exceeds 500 mm, while Form II, with a stiffening rib height of 100 mm, is used when the chord width or height falls outside this range.

To ensure that the members of the CFST composite truss fail before the joints, the joint bearing capacity should be higher than that of the members. The structural system must form "sectional plastic hinges" before joint failure. This "strong-joint, weak-member" design concept imposes requirements on the cross-section similar to those of compact sections in Eurocode 4 for composite steel-concrete beams (where a plastic hinge with sufficient rotation capacity forms before local buckling occurs in the plate elements).

In this design, PBL-stiffened CFRST joints with uniform width are adopted to significantly improve the mechanical performance of the joints. The chord and brace members of the joints are connected by welding, with most of the welding completed in the factory. Only minimal on-site welding and assembly are required, which greatly enhances construction efficiency and ensures welding quality.

Fatigue failure of joints is a critical consideration in truss design. According to Table C.0.7 in Appendix C of *Code for Design of Highway Steel Bridges* (JTG D64–2015), the fatigue detail category of rectangular hollow K-type joints is only 35 MPa, which is insufficient to meet the fatigue design requirements of bridges. To further improve the fatigue performance of joints, this design integrates high-fatigue-resistance joint plates into the PBL-stiffened CFRST joints, as shown in Fig. 9.11.

The edges of the joint plates are rounded, and full-penetration butt welding is used with adjacent plates. The ends of the welds are ground and

Table 9.3 Side-span main truss cross-section dimensions.

Span length/m	Top chord dimensions $b_0 \times h_0 \times t_0$/mm		Bottom chord dimensions $b_0 \times h_0 \times t_0$/mm		Brace member dimensions $b_1 \times h_1 \times t_1$/mm	
	Standard segment	Reinforced segment	Standard segment	Reinforced segment	Standard segment	Reinforced segment
50	400×400×32	600×400×40	400×400×24/32	600×400×36	400×250×12/16/20	600(32)×250(20)
60	500×500×30	700×500×44	500×500×24/36	700×500×36	500×300×16/20/22	700(32)×300(24)
70	500×500×28/36	900×500×44	500×600×28/36	900×600×40	500×300×16/20/22	900(36)×300(30)
80	600×600×20/36	1000×600×44	600×700×24/36	1000×700×40	600×400×16/20/22	1000(36)×400(30)

Table 9.4 Cross-section dimensions of main truss in central span.

Span length/m	Top chord dimensions $b_0 \times h_0 \times t_0$/mm		Bottom chord dimensions $b_0 \times h_0 \times t_0$/mm		Brace member dimensions $b_1 \times h_1 \times t_1$/mm	
	Standard segment	Reinforced segment	Standard segment	Reinforced segment	Standard segment	Reinforced segment
50	400×400×16/32	600×400×40	400×400×22/32	600×400×36	400×250×12/16	600(32)×250(20)
60	500×500×20/30	700×500×44	500×500×20/30	700×500×36	500×300×16/20	700(32)×300(24)
70	500×500×20/36	900×500×44	500×600×20/30	900×600×36	500×300×16/20	900(36)×300(30)
80	600×600×20/36	1000×600×44	600×700×20/36	1000×700×40	600×400×16/20	1000(36)×400(30)

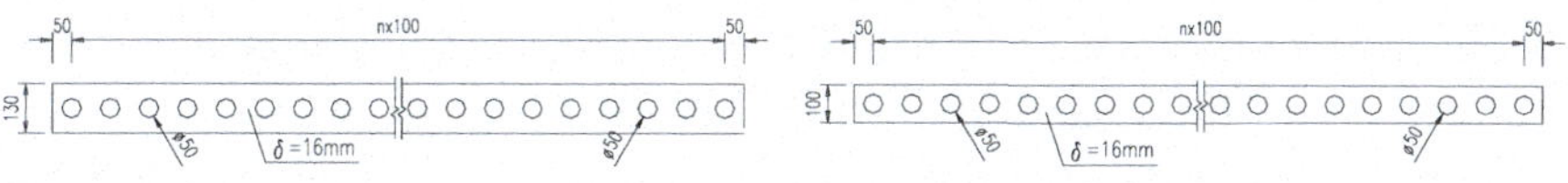

(A) (B)

Figure 9.10 PBL key size. (A) Form I. (B) Form II.

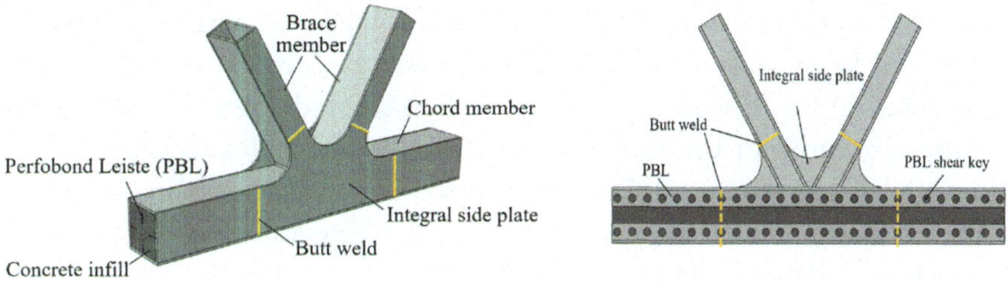

Figure 9.11 PBL-reinforced CFRST joint with gusset plate.

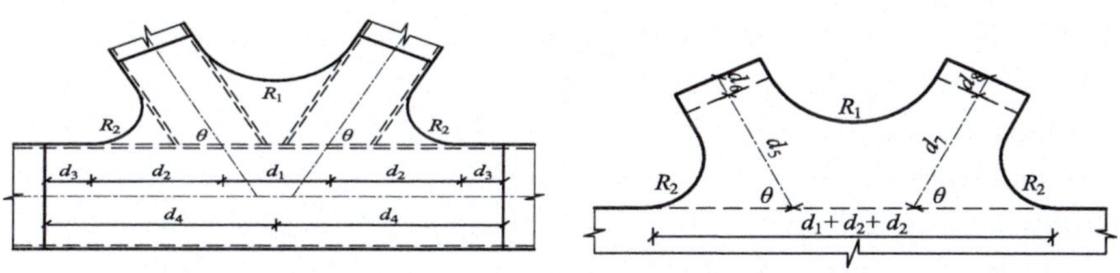

(A) (B)

Figure 9.12 Standard K-type joint for lower chord. (A) Standard joint. (B) Detailing of gusset plate.

hammer-treated to meet the fatigue performance requirements of the welds in this area. In addition, the overall joint plate design must meet the following requirements:①The radius of the rounded edge of the joint plate should be greater than half the height of the chord.②The distance between the butt weld position of the joint plate and the end of the arc should be at least 100 mm.

The K-type joint in the general drawing of the 50~80 m CFRST composite truss bridge is shown in Fig. 9.12. Within the joint range, the side plates of the chord members and the side plates of the brace members are made of a single integral steel plate. The joint eccentricities all meet the requirement of the specification that $-0.55 \le e/h_0 \le 0.25$.

Due to the adoption of the hierarchical design concept in this design, it can be seen from Fig. 9.9 that there are many types of joints and gusset plates. Only the standard joint dimensions of the lower chord members of the composite truss bridge are provided, as shown in Table 9.5.

When it is difficult to realize integral transportation and assembly due to poor road transportation conditions at the bridge site, insufficient on-site space, and large hoisting weight of the truss, the steel truss girder can be considered to be "divided into parts". The scattered assembly components are manufactured in the factory, and then welded and assembled on site. Since the integral gusset plate is adopted in this design, the division of scattered components shall be based on the integral gusset plate as the division boundary, as shown in Fig. 9.13.

When the bridge site is located in an area with convenient transportation and good transportation and hoisting capacity, the segmental construction method of "integrating parts into a whole" can be adopted. The on-site connection method between the main truss segments directly determines the overall forming construction quality of the main truss. The final selection of the chord joint is "built-in connecting plate + outer-wrapped steel pipe welding", and the brace member joint is "outer-wrapped steel pipe welding", as shown in Fig. 9.14. When connecting the chords, attention should be paid to welding the bottom plate first, then the side plates, and finally the top plate to ensure the welding quality.

9.4 Structural verification

9.4.1 Most unfavorable layout

The general drawing design of this CFRST composite truss bridge is applicable to 4 to 8 spans, adopting an equal-span design, and the standard span lengths are divided into two types: side spans and middle spans. Fig. 9.15 shows the bending moment diagram of a 4-to-8-span continuous beam with equal spans under uniformly distributed load, where the single-span length L=80 m and the uniformly distributed load q =1kN/m. The pier at the left-end joint is defined as Pier 0#, and the piers from left to right are numbered Pier 1#, Pier 2#, and so on. It can be seen from Fig. 9.15 that when the continuous beam is arranged with equal spans, the maximum positive and negative bending moments are located at the top of

Table 9.5 Standard joint size of bottom chord.

Span length layout/m	θ/°	R_1/mm	R_2/mm	d_1/mm	d_2/mm	d_3/mm	d_4/mm	d_5/mm	d_7/mm
4×50	59.3	200	200	341	497	133	800	426	426
4×60	62.1	300	200	397	502	199	900	424	424
4×70	56.8	400	200	433	550	234	1000	469	469
4×80	63.5	400	250	519	627	213	1100	623	623

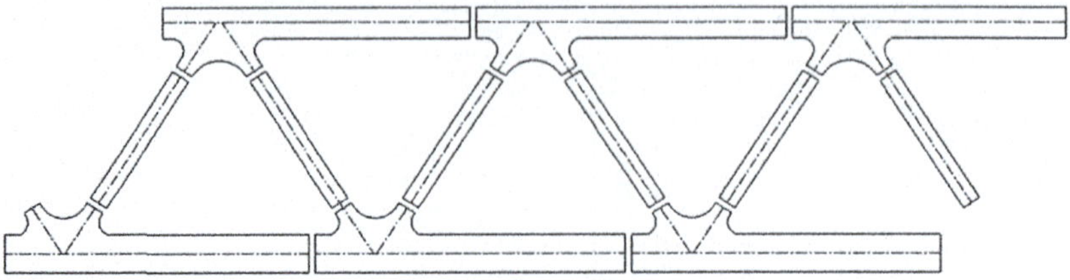

Figure 9.13 Division of main truss loose components.

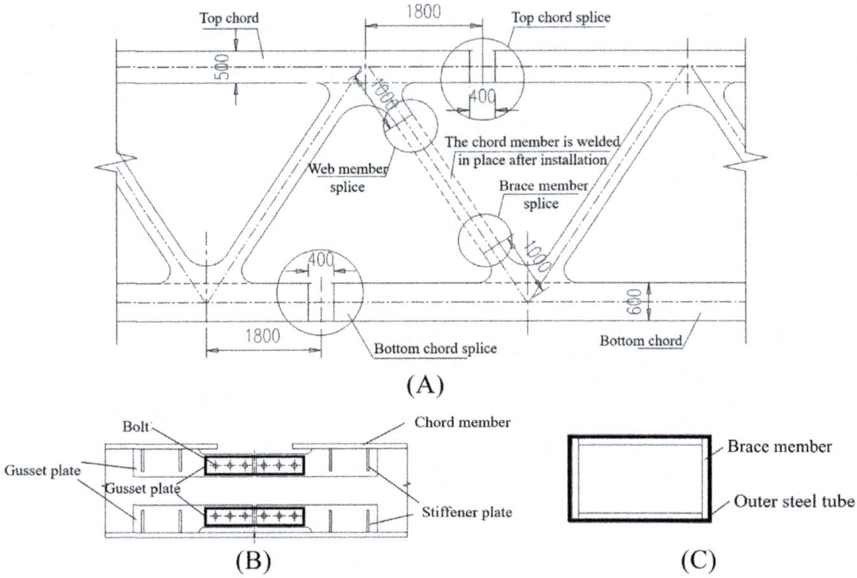

Figure 9.14 Main truss segment connection. (A) Location schematic of inter-segment connections. (B) Chord joint. (C) Brace member joint.

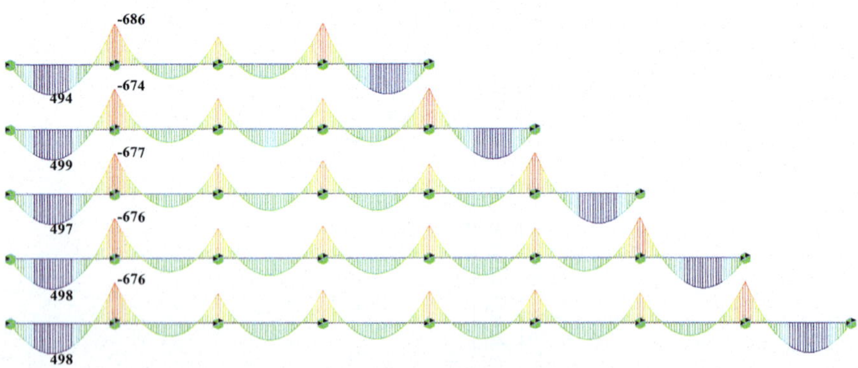

Figure 9.15 Bending moment of continuous beams with 4 to 8 spans (unit: kN·m).

Pier 1# and the mid-span of the side span respectively, and the bending moment levels of different arrangement forms differ slightly. The negative bending moment at the pier top is the largest when arranged with 4 spans, and the positive bending moment at the mid-span is the largest when arranged with 5 spans. In the mid-span positive bending moment zone, the upper chord members and bridge deck are in compression, while the lower chord members are in tension; in the pier-top negative bending moment zone, the upper chord members and bridge deck are in tension, while the lower chord members are in compression. Combined with the stress characteristics of steel and concrete, it can be concluded that the most unfavorable stress position is the top of Pier 1#. From the above analysis, the 4-span arrangement is the most unfavorable bridge span arrangement. Therefore, this section only calculates and analyzes the 4-span 50~80 m CFRST composite truss bridge.

9.4.2 Model calculation parameters and loads

9.4.2.1 Calculation parameters

The properties of concrete materials are shown in Table 9.6. C60 steel fiber concrete is used for the filled concrete inside the upper and lower chord members of the main truss; C50 concrete is used for the precast bridge deck concrete; C50 micro-expansive concrete is used for the cast-in-place bridge deck concrete and cast-in-place bottom slab concrete.

The material properties of the main truss steel plates are shown in Table 9.7. Q420qD steel is used for the rectangular cross-section members of the main truss, such as chord members, brace members, and end cross bracings. Q345qD steel is used for the I-shaped cross-section members of the main truss, such as upper and lower lateral bracings.

Table 9.6 Concrete material properties.

Strength grade	Elastic modulus E_c/MPa	Characteristic value/MPa		Design value/MPa	
		f_{ck}	f_{tk}	f_{cd}	f_{td}
C60	36000	38.5	2.85	26.5	1.96
C50	34500	32.4	2.65	22.4	1.83

Table 9.7 Steel plate material properties.

Steel plate		Elastic modulus	Design values for tensile, compressive and flexural strength	Design value for shear strength
Grade	Thickness/ mm	E_s/MPa	f_d/MPa	f_{vd}/MPa
Q345 steel	≤16	206000		160
	16~40		270	155
	40~63		260	150
Q420 steel	≤16		335	195
	16~40		320	185
	40~63		305	175

Prestressed Steel Bars: The nominal diameter of low-relaxation steel strands is d=15.2 mm, the characteristic value of tensile strength is fpk=1860MPa, and the elastic modulus is Ep=195000 MPa. The relaxation coefficient is 0.3; the deformation of anchors and retraction of steel bars is taken as 6 mm (per end); the pipeline friction coefficient is μ=0.25, and the pipeline deviation coefficient is k = 0.0015.

9.4.2.2 Calculate load

Self-weight: The unit weight of concrete is 26kN/m^3, and the unit weight of steel is 78.5kN/m^3.

Secondary permanent load: Including auxiliary facilities such as bridge deck pavement and guardrails. The unit weight of bridge deck pavement is 24kN/m^3, the load of bridge deck pavement on one side is 16 kN/m^3, and the load of guardrail on one side is 6 kN/m^3.

Vehicle load: Highway ClassI, with loading on 3 one-way lanes.

System temperature difference: Considered according to severe cold areas. The effective high temperature of the structural system is 39°C, and the effective low temperature is -32°C. Assuming the closure temperature is 10°C, the temperature rise of the structural system is 29°C, and the temperature drop is -42°C.

Temperature gradient: The positive temperature difference T_1 = 14°C, T_2 = 5.5°C, and the negative temperature difference is -0.5 times the positive temperature difference.

Shrinkage and creep of concrete: The annual average relative humidity is 55%.

9.4.3 Finite element model

Based on the structural arrangement, cross-sectional dimensions and construction stage division in the design drawings, the Midas Civil finite element software is used for simulation and analysis. Truss members are simulated using beam elements, among which PBL connectors are only used as safety reserves and not considered during modeling. The bridge deck is simulated with two longitudinal beam elements, and virtual beam elements (considering stiffness only, not weight) are used to simulate its transverse stiffness. The concrete bottom slab near the pier top is simulated with plate elements. The concrete inside the steel pipes of the lower chord members near the mid-span and the upper chord members near the pier top is in tension; during simulation, only the unit weight of concrete is considered, and its tensile contribution is not taken into account. Truss joints are spatial fully welded joints, which are approximately simulated using common joints. The finite element analysis model of the entire bridge is shown in Fig. 9.16.

9.4.3.1 Boundary condition simulation

General supports are adopted at the pier bottoms and abutments. The simulation methods for the connection between the bridge deck and the upper chord members are divided into two types: in the positive bending moment zone of the bridge deck, general elastic connections are used to simulate ordinary shear studs; in the negative bending moment zone of the bridge deck, elastic connections with released longitudinal and lateral stiffness are used to simulate uplift-resistant but shear-non-resistant shear studs.

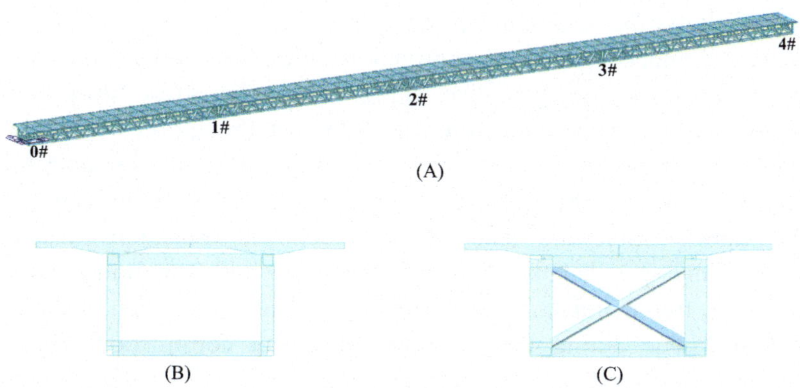

Figure 9.16 Full bridge finite element model. (A) Three-dimensional mode. (B) Mid-span cross-section. (C) Pier top cross-section.

9.4.3.2 Load conditions and load combinations

The structure is checked and verified in accordance with *Code for Design of Highway Steel Bridges* (JTG D64–2015), *Technical Specification for CFRST Structures* (CECS 159:2004), *Code for Design of Highway Reinforced Concrete and Prestressed Concrete Bridges and Culverts* (JTG 3362-2018), and *Code for Design and Construction of Highway Steel-Concrete Composite Bridges* (JTG/T D64-01-2015). *Load combinations are made in accordance with General Code for Design of Highway Bridges and Culverts* (JTG D60–2015).

9.4.3.3 Construction phase division

There are two types of construction methods applicable to CFRST composite truss bridges: when the height difference between the road alignment and the on-site ground is large (such as in loess gully terrain), the main truss can adopt the incremental launching construction method; when the height difference between the alignment and the existing ground is small, the main truss can adopt the hoisting construction method. This section only analyzes and verifies the structural strength and deformation during hoisting construction. The schematic diagram of the construction steps of the hoisting method is shown in Fig. 9.17, and the division of key construction stages in the finite element model is shown in Fig. 9.17 Guiding construction steps of the hoisting method Table 9.8.

9.4.4 Design verification

9.4.4.1 Structural verification during construction

Fig. 9.18 shows the maximum stress values of the main truss chord members and brace members during key construction stages. Among these, the concrete had not developed strength when taking the member stress values in stages CS2 to CS5. It can be seen from Fig. 9.18 that for the 50~80 m CFRST composite truss bridge, the variation trends of the maximum member stresses during key construction stages are consistent, and the maximum stress levels differ slightly—indicating that the cross-sectional dimensions match the corresponding span lengths. When hoisting the mid-span bridge decks of each span, the member stress increases suddenly, and then reaches the maximum value during the secondary pavement construction. Since the overall stress level of the main truss members is relatively low, there is no need to further check and verify the bearing capacity of chord members and the stability bearing capacity of brace members during the construction stage. Further analysis of Fig. 9.19 shows that the maximum tensile stress of chord

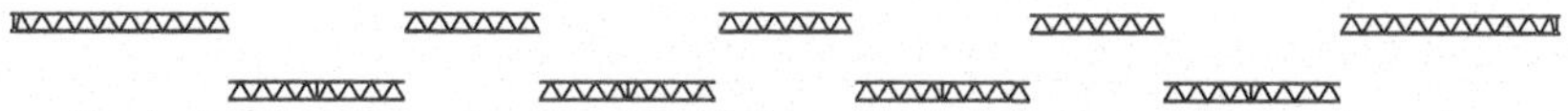

Step 1: Complete the construction of foundations, piers, and abutments; assemble the steel truss in the factory or on-site. The hoisting length shall take into account the lifting capacity of the hoisting machinery, and the segmental splicing position shall be preferably near the bending moment zero point of the steel truss girder.

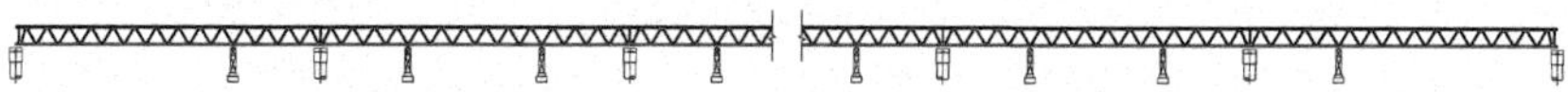

Step 2: Set up temporary facilities. After hoisting the steel truss into position, perform pre-assembly, and then carry out on-site connection of the member joints.

Step 3: Dismantle the temporary piers; at the same time, prefabricate the concrete bridge decks in the prefabrication yard. After the concrete reaches the curing strength, tension the transverse prestressed steel bars that do not pass through the shear grooves of the bridge decks.

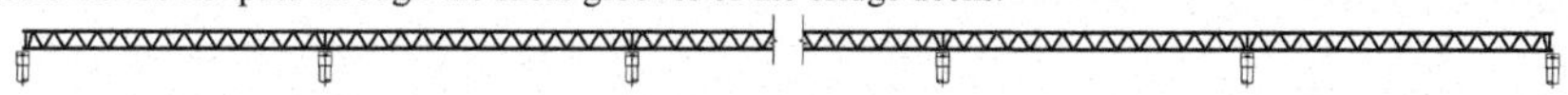

Step 4: Pour concrete into the lower chord members; bind steel bars and pour concrete for the bottom slab in the pier-top negative bending moment zone; pour concrete into the upper chord members.

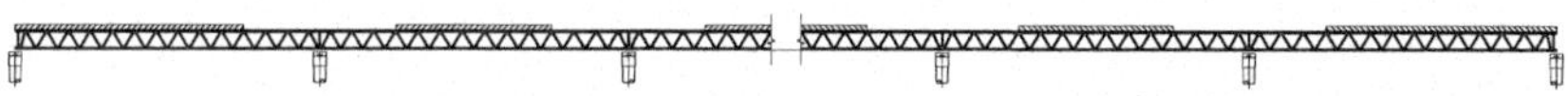

Step 5: Hoist the prefabricated concrete bridge decks at the mid-span of each span, and pour concrete for the wet joints and shear grooves of the bridge decks; at the same time, pour concrete at the expansion joints at the joint ends; tension the remaining transverse prestressed steel bars of the mid-span bridge decks of each span, and pour concrete for the post-cast strips on the side of the decks.

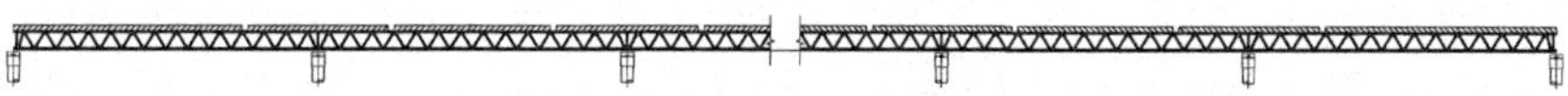

Step 6: Hoist the prefabricated concrete slabs near the pier tops, and cast-in-place the pier-top bridge decks in the negative bending moment zone, as well as the remaining concrete for wet joints (except for the transverse wet joints at both ends) and shear grooves.

Step 7: After the concrete reaches the specified strength and meets relevant requirements, tension the longitudinal prestressed steel bars.

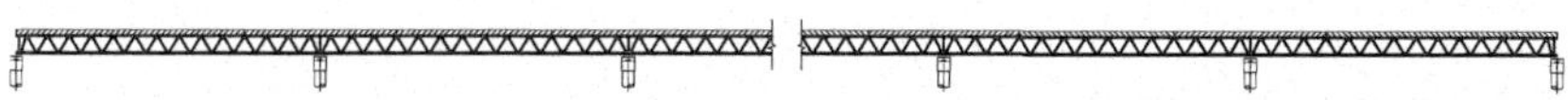

Step 8: Pour concrete for the transverse wet joints at both ends of the pier tops. After the concrete reaches the required strength, tension the remaining transverse prestressed steel bars and pour concrete for the remaining side post-cast strips of the slabs.

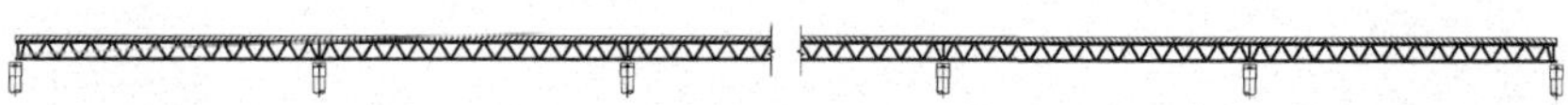

Step 9: Construct anti-collision guardrails and bridge deck pavement; complete the bridge construction.

Figure 9.17 Guiding construction steps of the hoisting method.

members is located at the top of Pier 1#, and the maximum compressive stress is located at the mid-span of the side span. At the pier top, the concrete bottom slab and the concrete inside the steel tube assist the lower chord steel

Table 9.8 Construction phase division of hoisting method.

Phase	Construction activities
CS1	Segment-wise hoisting of main truss (segment length: 20m–50 m) and positioning of truss members
CS2	Concrete placement in bottom chords and cast-in-place concrete base slab at pier tops
CS3	Concrete placement in top chords
CS4	Hoisting of precast concrete deck slabs at mid-span sections of each span
CS5	Hoisting of precast deck slabs near pier tops and cast-in-place deck slab at pier top zones
CS6	Tensioning of longitudinal prestressing tendons
CS7	Construction of bridge deck pavement and ancillary facilities

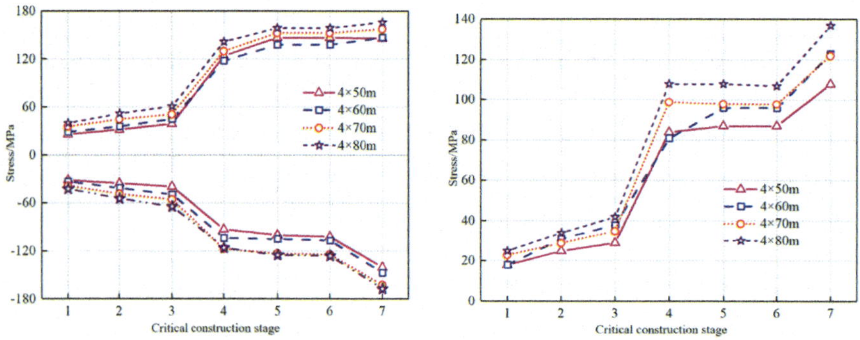

(A) (B)

Figure 9.18 Maximum stress of main truss during critical construction stage. (A) Chord member. (B) Brace member.

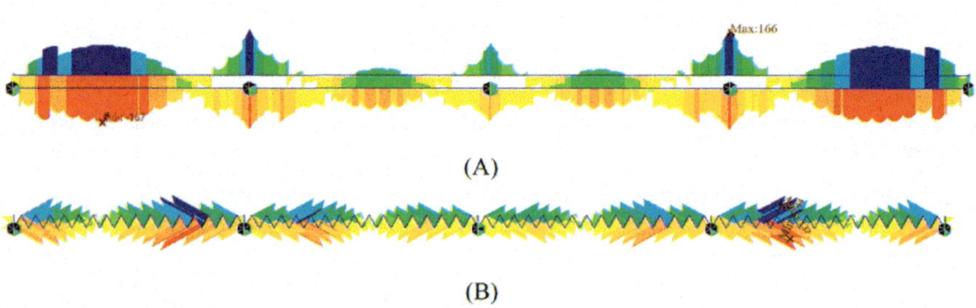

(A)

(B)

Figure 9.19 Maximum stress position of main truss during second phase paving. (A) Chord member. (B) Brace member.

tube in bearing loads, effectively controlling the stress level of the steel tube. The maximum stress of brace members is located near the inflection point, which is the transition position between the reinforced segment and the ordinary segment; the reduction in steel tube wall thickness leads to an increase in steel tube stress. Except for the pier top and inflection point areas, the steel tube stress distribution at other positions is relatively uniform—indicating that the hierarchical design concept of the main truss is relatively reasonable.

It can be seen from Fig. 9.20 that the minimum value of the overall stability safety factor during construction is 8.6, indicating that the overall structural stability meets relevant requirements. As the span increases, the overall stability safety factor during the construction stage decreases, which should be due to the increase in beam height. The overall stability safety factor of the structure is the smallest when the mid-span bridge deck has not developed strength; subsequently, the mid-span bridge deck combines with the main truss to form an integrated structure, and the stability safety factor increases significantly; when the bridge deck in the pier-top area develops strength, the stability safety factor reaches its maximum value.

9.4.4.2 Structural check and verification at the completed bridge stage

The natural vibration modes of the CFRST continuous composite truss bridge (taking the 4×80 m model as an example) are shown in Table 9.9.

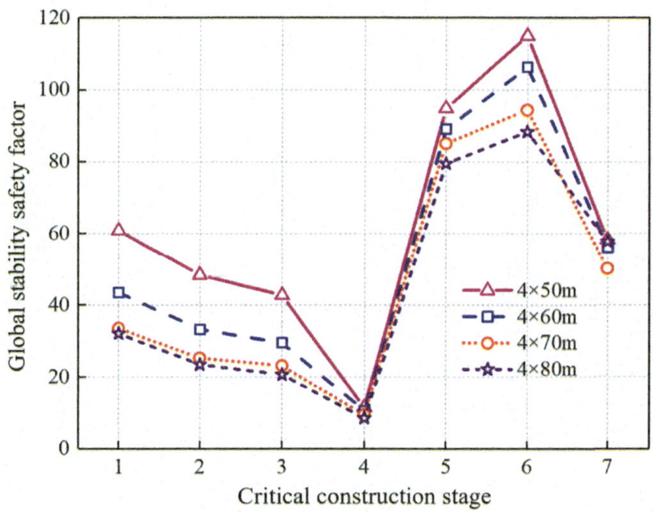

Figure 9.20 Overall stability safety factor.

Table 9.9 Natural vibration modes of CFRST composite truss bridge.

Mode	Frequency	Vibration shape	Description of mode shape
1	1.06	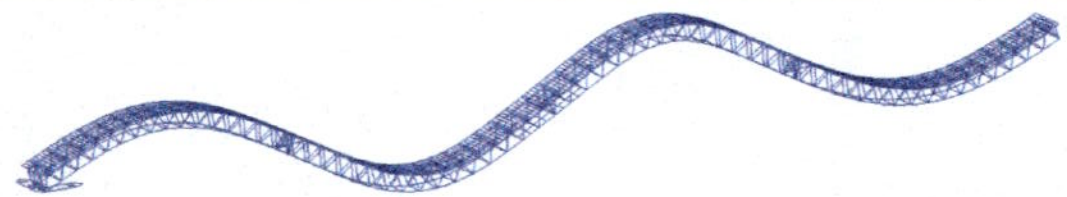	First-order antisymmetric vertical bending
2	1.37	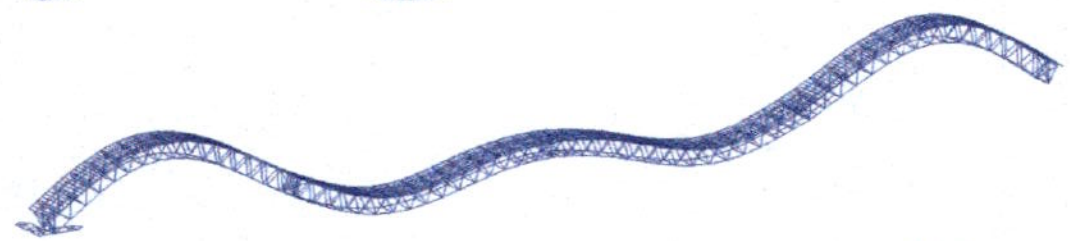	Second-order symmetric vertical bending
3	1.77	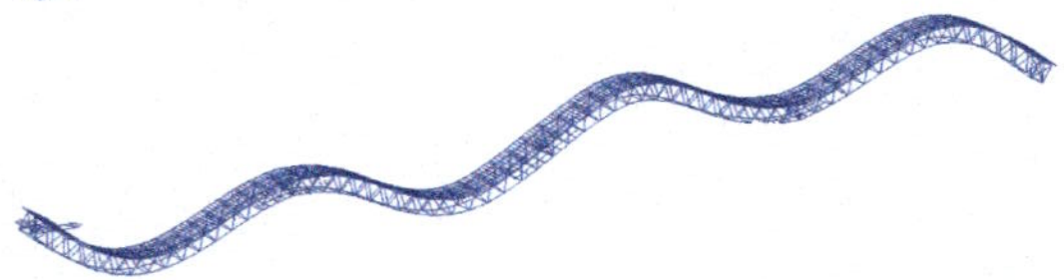	Third-order antisymmetric vertical bending
4	1.97	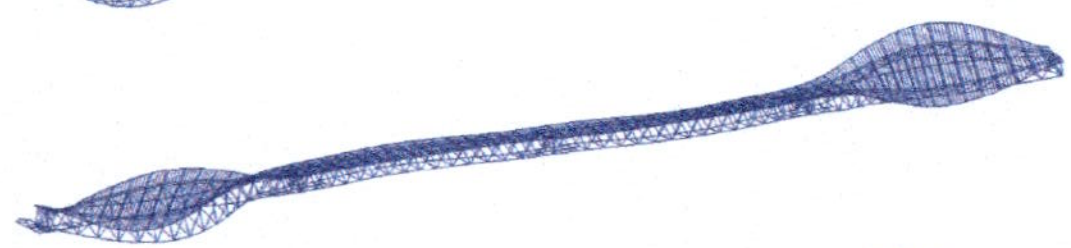	First-order antisymmetric torsion
5	1.99	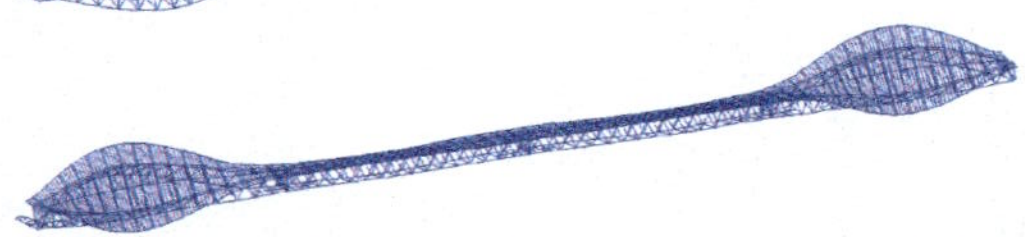	Second-order symmetric torsion

The first three vibration modes are all vertical bending, and the torsional vibration mode only appears in the fourth vibration mode—indicating that the composite effect of the truss girder and the bridge deck enhances the lateral stability of the main truss.

9.4.4.2.1 Structural stiffness check and verification

According to Article 4.2.3 of *Code for Design of Highway Steel Bridges* (JTG D64–2015): When the bridge structure is a simply supported or continuous truss, the vertical deflection caused by vehicle load (excluding impact force) shall not exceed L/500. It can be seen from Table 9.10 that the sum of the absolute values of the positive and negative displacements of the main truss under vehicle load is less than the deflection limit, indicating that the structural stiffness meets the requirements.

9.4.4.2.2 Main truss check and verification

The main load-bearing members of the main truss include upper chord members, lower chord members, and brace members. Among them, the upper and lower chord members are all PBL-stiffened CFRST members, and the brace members are hollow steel tube members. When chord members are under compression, they are checked as CFRST members; when under tension, the contribution of concrete is ignored (only considered as wet weight), and they are checked as rectangular hollow steel tube members.

To summarize, the check types of main truss members can be divided into three categories: CFRST compression members, hollow steel tube tension members, and hollow steel tube compression members.

It can be seen from Fig. 9.21 that under the ultimate limit state of bearing capacity, the upper chord members near the pier tops are under

Table 9.10 Deflection of 50~80 m CFRST composite truss beams under vehicle load.

Span arrangement/m	Positive deflection/mm	Negative deflection/mm	Total structural deflection/mm	Deflection limit/mm
4×50	32.8	-14.5	47.3	100
4×60	33	-15.1	48.1	120
4×70	77	-21	98	140
4×80	36.5	-18.9	55.4	160

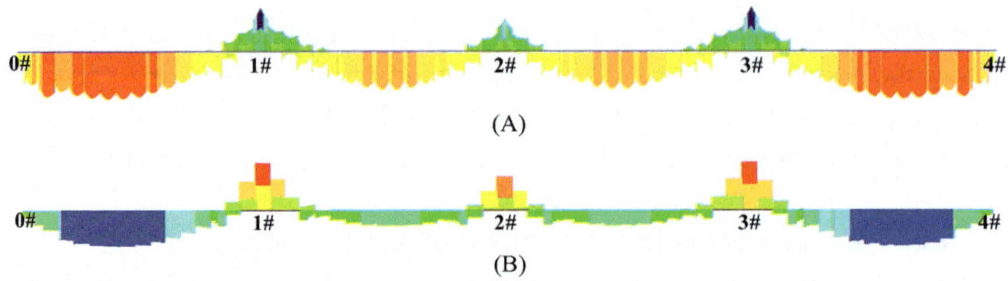

Figure 9.21 Upper chord stress and axial force envelope. (A) Stress. (B) Axial force.

tension, with the most unfavorable tension position at the top of Pier 1#; the upper chord members at the mid-span are under compression, with the most unfavorable compression position at the mid-span of the side span. Due to the equal-span arrangement of the structure (divided into side spans and middle spans), the cross-sectional dimensions of the members at the top of Pier 1# and Pier 2# must be consistent according to the requirements of standardized design, which results in low material utilization in the area at the top of Pier 2#.

The strength of the upper chord members is checked, as detailed in Table 9.11. The calculated stress and axial force values in the Table have all considered the structural importance coefficient γ_0, with $\gamma_0=1.1$ (the same below). The member efficiency coefficient is calculated based on the ratio of member internal force to resistance. It can be seen from Table 9.11 that the upper chord member checks of the 50~80 m CFRST composite truss bridges all pass. The chord member efficiency coefficients range from 0.88 to 0.90, indicating that the cross-sectional design dimensions corresponding to the four span combinations of truss bridges are relatively reasonable, with high material utilization.

It can be seen from Fig. 9.22 that under the ultimate limit state of bearing capacity, the lower chord members at the mid-span are under tension, with the most unfavorable tension position at the mid-span of the side span; the lower chord members in the pier-top area are under compression, with the most unfavorable compression position at the top of Pier 1#. The compressive stress of the lower chord members is relatively high at the transition position between the reinforced segment and the ordinary segment of the middle span. The main reason is that the cross-sectional size of the steel tube in this area is reduced and no concrete bottom slab is installed.

Table 9.11 Calculation results of upper chord bearing capacity.

Span arrangement/m	Stress/ MPa	Design strength /MPa	Axial force /kN	Stability coefficient φ	Member stability capacity /kN	Verification result	Member efficiency factor
4×50	273	305	9486	0.929	16782	Pass	0.90
4×60	268	305	9382	0.943	22827	Pass	0.88
4×70	275	305	15719	0.914	23979	Pass	0.90
4×80	270	305	8201	0.939	31341	Pass	0.89

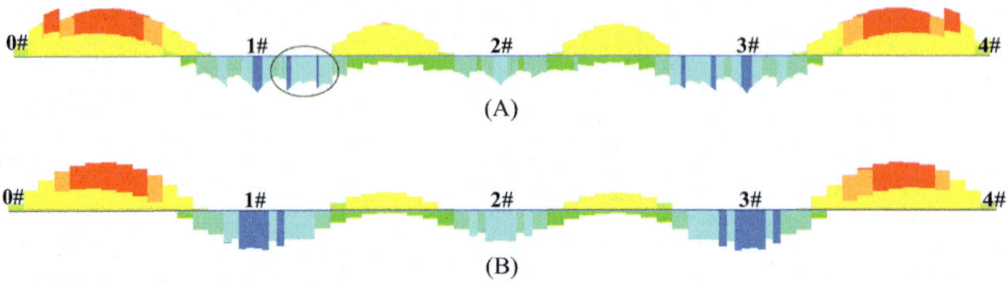

Figure 9.22 Lower chord stress and axial force envelope. (A) Stress. (B) Axial force.

The strength check of the lower chord members is shown in Table 9.12. It can be seen from Table 9.12 that the strength of the lower chord members meets the requirements. The stress law of tension and compression members is similar to that of the upper chord members, so it will not be repeated here.

It can be seen from Fig. 9.23 that under the ultimate limit state of bearing capacity, the axial force of the brace members in the pier-top area of Pier 1# is the largest. Due to the increased wall thickness of the steel tubes in the brace members of the reinforced segment and the transition brace members (the 5th diagonal brace members on both sides centered at the pier top, which are compression members), the positions with the maximum tensile and compressive stresses of the brace members are at the 6th and 7th diagonal brace members on the left side of Pier 1#.

The strength check of the brace members is shown in Table 9.13. It can be seen from Table 9.13 that the strength of the brace members meets the requirements, and both the tensile and compressive stresses of the brace members have a certain safety margin.

(c) Joint bearing capacity check and verification

The common drawing joints of composite truss bridges are classified and numbered based on the angle between the branch and main pipes of the joint, joint stress type, main pipe diameter, etc., as shown in Fig. 9.24. The joints of CFRST composite trusses have high compressive bearing capacity and generally do not suffer from local buckling failure of the steel tube side wall. However, the main pipe steel plate connected to the tension branch pipe may be the first to fail. This section focuses on checking K-type joints, and combines the joint bearing capacity formula to merge and delete the joints that need to be checked. For example, for joints JD4 and JD9, only the joint with a smaller main pipe wall thickness (JD9) is checked conservatively, as shown in Fig. 9.25.

Table 9.12 Calculation results of the bearing capacity of the bottom chord.

Span arrangement/m	Stress/ MPa	Design strength/ MPa	Axial force/ kN	Stability coefficient φ	Member stability capacity /kN	Verification result	Member efficiency factor
4×50	293	320	9317	0.963	25010	Pass	0.92
4×60	276	320	11229	0.967	32019	Pass	0.86
4×70	292	320	11576	0.943	44935	Pass	0.91
4×80	285	320	18006	0.957	54154	Pass	0.89

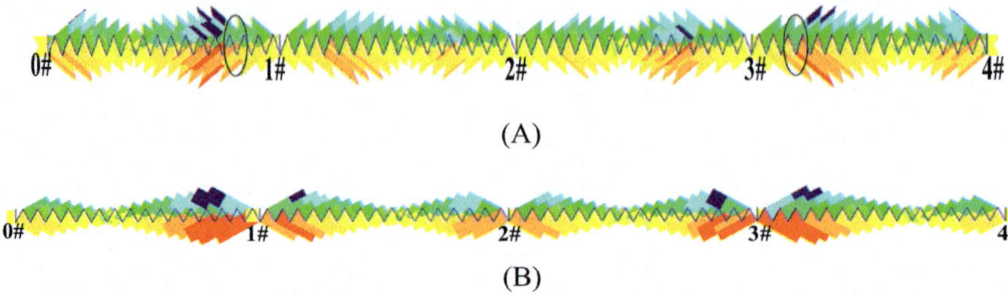

Figure 9.23 Brace member stress and axial force envelope. (A) Stress. (B) Axial force.

The joint bearing capacity check results for the common drawings of 50~80 m CFRST composite truss bridges are shown in Tables 9.14 to 9.17. The member bearing capacity is determined when the full cross-section of the branch pipe reaches the design strength, and the joint bearing capacity efficiency coefficient is the ratio of the joint bearing capacity to the corresponding branch pipe member bearing capacity.

9.4.4.3 Joint fatigue check and verification

The same joint fatigue check method as in Section 8.3.2.5 of this book is used to check the common drawing joints of 50~80 m CFRST composite truss bridges. The results are shown in Table 9.18. It can be seen from Table 9.18 that the fatigue resistance of the joints meets the requirements.

9.4.4.4 Bridge deck crack resistance check and verification

9.4.4.4.1 Longitudinal crack resistance check of bridge deck

In accordance with the requirements of Article 7.5.1 in *Code for Design and Construction of Highway Steel-Concrete Composite Bridges* (JTG/T D64-01-2015), the maximum crack width Wcr of the bridge deck in the negative moment zone of the composite truss girder under the serviceability limit state shall be checked in accordance with the relevant provisions of Articles 6.4.3 and 6.4.4 in *Code for Design of Highway Reinforced Concrete and Prestressed Concrete Bridges and Culverts* (JTG 3362-2018). The limit value of the maximum crack width of the bridge deck is 0.2 mm. The axial force of the bridge deck under the frequent effect combination of the serviceability limit state is shown in Fig. 9.26. The adoption of prestressed steel bars and shear studs that resist pulling but not shearing effectively reduces the tensile stress of the bridge deck in the negative moment zone, significantly improving the crack resistance of the bridge deck.

Table 9.13 Calculation results of brace member bearing capacity.

Span arrangement/m	Stress/ MPa	Design strength /MPa	Axial force/ kN	Stability coefficient φ	Member stability capacity /kN	Verification result	Member efficiency factor
4×50	232	320	275	0.97	310	Pass	0.89
4×60	227	320	252	0.973	311	Pass	0.81
4×70	257	320	282	0.963	308	Pass	0.92
4×80	262	320	288	0.967	309	Pass	0.93

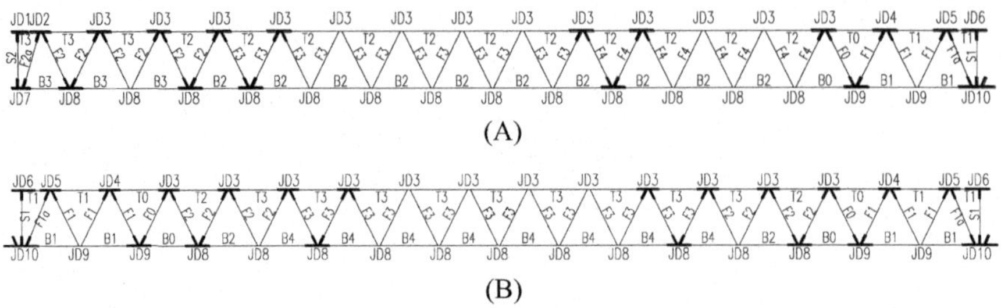

Figure 9.24 Classification and numbering of steel truss joints. (A) Side span. (B) Central span.

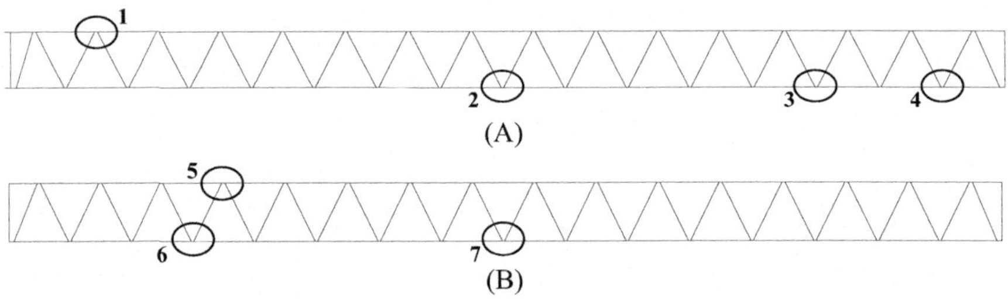

Figure 9.25 Joint numbers to be verified for steel trusses. (A) Side span. (B) Central span.

The bridge deck is checked as an ordinary reinforced concrete member in the longitudinal direction, and the check results are shown in Table 9.19. The CFRST composite truss bridges with spans of 50 m and 60 m adopt a consistent reinforcement style for the negative moment zone, while those with spans of 70 m and 80 m also share a uniform reinforcement style. The crack width of all these bridges meets the requirements of the code.

9.4.4.4.2 Transverse crack resistance check of bridge deck

For the transverse calculation model of the bridge deck, a 1m-wide longitudinal strip of the bridge deck is selected, as shown in Fig. 9.27. By applying a unit load at the mid-span position in the overall calculation model of the entire bridge, the vertical support stiffness at the upper chord of the bridge deck can be obtained. The vertical support stiffness of the bridge deck for composite truss bridges with different span combinations is basically consistent, approximately 43,160 kN/m.

Table 9.14 Verification results of joints for general drawings of 50 m composite truss girder bridge.

Joint No.	Chord dimension $b_0 \times h_0 \times t_0$/mm	Brace dimensions $b_{1,2} \times h_{1,2} \times t_{1,2}$/mm	Joint capacity/kN		Calculated axial force in member/kN		Verification result	Member capa-city/kN	Joint capacity efficiency factor
			Tensile brace	Compressive brace	Tensile brace	Compressive brace			
1&6	400×400×32	400×250×16	6328	-10353	4153	-4806	Pass	6625	0.96
2	400×400×32	400×250×12	4808	-10353	2509	-3080	Pass	5033	0.96
3	400×400×32	400×250×20	7808	-10353	5124	-6100	Pass	7808	1.00
4	600×400×36	600(32)×250(20)	11890	-24797	9076	-9306	Pass	14669	0.81
5	400×400×22	400×250×16	5662	-7618	4173	-4825	Pass	6625	0.85
7	400×400×22	400×250×12	4808	-7618	2118	-2975	Pass	5033	0.96

Table 9.15 Verification results of joints for general drawings of 60 m composite truss girder bridge.

Joint No.	Chord dimension $b_0 \times h_0 \times t_0$/mm	Brace dimensions $b_{1,2} \times h_{1,2} \times t_{1,2}$/mm	Joint capacity/kN		Calculated axial force in member/kN		Verification result	Member capa-city/kN	Joint capacity efficiency factor
			Tensile brace	Compressive brace	Tensile brace	Compressive brace			
1&6	500×500×30	500×300×20	9275	-11975	5594	-6354	Pass	9728	0.95
2	500×500×30	500×300×16	7864	-11975	4227	-4964	Pass	8233	0.96
3	500×500×30	500×300×22	9275	-11975	6852	-7617	Pass	10644	0.87
4	700×500×36	700(32)×300(24)	13087	-24980	11226	-11562	Pass	17961	0.73
5	500×500×20	500×300×20	5765	-9624	4773	-5259	Pass	9728	0.59
7	500×500×20	500×300×16	5765	-9624	2949	-3932	Pass	8233	0.70

Table 9.16 Verification results of joints for general drawings of 70 m composite truss girder bridge.

Joint No.	Chord dimension $b_0 \times h_0 \times t_0$/mm	Brace dimensions $b_{1,2} \times h_{1,2} \times t_{1,2}$/mm	Joint capacity/kN		Calculated axial force in member/kN		Verification result	Member capacity/kN	Joint capacity efficiency factor
			Tensile brace	Compressive brace	Tensile brace	Compressive brace			
1	500×500×28	500×300×20	9037	-11678	6141	-6232	Pass	9728	0.93
2	500×600×36	500×300×16	7864	-10935	3342	-4252	Pass	8233	0.96
3	500×600×36	500×300×22	10644	-10935	7784	-8649	Pass	10644	1.00
4	900×600×40	900(36)×300(30)	17814	-31184	12588	-12101	Pass	25114	0.71
5	500×500×20	500×300×20	6257	-9884	5320	-6992	Pass	9728	0.64
6	500×600×36	500×300×20	9728	-10935	6293	-6992	Pass	9728	1.00
7	500×600×20	500×300×16	6257	-8916	3858	-5358	Pass	8233	0.76

Table 9.17 Verification results of joints for general drawings of 80 m composite truss girder bridge.

Joint No.	Chord dimension $b_0 \times h_0 \times t_0$/ mm	Brace dimensions $b_{1,2} \times h_{1,2} \times t_{1,2}$/ mm	Joint capacity/kN		Calculated axial force in member/kN		Verification result	Member capa-city/kN	Joint capacity efficiency factor
			Tensile brace	Compressive brace	Tensile brace	Compressive brace			
1&5	600×600×20	600×400×20	6994	-14122	6605	-7680	Pass	12288	0.57
2	600×700×36	600×400×16	9912	-15844	5049	-5111	Pass	10377	0.96
3	600×700×36	600×400×22	13384	-15844	10011	-11021	Pass	13460	0.99
4	1000×700×40	1000(36)×400(30)	18942	-39003	14669	-14485	Pass	29338	0.65
6	600×700×36	600×400×20	12288	-15844	8141	-8986	Pass	12288	1.00
7	600×700×20	600×400×16	6994	-13064	5390	-6009	Pass	10377	0.67

Table 9.18 Verification results of joints for general drawings of composite truss girder bridge.

Span arrangement	Joint location		Normal stress $\triangle\sigma_P$ (MPa)	Bending stress $\triangle\sigma_M$ (MPa)	Hot-spot stress concentration factor (SCF)				Hot-spot stress range S_h(MPa)	Hot-spot stress range $\gamma_{Mf}S_h$(MPa)	Fatigue stress range limit $\triangle\sigma_C$(MPa)	Verification result
					$SCF_{brace,I}$	$SCF_{chord,I}$	$SCF_{chord,II}$	$SCF_{chord,III}$				
4×50 m	Mid-span joint of central span	Chord	19.5	8.8	1.29	0.40	1.31	1.53	47.65	64.33	90.5	Pass
		Brace	16.6	5.0					27.86	37.62	113.8	Pass
4×60 m	Mid-span joint of central span	Chord	8.5	3.4	2.07	0.56	1.36	1.80	25.74	34.75	93.8	Pass
		Brace	9.0	5.4					29.81	40.24	102.1	Pass
4×70 m	Mid-span joint of central span	Chord	17.1	4.0	1.95	0.57	1.38	1.72	39.48	53.30	93.8	Pass
		Brace	9.2	6.6					30.81	41.59	102.1	Pass
4×80 m	Mid-span joint of central span	Chord	16.1	3.8	2.44	0.59	1.40	1.97	38.11	51.45	93.8	Pass
		Brace	7.5	6.2					33.43	45.13	102.1	Pass

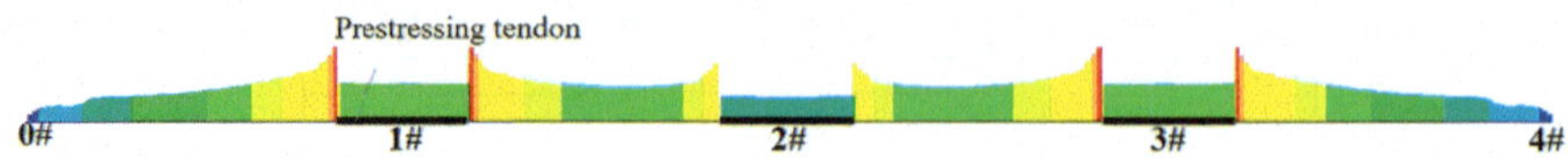

Figure 9.26 Xial force of bridge deck under combined action of short-term effects.

The transverse crack resistance of the bridge deck is checked as a Class B prestressed concrete member, and the check results are shown in Table 9.20. The transverse crack resistance meets the requirements of the code.

9.4.4.5 Connector check and verification

In accordance with the requirements of Article 9.3.4 in *Code for Design and Construction of Highway Steel-Concrete Composite Bridges* (JTG/T D64-01-2015), the shear stud fracture failure mode and concrete local crushing failure mode correspond to different design values of bearing capacity for shear connectors. The design value of the shear bearing capacity of a single shear stud is taken as the smaller of the two. Calculations show that the design value of bearing capacity corresponding to the fracture failure mode of a single shear stud is smaller, at 106.4 kN.

In accordance with the requirements of Articles 9.3.3 and 9.3.5 in *Code for Design and Construction of Highway Steel-Concrete Composite Bridges* (JTG/T D64-01-2015), shear connectors shall undergo bearing capacity check under the ultimate limit state of bearing capacity and slip amount check under the serviceability limit state. As shown in Table 9.21, the shears $\gamma 0 V_d$ and V_{sd} respectively represent the maximum shear forces at the steel-concrete interface under the ultimate limit state of bearing capacity and the serviceability limit state. It can be seen from Table 9.21 that both the bearing capacity check and slip amount check of the shear connectors meet the requirements of the code.

9.5 Economic analysis

Tables 9.20 and 9.21 present statistics on the consumption of main materials for the side spans and middle spans of composite truss bridges with spans ranging from 50 m to 80 m, where the numbers in parentheses represent the consumption of PBL connector steel. As can be seen from the tables, the consumption of each material in the middle spans is slightly higher than that in the side spans. The reason for this is that the mid-span of the side spans and the top of Pier 1# (Fig. 9.16) are the most unfavorable stress positions. In accordance with the requirements of standardized

Table 9.19 Calculation results of longitudinal crack resistance of bridge deck.

Span arrangement/m	Reinforcement layout type/mm	Reinforcement ratio/%	Axial force N_s/kN	Reinforcement stress/MPa	Crack width/mm	Crack width limit/mm
4×50	Φ25@125	2.19	17376	183	0.168	0.2
4×60			16831	179	0.173	
4×70	Φ28@125	2.22	20122	170	0.122	
4×80			26084	221	0.192	

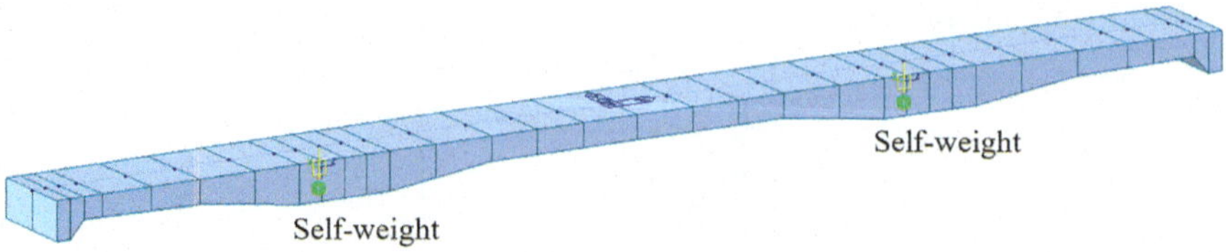

Figure 9.27 Bridge deck transverse finite element model.

Table 9.20 Calculation results of transverse crack resistance of bridge deck.

Design location	Load combination	Maximum stress Sig_{MAX}/MPa	Allowable stress Sig_{ALW}/MPa
At mid-span in transverse direction	Short-term effect	-1.219	1.855
	Long-term effect	-3.420	0
At support in transverse direction	Short-term effect	-1.303	1.855
	Long-term effect	-1.555	0

Note: Tensile stress is positive and compressive stress is negative.

design, the cross-sectional dimensions of the members at the top of Pier 1# and Pier 2# must be consistent, which consequently reduces the material utilization rate in the area at the top of Pier 2# (Table 9.22).

The steel consumption of composite truss girders with spans of 50–80 m and the self-weight of the superstructure (average value of side spans and middle spans) are shown in Table 9.23 and Fig. 9.28. It can be seen from Fig. 9.28 that as the span increases, the steel consumption index per square meter also increases accordingly.

Compared with prestressed concrete box girder bridges, CFRST composite truss bridges have a slight increase in steel consumption but a significant reduction in concrete consumption. This leads to a substantial decrease in the self-weight of the superstructure, which effectively reduces the engineering quantities of piers, foundations, and other components. Meanwhile, the lighter self-weight of the superstructure is more favorable for the seismic fortification of the substructure. To intuitively compare the material consumption of the two bridge types, a 4×70 m CFRST composite truss bridge and a 50+90+90+50 m four-span variable-section prestressed concrete continuous box girder bridge (with a bridge width of

Table 9.21 Verification results of shear connector bearing capacity and slip value.

Span arrangement/m	Top chord width/mm	Shear force $\gamma_0 V_d$/kN	Required number of shear studs/unit	Actually arranged shear studs/unit	Shear force V_{sd}/kN	Maximum slip value/mm	Slip limit/mm
4×50	400	2945	28	30	1939	0.13	0.2
4×60	500	2510	24	48	1568	0.11	
4×70	500	2745	24	48	1672	0.12	
4×80	600	2582	25	60	1673	0.1	

Table 9.22 Main material consumption of side span.

Span arrangement/m	Steel for main truss kg/m^2	Ordinary reinforcement kg/m^2	Prestressing steel kg/m^2	Total steel kg/m^2	Concrete m^3/m^2	Self-weight of superstructure kg/m^2
4×50	253(15.8)	119	12.5	385	0.38	1297
4×60	313(15.8)	118	12.5	444	0.4	1404
4×70	344(18.1)	129	10.9	484	0.47	1612
4×80	416(20.1)	130	10.5	557	0.5	1757

Table 9.23 Main material consumption of mid-span.

Span arrangement/m	Steel for main truss kg/m^2	Ordinary reinforcement kg/m^2	Prestressing steel kg/m^2	Total steel kg/m^2	Concrete m^3/m^2	Self-weight of superstructure kg/m^2
4×50	261(15.8)	127	13.6	402	0.42	1410
4×60	313(15.8)	125	13.7	452	0.45	1532
4×70	354(18.1)	137	11.6	503	0.53	1775
4×80	416(20.1)	138	11.3	565	0.56	1909

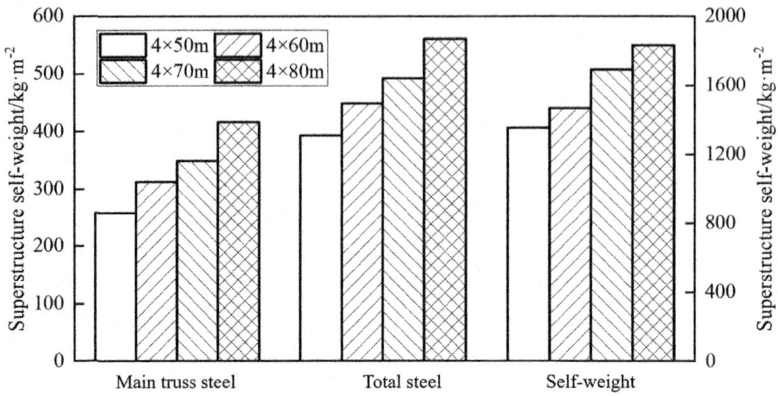

Figure 9.28 Structural steel consumption and deadweight.

12.75 m and a main girder cross-section in the form of a single-box single-cell) are selected as comparison cases. As shown in Table 9.24, compared with the prestressed concrete box girder bridge, the CFRST composite truss bridge has a 36.3% increase in steel consumption, a 53.7% decrease in concrete consumption, and a 42.8% decrease in the self-weight of the superstructure.

9.6 Summary

This chapter conducts a design study on prefabricated CFRST composite truss bridges, and develops general drawings for continuous beam bridges of CFRST composite trusses with four span lengths: 50 m, 60 m, 70 m, and 80 m. The main truss in the general drawings adopts segmental steel main trusses and prefabricated bridge decks. The bridge decks are placed on the chord members of the main truss and connected by shear studs, which fully utilizes the characteristics of rectangular steel tube sections and enables rapid bridge construction. The tension-compression member system of the main truss of the composite truss bridge can give full play to the advantages of material strength; the use of shear studs that resist pulling but not shearing in the negative moment zone can effectively improve the crack resistance of the bridge deck. Considering the overall cost of the superstructure and substructure as well as the convenience of construction, the technical and economic performance of the composite truss bridge is significantly better than that of the prestressed concrete box girder bridge.

Table 9.24 Comparison of material usage of two types of bridges.

Bridge type	Steel		Concrete		Superstructure self-weight	
	Quantity kg/m^2	Variation range/%	Quantity m^3/m^2	Variation range /%	Self-weight kg/m^2	Variation range /%
CFRST composite truss girder bridge	493	36.3	0.5	-53.7	1693	-42.8
Prestressed concrete box girder bridge	366		1.08		2958	

Extended reading

Gao YM, Liu YJ, Zhou XH, et al. High-performance CFST Composite Truss Bridge[J]. China Journal of Highway and Transport 2018;31(12):174–87.
Liu Y.J., Liu B., Dai L.X., et al. Fabricated CFRST Composite Truss Bridge and Construction Method[P]. Shaanxi: CN106223183A [P]. 2016-12-14.
Liu B, Liu YJ, Zhou XH, et al. Design of Mid-span Fabricated RCFST Composite Truss Bridge[J]. Journal of Traffic and Transportation Engineering 2017;17(04):20–31.
Liu B, Liu YJ, Yang YH, et al. Design of PBL-Stiffened Steel Truss-Concrete Composite Girder Bridge with Truss Made of Rectangular Steel Tubes[J]. Bridge Construction 2019;49(05):97–102.

Index

Note: Page numbers followed by "*f*" and "*t*" refer to figures and tables, respectively.

B

Bearing capacity, 183–193, 333–346
  calculation method, 333–346
    basic load cases, 343*f*
    concrete-filled rectangular steel tubular joints, 333–337, 336*t*, 338*f*
    fatigue calculation method, 338–346
    fatigue checking, 345–346
    fatigue resistance S-N curve, 344–345
    fatigue strength curve, 344*f*, 345*f*
    flexural capacity of concrete-filled steel tubular composite truss girders, 338
    hot-spot stress method, 339–344
    parameter formulas, 341*t*, 342*t*, 343*t*
    rectangular hollow section joints, 334*t*
  code formulas, 190–193
    base area, 189*f*
    compressive bearing capacity, 192–193
    design formula, 190–192
    effective width failure mode, 191*f*, 194*f*
    flexural bearing capacity, 193
    punching-shear failure mode, 192*f*
  concrete-filled rectangular steel tubular truss joints, 187–190
    formula for compression joints, 188–190
    formula for tension joints, 187–188
  failure mode criteria, 339*t*
  flexural capacity, 340*t*
  gap determination formula, 183–187
  internal work, 185–186
  yield-line failure model, 186*f*, 190*f*
Bending joint, 77–78
  failure modes, 77*f*
    concrete-filled rectangular hollow section joints under bending, 78*f*
Bernoulli-Euler beam theory, 47
Box girder bridges, 422–426
Bracing system, 28

C

Calculation methods, 325
Combined punching-shear and yield-line failure model, 64–72
  bearing capacity, 71*f*
  chord failure capacity, 65*f*
  formula verification, 68–72
  rectangular hollow section joints, 66*f*
  results, 71*f*
  test data, 69*t*
Compression joints, 72–75
  load transfer mechanism, 72*f*
  local compression, 74*f*
  T-type, 72
  X-type, 73, 73*f*
  Y-type, 74
Compressive performance, CFRST, 93–119
  analysis, 97–115
  calculation method, 103*f*
  design, 93–94, 94*f*
  displacement measurement points, 97*f*
  experimental method, 96–97, 101*t*
  fabrication, 93–94
  infill concrete failure, 99*f*
  local compression enhancement factors, 104*f*
  Packer's method, 105*f*
  specimen failure condition, 99*f*
  specimen parameters, 95*t*
  strain measurement points, 98*f*
  tensile and bending performance, 119–145
  test results, 97–115
  test setup, 96*f*
  theoretical calculation, 101*t*
  transverse local bearing strength, 115–119
  transverse local compression strength, 103*f*
Concrete box girders, 1

Concrete-filled rectangular steel tubular (CFRST), 45, 91, 93, 193–194, 197, 278, 349
compressive performance, 93–119
analysis, 97–115
calculation method, 103*f*
design, 93–94, 94*f*
displacement measurement points, 97*f*
experimental method, 96–97, 101*t*
fabrication, 93–94
infill concrete failure, 99*f*
local compression enhancement factors, 104*f*
Packer's method, 105*f*
specimen failure condition, 99*f*
specimen parameters, 95*t*
strain measurement points, 98*f*
tensile and bending performance, 119–145
test results, 97–115
test setup, 96*f*
theoretical calculation, 101*t*
transverse local bearing strength, 115–119
transverse local compression strength, 103*f*
Concrete-filled steel tube (CFST), 1, 11, 15
Concrete-filled steel tube composite truss girder bridge, 1, 2*f*
characteristics, 2–11
bridge deck *vs.* main truss connection, 9*f*
node force transmission paths, 8*f*
node structure comparison, 8*f*
storage comparison, 9*f*
structural characteristics, 7–9
components, 9–11
rectangular steel tube concrete composite truss bridge, 10*f*
construction methods, 3*f*
force transmission, 2*f*
high construction performance, 3–4
high economic performance, 4–6
Ganhaizi Bridge, 6
material cost and usage, 6*t*
material quantity, 4*t*
Nanhai Zidong Bridge, 4–5, 5*f*
Xiangjiaba Bridge, 5–6, 5*f*
structural performance, 2–3
Construction methods, 32–44
infilled concrete, 40–41
prefabricated deck panels, 41–44, 42*f*
steel trusses, 33–40, 34*f*
cantilever erection, 39–40
Dongjiang Bridge, 40*f*
girder jacking scheme, 35*f*
Hong Kong-Zhuhai-Macao Bridge, 43*f*
Huangyan Expressway, 37*f*
jacking and sliding device, 36*f*
Øresund Bridge, 38*f*
Padma Bridge, 38*f*
schematic diagram, 35*f*
segmental erection, 36–38
single-sided cantilever assembly, 39*f*
symmetrical cantilever assembly, 39*f*
Ulla Bridge, 41*f*
Wenchuan Keku Bridge, 43*f*
wet joints, 32*f*

D

Deck system, 29–31
cast-in-place concrete bridge, 30*f*
common cross-link structures, 29*f*
plate truss connection method, 30*f*
precast concrete bridge deck, 31*f*
shear trough structure, 32*f*
vertical connection form, 28*f*, 29*f*
crisscross configuration, 28*f*
diamond configuration, 28*f*
K-shaped configuration, 28*f*
Design verification, 402–420
bottom chord, bearing capacity of, 411*t*
brace member bearing capacity, 413*t*
brace member stress, 412*f*
during construction, 402–405
deflection, 407*t*
joint fatigue check, 412
lower chord stress, 410*f*
main truss check, 407–412
main truss during second phase paving, 404*f*
maximum compressive stress, 402–405, 404*f*

natural vibration modes, 406*t*
numbering, 414*f*
safety factor, 405, 405*f*
steel trusses, 414*f*
stiffness check, 407
upper chord bearing capacity, 409*t*
upper chord stress, 408*f*
Detailing requirements, 325–328
concrete-filled rectangular steel tubular members, 328
joint detailing, 325–327, 326*f*
continuous truss with concrete-filled rectangular steel tubular, 326*f*, 327*f*
eccentricity, 327*f*
joint dimensions, 326–327
T-type joint, 325–326

E

Economic analysis, 420–426
Elastoplastic stage, 295
European Code for Concrete Structures (CEB-FIP), 100

F

Fatigue calculation method, 338–346
fatigue load model, 338–339, 343*f*
Fatigue test, 227–246
arrangement, 230*f*
crack initiation, 239–244
crack propagation, 239–244, 242*f*, 243*f*
geometry, 229*t*
internal force distribution, 232*f*
joint configuration, 228*f*
loading parameters, 231*t*
modes, 241*f*
nominal stress point layout, 231*f*
point layout, 230*f*
procedure, 228–231
results, 240*t*
Shs-N curve, 244–246
curve analysis, 246*f*
data analysis, 244–245
fatigue design, 245–246
specimen design, 227–231
static test procedure, 228
Finite element model, 247–250, 329, 401–402
boundary conditions, 249
boundary condition simulation, 401
CFRST Y-type joint, 248*f*
construction phase division, 402, 404*t*
element type, 247–248
finite element simulation method, 249–250
full bridge finite element model, 401*f*
hoisting method, 403*f*
load conditions and combinations, 402
material properties, 248
mesh size, 247–248
steel-concrete interface relationship, 249
weld configuration, 248, 249*f*
Flexural capacity, 312–322
bearing capacity calculation, 312–316
calculation model, 313*f*
comparative analysis, 319–322
comparison, 316–319
compressed concrete deck slab, 314
cross-sectional area, 314
failure mode determination, 312–316, 320*t*, 322*f*
formula verification, 316–319
under negative bending moment, 317*t*
under positive bending moment, 316*t*
test data, 318*t*
width coefficient, 312*f*
Flexural performance, 281
concrete slab cracking, 289–292, 290*f*
failure mode, 286*f*
specimen CTB2, 288*f*
load-displacement curves, 288–289, 289*f*
maximum crack width with load, 291*f*
maximum crack width with reinforcement strain, 292*f*
negative moment region, 281–296
loading scheme, 285
material properties, 282, 283*t*
specimen design, 282
strain distribution, 292–294
brace members, 295*f*
concrete slab, 292–294
load-slip curve, 296*f*
load-strain curves, 294–296, 294*f*
longitudinal strain distribution, 293*f*
test results, 285–296
geometric dimensions, 283*f*

specimen configuration, 283*f*
steel material properties, 284*t*
test setup, 285*f*
Full-scale test, 297–311
bridge deck, 310–311
effective width, 310–311
bridge test load cases, 301*t*
global mechanical performance, 304–306
equivalent mid-span displacement, 305
schematic diagram, 303*f*
sensor arrangement, 303*f*, 304*f*
instruments, 300–302
load cases, 297–300
loading method, 297–300
loading vehicle axle loads, 298*f*
measured loading vehicle axle loads, 299*t*
mechanical performance, 306–310
axial force *vs.* measured strain curves, 305*f*
bottom chord axial stiffness, 310*t*
cross-sectional area, 307–308
elastic modulus, 307–308
equivalent mid-span load *vs.* displacement curves, 306*f*
load distribution ratio, 306–308, 308*t*, 309*f*
measured member, 307*f*
stiffness, 308–310, 309*f*
overview, 297–302
load-maximum slip curves, 297*f*
photos, 302*f*
schematic layout of, 298*f*
sensor arrangement, 300–302

G

Ganhaizi Bridge, 6
full-scale bridge, 7*f*
material indicators, 7*t*
schematic cross section, 7*f*
Global mechanical performance, 304–306
equivalent mid-span displacement, 305
schematic diagram, 303*f*
sensor arrangement, 303*f*, 304*f*

H

Hot-spot stress method, 197, 201–204, 205*f*
CIDECT code, 201–203
concrete constituents, 202*t*
concrete properties, 202*t*
extrapolation method, 203*f*
extrapolation region, 204*t*
geometric parameters, 199*t*
intersection line, 213*f*
nominal stress measurement scheme, 204–205, 205*f*
specimen fabrication, 200*f*
steel properties, 201*t*
test setup, 203*f*
Huangyan Expressway, 349, 381
economic analysis, 378–381
first-order buckling mode, 379*f*
joint fatigue verification, 377*t*
material consumption comparison, 380*t*
photo of, 379*f*
project overview, 349–350
bridge location plan, 350*f*
bridge scheme 1, 351*f*
bridge scheme 2, 352*f*
bridge scheme 3, 353*f*
bridge scheme 4, 354*f*
structural design, 350–361
bearing capacity verification, 365*t*
bottom chord, 367*t*
brace members during construction stages, 366*t*, 369*t*
calculation principles, 357–358, 359*f*
cross-section layout, 355, 357*f*
deck slab structural design, 357–358, 358*f*
deflection curve, 370*f*
double-legged "Y"-shaped Perforated Bond Link, 360*f*
elevation layout, 357, 357*f*
envelope diagram, 364*f*, 368*f*
material consumption, 356*t*
mechanical analysis, 363*f*
mechanical performance, 362*t*
pier configuration, 358–361
pier-girder connection form, 358–361, 360*f*
prestressing tendons, 359*f*
scheme comparison, 350–361
scheme selection, 350–361
staged composite section, 363

structural verification, 361–378
completed bridge stage, 368–378
during construction stages, 361–364
crack resistance, 373*t*
deck slab bearing capacity, 368–371
global stiffness, 371
internal force distribution, 376*f*
internal force verification, 372*t*
joint bearing capacity verification, 371–374
joint fatigue resistance, 374–378
joint internal forces, 376*f*
main truss, 368
stress envelope diagram, 370*f*
structural deflection, 374*f*
vertical bending vibration, 378*f*
Huangyan Expressway Overpass Bridge, 297

I

Influencing factors, 78–79
Internal forces, 328–333
finite element model, 329
multiscale finite element calculation method, 330–333, 331*f*
beam elements, 329*f*
characteristics, 330
comparative analysis, 332, 332*t*, 333*f*
"Embedded" constraint, 330–331
establishment, 330–333
simplified calculation method, 328

J

Joint test, 206–216, 215*f*, 216*f*
axial force, 220, 222*f*
bending moment, 222*f*
brace-chord angle, 214, 214*f*, 215*f*
brace-chord thickness, 216
chord width-to-thickness ratio, 215
concrete filling, 210–214
configuration and dimensions, 217*f*
CTB1 *vs.* CTB2, 226*f*
geometric dimensions, 223
internal force distribution, 221*f*, 224*f*
measurement points, 216–219
nominal stress measurement, 218*f*
Perfobond Leister ribs stiffeners, 210–214
points, 218*f*
specimen design, 216
specimens YH, YC, and YCP-1, 213*f*
strain distribution, 220
tension and compression, 211*t*

K

K-type joints, 145–183
analysis model, 158*f*
brace measurement points, 179*f*, 180*f*
branch load, 167
chord surface strains, 177
configuration, 145*f*
distribution of longitudinal strain, 182*f*, 183*f*
chord member, 184*f*
eccentric compression, 159*f*
fabrication, 145–148
failure load, 154*t*
failure modes, 149–168, 154*t*
flexural capacity, 155*f*
force analysis, 161*f*
geometric parameters, 147*t*
joint deformation, 159–168
load-average strain curves, 179*f*
load-displacement curves, 156–159, 174–175, 176*f*
load-strain curves, 159–168, 175–183, 178*f*
load *vs.* average longitudinal strain, 181*f*
neutral axis, 164–165
schematic diagram, 167*f*
specimen design, 145–148
tension zone, 164
test method, 148–149
failure condition, 153*f*
joint displacement measurement points, 150*f*
measurements, 148–149
setup, 148, 149*f*
strain measurement point layout, 151*f*, 152*f*
test results, 169–183
bending deformation, 174
chord brace measurement points, 168*f*
failure load, 171*t*
failure modes, 170*f*, 171*t*

loading pattern, 173
member bearing capacity, 172
schematic diagram, 175f, 182f
transverse strain, 184f, 185f

L

Load-deformation relationship, 104–106
CFST, characteristics of, 106
specimens, curves of, 105f
Load-longitudinal strain relationship, 106–115
brace measurement points, 110f
far-from-bearing-plate section, 113–114
flange edge measurement points, 109f
infilled concrete, 115
load-transfer mechanism, 115
measurement points, 107f, 112f, 113f, 114f
mid-flange measurement points, 111f

M

Main truss, 22–28
beam-arch combination system, 21f
brace arrangement, 23f
diamond brace member truss, 23f
K-braced truss, 23f
modified triangular truss, 23f
single-diagonal truss, 23f
triangle truss, 23f
brace cross section, 26f
brace member arrangement, 22–24
chord cross section, 26f
common cross-sectional forms, 21f
four-limb cross-section, 22t
joints, 27–28
rectangular CFST trusses, 27f
rectangular steel pipe processing method, 24f
welding four plates, 25f
three-limb cross section, 22t
types, 24–26
Mechanical performance, full-scale test, 306–310
axial force *vs.* measured strain curves, 305f
bottom chord axial stiffness, 310t
cross-sectional area, 307–308
elastic modulus, 307–308
equivalent mid-span load *vs.* displacement curves, 306f
load distribution ratio, 306–308, 308t, 309f
measured member, 307f
stiffness, 308–310, 309f
Mechanical properties, 53–79
combined punching-shear and yield-line failure model, 64–72
bearing capacity, 71f
chord failure capacity, 65f
formula verification, 68–72
rectangular hollow section joints, 66f
results, 71f
test data, 69t
failure modes, 58–64
axial force in branch members, 58–63, 59f
brace bending, 63f
effective width failure, 61f
equivalent column, 62f
flexure in branch members, 63–64
punching-shear model, 61f
yield-line model, 60f
joint configuration, 55–58
truss joint types, 53–55
cross-sectional dimensions, 57–58
geometric parameters, 56t
joint notation convention, 55f
K-type, 53, 54f
N-type, 53
strengthening, 57f, 58
T-type, 53f
Y-type, 53f, 54f
Midas Civil, 299
Multiscale finite element calculation method, 330–333

N

Nanhai Zidong Bridge, 4–5

P

Parametric analysis, 247–277
calculation formulas, 260–268
brace in-plane bending moment, 266
chord axial force, 266

chord in-plane bending moment, 266
FE calculation results *vs.* fitting formula calculation results, 259*f*
fitted formulas, 267*f*
joint deformation, 258*f*
rectangular steel tube joint *vs.* rectangular steel tube concrete joint, 266–268, 267*f*
SCF chart method, 263*f*, 264*f*, 265*f*
SCF graphology, 261*t*, 263–266
stress concentration factor parameter, 260
stress distribution, 258*f*
FE calculation results, 252*t*
finite element modeling method, 247–250, 257*f*
boundary conditions, 249
CFRST Y-type joint, 248*f*
element type, 247–248
finite element simulation method, 249–250
material properties, 248
mesh size, 247–248
steel-concrete interface relationship, 249
weld configuration, 248, 249*f*
K-type joints, 268–270, 270*f*
boundary conditions, 270
CFRST K-type joint, 268*f*, 271*f*, 276*t*
load cases, 270
parametric design, 269*t*
SCF parameters, 272*t*
load cases, 256*f*
stress concentration factor parameters, 270–277
fitted formulas, 274*f*
fitting, 270–275
multiple regression analysis, 270–271
rectangular steel tube joint *vs.* rectangular steel tube concrete joint, 275–277, 277*f*, 278*f*
validation, 270–275
T-type and X-type joints, 250–259, 251*f*, 251*t*
boundary condition, 256
loads, 256
parameter design, 250–255
verification, 247–250
Perfobond Leister ribs (PBL), 282
Prefabricated concrete-filled rectangular steel tubular composite truss bridge, 383
bending moment, 390*f*
CFRST joint with gusset plate, 395*f*
hierarchical design concept, 396
integral transportation, 396
internal force distribution, 391*f*
joint selection, 387–396
main truss connection structure, 387–396
main truss in central span, 394*t*
main truss loose components, 398*f*
overview, 383–384
design specifications and standards, 384
material parameters, 383
technical standards, 383
PBL key size, 395*f*
shear stud arrangement, 390*f*
shear stud shape, 389*f*
side-span main truss cross-section dimensions, 393*t*
standard K-type joint, 395*f*, 397*t*
structural design, 384–396
80 m elevation layout, 385*f*
arrangement, 385*t*
bridge deck selection, 384–387, 386*f*, 388*t*
cross-sectional layout, 384–387, 385*f*
layout, 384
main truss selection, 384
plate-truss connection structure, 384–387
transverse layout, 389*f*
truss grading design, 391*f*
truss members, 387–396

R

Rectangular concrete-filled steel tubular truss tests, 79–91
failure mechanism, 88–91
concrete cracking patterns, 88*f*
concrete-filled steel tubular (CFST) trusses, 90*f*
specimen failure modes, 87*f*
failure modes comparison, 85*f*, 86*f*

overview, 80
parameters, 81*t*
punching-shear failure, 84*f*
specimens, 80–84
specimens R0-R2 and C0-C2, 84–85
test loading setup, 83*f*
test results, 86–88
upper chord hollow section joint in compression, 83*f*
specimen A0, 83*f*
specimen A1, 83*f*
specimen C0, 84*f*
specimen R0, 84*f*
Rectangular steel tubular (RST), 45
concrete-filled steel tubular trusses, 48–52
calculation model, 48*f*
truss loading failure modes, 51*t*, 52
deformation schematics, 46*f*
failure modes, 46*f*
flexural capacity, 45–52
flexural stiffness, 47*f*
mechanical behavior, 45–52

S

Simplified calculation method, 328
Static test, 232–239, 236*t*
hot-spot stress results, 235–239
joint under brace tension, 234*t*
nominal stress results, 232–235
SCF *vs*. RST joints, 238*f*
stress distribution, 233*f*
Steel trusses, 33–40, 34*f*
cantilever erection, 39–40
Dongjiang Bridge, 40*f*
girder jacking scheme, 35*f*
Hong Kong-Zhuhai-Macao Bridge, 43*f*
Huangyan Expressway, 37*f*
jacking and sliding device, 36*f*
Øresund Bridge, 38*f*
Padma Bridge, 38*f*
schematic diagram, 35*f*
segmental erection, 36–38
single-sided cantilever assembly, 39*f*
symmetrical cantilever assembly, 39*f*
Ulla Bridge, 41*f*
Wenchuan Keku Bridge, 43*f*
wet joints, 32*f*
Stress concentration, 197–227
hot-spot stress method, 201–204, 205*f*
CIDECT code, 201–203
concrete constituents, 202*t*
concrete properties, 202*t*
extrapolation method, 203*f*
extrapolation region, 204*t*
geometric parameters, 199*t*
nominal stress measurement scheme, 204–205, 205*f*
specimen fabrication, 200*f*
steel properties, 201*t*
test setup, 203*f*
joint force analysis, 206*f*
joint test results, 206–216, 215*f*, 216*f*
axial force, 220, 222*f*
bending moment, 222*f*
brace-chord angle, 214, 214*f*, 215*f*
brace-chord thickness, 216
chord width-to-thickness ratio, 215
concrete filling, 210–214
configuration and dimensions, 217*f*
CTB1 *vs*. CTB2, 226*f*
geometric dimensions, 223
internal force distribution, 221*f*, 224*f*
measurement points, 216–219
nominal stress measurement, 218*f*
Perfobond Leister ribs stiffeners, 210–214
points, 218*f*
specimen design, 216
specimens YH, YC, and YCP-1, 213*f*
strain distribution, 220
tension and compression, 211*t*
node trials, 197–205
static testing, 201
typical joint configuration, 198*f*
weld toe side *vs*. weld heel side, 208*t*
Stress concentration factor (SCF), 197
Structural system, 15–22
classification, 15
forms, 15, 16*f*
general layout, 15–22
continuous truss bridges, 17*f*
cross-sectional layout, 20–22
curved chord truss bridge, 20*f*
elevation configuration, 19–20

height-to-span ratio, 18*f*
parallel chord girder bridge, 19*f*
span arrangement, 16–19
steel consumption, 18*f*
truss bridge with stiffened chords, 20*f*
Structural verification, 396–420
bending moment, 398*f*
calculation parameters, 399–400
concrete material properties, 399*t*
connector check, 420
deadweight, 426*f*
design verification, 402–420
bottom chord, bearing capacity of, 411*t*
brace member bearing capacity, 413*t*
brace member stress, 412*f*
during construction, 402–405
deflection, 407*t*
joint fatigue check, 412
lower chord stress, 410*f*
main truss check, 407–412
main truss during second phase paving, 404*f*
maximum compressive stress, 402–405, 404*f*
natural vibration modes, 406*t*
numbering, 414*f*
safety factor, 405, 405*f*
steel trusses, 414*f*
stiffness check, 407
upper chord bearing capacity, 409*t*
upper chord stress, 408*f*
finite element model, 401–402
boundary condition simulation, 401
construction phase division, 402, 404*t*
full bridge finite element model, 401*f*
hoisting method, 403*f*
load conditions and combinations, 402
loads, 399–400
longitudinal crack resistance check, 412–414, 421*t*
material consumption, 419*t*, 424*t*, 425*t*, 427*t*
shear connector bearing capacity, 423*t*
steel plate material properties, 400*t*
transverse crack resistance check, 414–420
unfavorable layout, 396–399
xial force, 420*f*

T

Tensile and bending performance, 119–145
fabrication, 121–122
geometric parameters, 123*t*
joint configuration, 121*f*
load-deformation relationship, 131–133
curves, 132*f*
load-strain relationship, 133–137
brace measuring points, 133*f*
chord flange measuring points, 134*f*, 135*f*
mechanical properties, 124*t*
packer mechanical properties, 124*t*
test design, 121–122, 125*f*
test method, 124–125
displacement measuring points, 126*f*
loading scheme, 124
measurements, 125
setup, 124
strain measuring points, 126*f*
theoretical bearing capacity calculations, 129*t*
X-type bending joints, 137–145
analysis, 137–145
capacity calculation, 143–145
chord brace under load, 143*f*
chord flange, 142*f*
failure loads, 137–139, 138*t*
failure modes, 137–139, 137*f*, 138*t*
load-deformation relationship, 139–140
load-displacement curve, 139*f*
load-strain curve, 140*f*, 141*f*, 142*f*
load-strain relationship, 140–143
principal tensile strain angle, 144*f*
principal tensile strain curves, 136*f*
results, 137–145
X-type tensile joint, 127–137, 127*f*
analysis, 127–137
failure loads, 127–131
failure modes, 127–131, 128*f*
results, 127–137
Tension joints, 75–77
deformation comparison, 75*f*

relationship curve, 76, 76*f*
T-type, 75
X-type, 75
Y-type, 75
Transverse finite element model, 422*f*
Transverse local bearing strength, 115–119
calculation formula, 118–119
experimental *vs.* calculated results, 120*t*, 121*f*
influence coefficient, 118
participation coefficient, 116–118
simplified model, 117*f*
Truss joint types, 53–55
cross-sectional dimensions, 57–58
geometric parameters, 56*t*
joint notation convention, 55*f*
K-type, 53, 54*f*
N-type, 53
strengthening, 57*f*, 58
T-type, 53*f*
Y-type, 53*f*, 54*f*
Truss segment test, 216–219
T-type joints, 145–183
analysis model, 158*f*
brace member for T-type joint, 157*f*
branch load, 167
configuration, 145*f*
eccentric compression, 159*f*
fabrication, 145–148
failure load, 154*t*
failure modes, 149–168, 154*t*
flexural capacity, 155*f*
force analysis, 161*f*
geometric parameters, 147*t*
joint deformation, 159–168
joint Tc-c16, 163*f*, 164*f*, 165*f*, 166*f*
load-displacement curves, 156–159
load-strain curves, 159–168
neutral axis, 164–165
schematic diagram, 167*f*
specimen design, 145–148
tension zone, 164
test method, 148–149
failure condition, 153*f*
joint displacement measurement points, 150*f*
measurements, 148–149
setup, 148, 149*f*
strain measurement point layout, 151*f*, 152*f*

W

Wedging effect, 116*f*
Weld heel side, 243–244
Weld toe side, 239–243

X

Xiangjiaba Bridge, 5–6
X-type bending joints
analysis, 137–145
capacity calculation, 143–145
chord brace under load, 143*f*
chord flange, 142*f*
failure loads, 137–139, 138*t*
failure modes, 137–139, 137*f*, 138*t*
load-deformation relationship, 139–140
load-displacement curve, 139*f*
load-strain curve, 140*f*, 141*f*, 142*f*
load-strain relationship, 140–143
principal tensile strain angle, 144*f*
principal tensile strain curves, 136*f*
results, 137–145

Y

Y-type joints, 145–183
analysis model, 158*f*
brace member for Y-type joint, 160*f*
branch load, 167
configuration, 145*f*
eccentric compression, 159*f*
fabrication, 145–148
failure load, 154*t*
failure modes, 149–168, 154*t*
flexural capacity, 155*f*
force analysis, 161*f*
geometric parameters, 147*t*
joint deformation, 159–168
joint Yc-018, 162*f*
joint Yc-c16, 162*f*
load-displacement curves, 156–159
load-strain curves, 159–168
neutral axis, 164–165

schematic diagram, 167*f*
specimen design, 145–148
steel Y-type joint, 160*f*
tension zone, 164
test method, 148–149
  failure condition, 153*f*
  joint displacement measurement points, 150*f*
  measurements, 148–149
  setup, 148, 149*f*
  strain measurement point layout, 151*f*, 152*f*

Printed and bound by CPI Group (UK) Ltd, Croydon, CR0 4YY

14/09/2026

02216408-0007